Ralf Kilian, Sara Saba, Caroline Gietz (Hrsg.)

**EEHB 2022 Post Prints**



The 4th International Conference on
Energy Efficiency in Historic Buildings

4th and 5th May 2022
Benediktbeuern, Germany

Fraunhofer IRB

Fraunhofer IBP

Fraunhofer Centre Benediktbeuern

Ralf Kilian, Sara Saba, Caroline Gietz (Hrsg.)

# EEHB 2022 Post Prints

## The 4th International Conference on Energy Efficiency in Historic Buildings

4th and 5th May 2022
Benediktbeuern, Germany

3., erweiterte Auflage

with the kind support of

bezirk oberbayern

sponsored by

DBU Deutsche Bundesstiftung Umwelt

www.dbu.de

Fraunhofer IRB | Verlag

Bibliographic information published by Die Deutsche Nationalbibliothek

Die Deutsche Nationalbibliothek lists this publication in the Deutsche Nationalbibliografie; detailed bibliografic data is available in the Internet at www.dnb.de.

ISBN (Print): 978-3-7388-0967-1
ISBN (E-Book): 978-3-7388-0966-4

Herstellung, Umschlagentwurf: Gabriele Wicker, Fraunhofer IRB
Druck: Books on Demand GmbH, Norderstedt


Fraunhofer Information-Centre for Regional Planning and Building Construction IRB
P.O. Box 80 04 69, D-70504 Stuttgart
Nobelstrasse 12, D-70569 Stuttgart
Phone +49 711 970-2500
Fax +49 711 970-2508
E-Mail irb@irb.fraunhofer.de
URL www.baufachinformation.de

## New life within old walls

Churches, chapels and monasteries, castles and country estates, farmhouses and townhouses, industrial monuments and historic memorials: Few things shape the appearance of Upper Bavaria more than its tens of thousands of monuments. The protection of historic buildings and ensembles is a very high priority for the district of Upper Bavaria. We support all owners of listed buildings – whether private, ecclesiastical or municipal. In this way, we want to contribute to the preservation of our unique cultural landscape. Since 2004, the district of Upper Bavaria has spent over 40 million in grants for the restoration of private, ecclesiastical and municipal monuments.

From this point of view, it is a pleasure for me to congratulate the Fraunhofer-Gesellschaft on its conference "4th International Conference on Energy Efficiency in Historic Buildings". The District of Upper Bavarias own "Zentrum für Trachtengewand" and the Fraunhofer-Gesellschaft are not only located in good proximity to each other in the Benediktbeuern monastery area but are also linked in numerous projects in the field of historic preservation.

The idea of monument protection has a long and, especially in Bavaria, royal history. That the architectural heritage must be protected is a fundamental conviction of the Romanticism of the early 19th century. In Bavaria, it was King Ludwig I who helped these efforts achieve a breakthrough. Following the French model, he created the office of "Inspector General of the Plastic Monuments of the Middle Ages" in 1835. Soon this office became an independent authority for the care of all monuments in Bavaria, called "Landesamt für Denkmalpflege" in 1908.

Today, the Fraunhofer-Gesellschaft, especially through its institute in Benediktbeuern, is also one of those forces that strengthen and support the preservation of historical monuments through technical expertise. It is our common concern to enable "new life in old walls", thus modern living and working in listed buildings.

I wish the conference every success and look forward to continued cooperation in a spirit of trust.

Yours

Josef Mederer

Bezirkstagspräsident von Oberbayern
(District President of Upper Bavaria)

## Foreword to EEHB 2022

In a time of crisis, the Energy Efficiency in Historic Buildings conference EEHB also had to be postponed several times: the Corona pandemic did not allow us to come together safely for a long time. Finally, we are very pleased that it could take place in 2022, in the form of a hybrid event with participants from all over the world, both virtual and in presence.

The corona crisis, the crisis in Ukraine and the related international energy crisis currently underway are leading to immense social and economic changes. However, the climate crisis continues to progress and threatens to lose focus in the face of multiple overlapping crises. The EEHB has brought this challenge back to the forefront and made us all aware of our responsibility to act swiftly and purposefully.

Consequently, the EU also had to react with emergency measures that lead to a tightening of the energy saving targets as well as the stronger compulsion to expand renewable energies in the EU directives, which are currently being negotiated. These crises offer the opportunity for an ambitious approach to decarbonization in all sectors.

However, the obstacle here is the affordability of energy and the associated social impact. Saving energy, now famously referred to as "Efficiency First", is the essential starting point in further action. The operationalization of the "Efficiency First" principle includes various measures such as the exemplary role of the public sector, targeted planning of heating and cooling supply and energy efficiency obligation systems. This also includes mandatory renovation measures for buildings. The reform of the EPBD (the directive on the energy performance of buildings) and the "renovation wave" aim at reducing building-related greenhouse gas emissions.

Currently, limiting factors are especially the capacities of planners and executors as well as the availability of building materials and building services equipment. Thus, also the preservation or modernization of historic buildings faces special challenges. The historically valuable buildings have a high planning demand in order to ensure their sustainable preservation. Also, the energy demand of these buildings cannot be reduced as easily as in a completely new building. Nevertheless, to do nothing and to wait for others to act is not an option – thus reasonable measures for energy efficiency in historic buildings are to be taken.

Meanwhile, the value-giving characteristics of the historic building stock must not be forgotten: They are characteristic for the cityscape or townscape and it is impossible to imagine our culture without them. Due to their long service life, these buildings can also be regarded as long-term $CO_2$ storages, and the environmental impact caused by the manufacture of building products is spread over precisely this long service life. This is a clear advantage of historic buildings. On the other hand, the high energy demand due to "modern" use can leads to high costs and corresponding $CO_2$ emissions if comfort criteria of the users are to be met. However, it is precisely this proportion of the “worst” buildings that also offers a high savings potential and an effective approach to reducing emissions.

Digital planning methods and workflows, the special topic of this year’s EEHB conference, enable effective and sustainable optimization with regard to the relevant influencing parameters for the long-term preservation of our valuable historic buildings in an age of upheavals like ours.

Prof. Dr. Gunnar Grün

Deputy Institute Director of the Fraunhofer Institute for Building Physics IBP

## Foreword to EEHB 2022

Improving the energy performance of historic buildings has become a crucial task for achieving $CO_2$ reduction targets. But while great progress has been achieved in optimising the energy consumption of new buildings, there is still an urgent need for action when it comes to historic buildings, despite their great ecological and economic potential. For many owners, energetic renovation of historic buildings is still a challenge because there are no standard solutions that can be applied – unlike with new constructions.

The International Conference on Energy Efficiency in Historic Buildings made an important contribution to tackling this challenge by highlighting innovative digital approaches: Energy and hygrothermal building simulations are important planning tools when it comes to energy-efficient renovation. These simulations can be enhanced with the use of 3D models. While both technologies – building simulation and 3D modelling – have been available for some time. However, there are no proven approaches to combining the two in an automated way. More research is needed to formulate answers to many technical and practical questions at hand.

Here, the German Federal Environmental Foundation DBU would like to make an important contribution. We see digital technologies as key to environmental protection and sustainable development. We therefore support projects in the field of cultural heritage that make innovative use of digital solutions, for example automation, sensor technology, 3D modelling or AI-supported analyses for damage detection. It was therefore important to us to support the conference workshop "Recording Historic Buildings using Digital Workflows" that leads to valuable knowledge in digital building models and simulation of historic buildings.

The DBU expresses its gratitude towards all partners and participating experts and researchers for their high level of personal commitment to the project, especially the Fraunhofer Centre in Benediktbeuern with the Fraunhofer Institute for Building Physics IBP, the Belgian Building Research Institute BBRI, the University of Bamberg and its programme "Digital Technologies in Heritage Conservation", the International Committee for Documentation of Cultural Heritage as well as the "Competence Center 4.0".

Sincerely yours,

Alexander Bonde

Secretary General, German Federal Environmental Foundation (DBU)

# Table of Content

An integrated H-BIM approach for energy retrofit of built heritage 13
*Thravalou Stavroula, Alexandrou Kristis, Artopoulos Georgios*

Reality capture technologies as a support for efficient energy diagnosis and simulation of heritage buildings: the case of the 'Alte Schäfflerei' in Benediktbeuern 23
*S. Dubois, J. Desarnaud, M. de Bouw*

Seismic Upgrade as an Opportunity for Energy Retrofitting Historic Unreinforced Masonry Buildings in Aotearoa New Zealand 35
*P. Besen, P. Boarin*

Legal and Social Challenges to a Fossil-Free Historic Built Environment: A Comparative Review of Laws and Policies from Sweden and India 43
*H.V. Bhati, M. Geijer, T. Broström, Y. Epstein*

Damage-free energetic window renovation in old and listed buildings – Hygrothermal aspects of box-type windows 55
*S. Bichlmair, M. Krus, D. Stiegler, Ch. Milch*

Transparent internal wall insulation 63
*S. Bichlmair, M. Krus, R. Kilian*

Improving the (energy) performance of (historic) buildings and communities: towards "neighbourhood Energy Performance Certification" (nEPC) 73
*N.S. Brito*

Building stock categorization for energy retrofit of historic buildings: A literature review 81
*L. Briz*

Investigating stakeholders' opinion on energy retrofit of historic buildings, grasping their needs and experiences – An overview on the use of retrofit guidelines and decisional tools 93
*A. Buda*

Identification and thermal behaviour of traditional flat roofs in the Mediterranean using remote and in situ sensing technologies – an innovative approach 105
*J. Cassar, C. Galdies, E. Muscat Azzopardi*

Internal insulation – a preliminary assessment tool based on probabilistic simulations 113
*E. J. de Place Hansen, E. B. Møller, T. K. Hansen*

Does age matter? How building age influences energy use in the Swedish residential building stock 121
*A. Donarelli, T. Broström*

"Mechanization Takes Command" – Indoor Environmental Control in 20th-Century Historic Buildings 129
*Anna Bonora, Kristian Fabbri, Giulia Favaretto, Marco Pretelli*

Hygrothermal building simulation from a 3D point cloud – the Edo-Wiemken-Monument in Jever 137
*L. Feldmann, S. Bichlmair, R. Kilian, M. Winkler*

Preserving and Improving Energy Efficiency in Modern Architecture – Efficacy of Digital Tools 147
*G. Franco, S. Bergero, D. Desideri, B. Filograna*

Future energy performance and overheating risks in retrofitted historic buildings of South Tyrol 155
*L. Hao, D. Herrera-Avellanosa, Alexandra Troi*

Launching an experimental label on both energy retrofitting and heritage preservation: early lessons 163
*E. Heberle, J. Borderon, Y. Dervyn, A. Sage, H. Arnichand*

Retrofit plan for energy performance improvement of masonry historic buildings on campus: cultural heritage building cases 171
*Hyun Mi Cho, Hyeonseong Yuk, Ho Hyeon Jo, Beom Yeol Yun, Sungwoong Yang, Jihee Nam, Sumin Kim*

Balancing energy efficiency and conservation aspects of Terragni's Casa del Fascio in Como: thermal analysis, energy modelling and intervention proposals 181
*S. Mauri, V. Pracchi*

Traditional and innovative materials and solutions to improve the energy efficiency of historic windows: a literature review 189
*S. Mauri, V. Pracchi*

A Life Cycle Assessment Methodology to Compare Carbon Emissions from Building Refurbishments to Demolition and New Build 197
*A. Duffy, A. Nerguti, C. Engel Purcell*

Historic buildings and energy efficiency in New Zealand: A theoretical study of a holistic refurbishment of a 1930s heritage building's windows 205
*R. Paschoalin, N. Isaacs*

The Carbon Value of the U.K.'s Historic Housing Stock 215
*James Ritson*

Energy Efficiency Intervention in Brutalist buildings: New Challenges in preservation of 20th century built heritage 223
*E. J. Uranga, L. Azcona, L. Etxepare, I. Lizundia, M. Sagarna*

Action within the Flemish climate fund to reduce $CO_2$ emission of protected buildings in Flanders 231
*N. Vernimme, Y. Vanhellemont, M. de Bouw*

Barriers to reducing carbon from heritage buildings: residents' views from Cumbria and the English Lake District World Heritage Site 239
*F. Wise, A. Moncaster, D. Jones*

Link Historic Buildings to Cloud with Internet of Things and Digital Twins 249
*Zhongjun Ni, Yu Liu, Magnus Karlsson, Shaofang Gong*

# An integrated H-BIM approach for energy retrofit of built heritage

Thravalou Stavroula[1], Alexandrou Kristis[1], Artopoulos Georgios[2]

[1] Energy, Environment and Water Research Center (EEWRC), The Cyprus Institute (CyI), Nicosia, Cyprus. s.thravalou@cyi.ac.cy, k.alexandrou@cyi.ac.cy

[2] Science and Technology in Archaeology Research Center (STARC), The Cyprus Institute (CyI), Nicosia, Cyprus. g.artopoulos@cyi.ac.cy

**Keywords** – Built Heritage, Energy Retrofit, Energy-Efficiency, Heritage Building Information Modelling (EE-HBIM)

## 1. INTRODUCTION

The latest Green Deal policy, released by the EU, prioritises energy efficiency in the building sector and highlights the importance of digitalisation of the building retrofitting process [1]. Historic buildings are usually excluded from legislation regarding minimum energy performance requirements, yet there is a great potential of energy consumption and greenhouse gas emission reductions through the energy retrofit of the particular building stock [2]. Over the last decades, numerous guidelines and methodologies have been developed, outlining the procedure of decision-making for historic buildings refurbishment [3]. The stages of the process that are typically outlined in literature are: building survey and analysis including historical significance assessment, indoor environmental monitoring, energy auditing and dynamic simulation. Energy retrofits are often described in the literature as an act of balancing multiple criteria, among which conservation and energy consumption prevail. The criterion of economic viability is emerging, yet, the accessibility to funds is not covered by most methodology approaches, omitting a decisive factor in the implementation of the project.

Building Information Modelling (BIM) is an emerging and promising building asset management technology, able to integrate a broad spectrum of building information, such as object attributes and construction processes, that take place from the building's planning stage to its demolition. BIM supports a holistic modelling and analysis process by simultaneously assigning additional dimensions of information to the model objects, i.e., cost (4D), time (5D) and energy performance results (6D) [4]. The centralised digital platform of information management offered by native BIM software, ensures the minimization of duplicate modelling processes, provides a workflow less sensitive to human errors and eliminates accidental information neglection during the entire building development [4]. Despite the comparative advantages of BIM, its application for heritage refurbishments (HBIM) is rare. This is mainly attributed to the emerging complexities and the absence of standardised processes, namely, the scan-to-BIM intensive modelling process, insufficient software interoperability with third party numerical simulation engines and the inadequate data exchange between native BIM software [5]. Moreover, the lack of sufficient geometrical, historical and conservation state documentation data complicates the modelling and alphanumerical data collection for heritage buildings, since most of the heritage objects' geometry and data complexity impede standardisation and automation [6].

In order to tackle the challenges of an integrated H-BIM approach, the research project "BIM for Energy Efficiency in the Public sector" (BEEP) was launched in 2019, under the framework of ENI CBC MED [7]. BEEP main objective is to create a comprehensive methodology for Energy Efficiency Heritage BIM (EE-

HBIM), while supporting the financial decision-making through the enhancement of the Energy Performance Contracting process. The later mechanism promotes the involvement of private funds through the guaranteed energy savings deriving from the energy retrofit of the building [8]. The research findings and workflow guidelines will be implemented in seven pilot-cases from all the involved partner countries (Italy, Cyprus, Spain, Jordan, Palestine, Lebanon and Egypt). Through these pilot actions in public historic buildings, BEEP objective is to demonstrate the applicability of BIM technology in heritage buildings. This paper discusses the methodology of BEEP, along with its implementation stages in the pilot case-study in Nicosia, Cyprus.

## 2. AN INTEGRATED METHODOLOGY FOR EE-HBIM

BEEP project promotes two key elements for enriching the existing approaches regarding the energy retrofit of historic buildings, namely, the integration of digital technologies and the accessibility to funds. With regards to the latter, the objective of BEEP is to promote BIM use as trustworthy tool for financial evaluation of the return of investment (ROI) of the refurbishment project, and to foster the access of public administration to credit from private financial services. The project will provide guidelines, contributing towards new public administration practices by developing a single common approach for a) documenting and archiving heritage building's geometry and related construction attributes, b) performing dynamic energy performance analysis and selecting the energy retrofit measures, and finally c) preparing necessary documentation for pursuing private investments for retrofit implementation. Figure 1 presents the general outline of the concept and workflow of BEEP project.

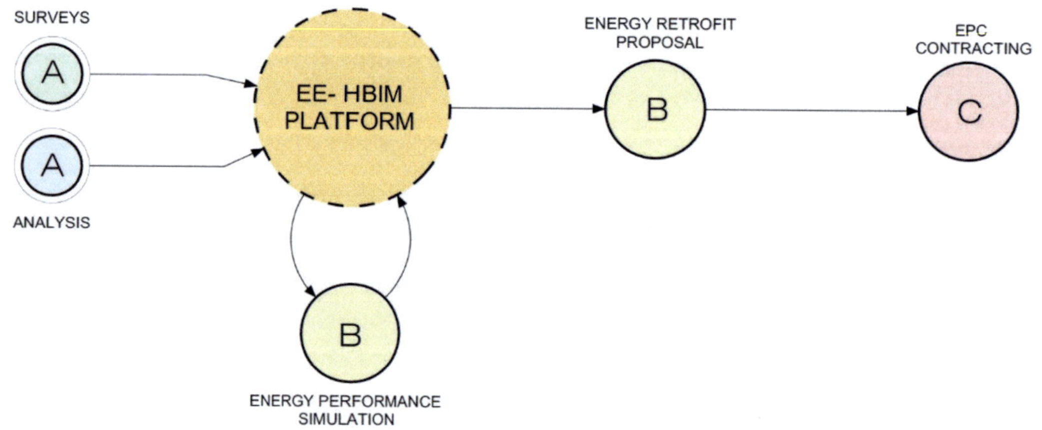


**Figure 1. Schematic representation of BEEP's project workflow.**

In addition, Figure 2 explains the BEEP methodology for achieving these objectives. The first step in the development of the Heritage Building Information Model (H-BIM) is the historic, geometric, and environmental data acquisition, i.e. *Building Analysis and Documentation* (*A*). In the consecutive Energy-Efficiency H-BIM model (EE-HBIM) preparation phase, additional integration of alphanumerical information of energy related analyses, is required, for passing on to the second step of *Energy Performance Assessment (B)*. Building Performance Simulations (BPS) enable the estimation of the environmental performance of the existing building and the selection of suitable energy retrofit measures. In the framework of the proposed pilot actions, three retrofit scenarios will be examined based on financial and energy consumption criteria, while accounting for compatibility with heritage building conservation constraints. The integration of the 4D and 5D dimensions (time and cost related metadata) in the subsequent enriched EE-HBIM model stage, focus on providing additional assessment features for the selection of the most cost-efficient rehabilitation strategy (cost-analysis indicators). Finally, having concrete results regarding the guaranteed energy savings, the respective payback period and the projects' implementation timeframe, enhances the approach of Energy Service Providers (ESPs). The final stage, i.e. *Design and deployment of financing mechanisms (C)*, entails the respective actions for the preparation of the legal and technical framework documentation for proceeding with a potential Energy Performance Contract.

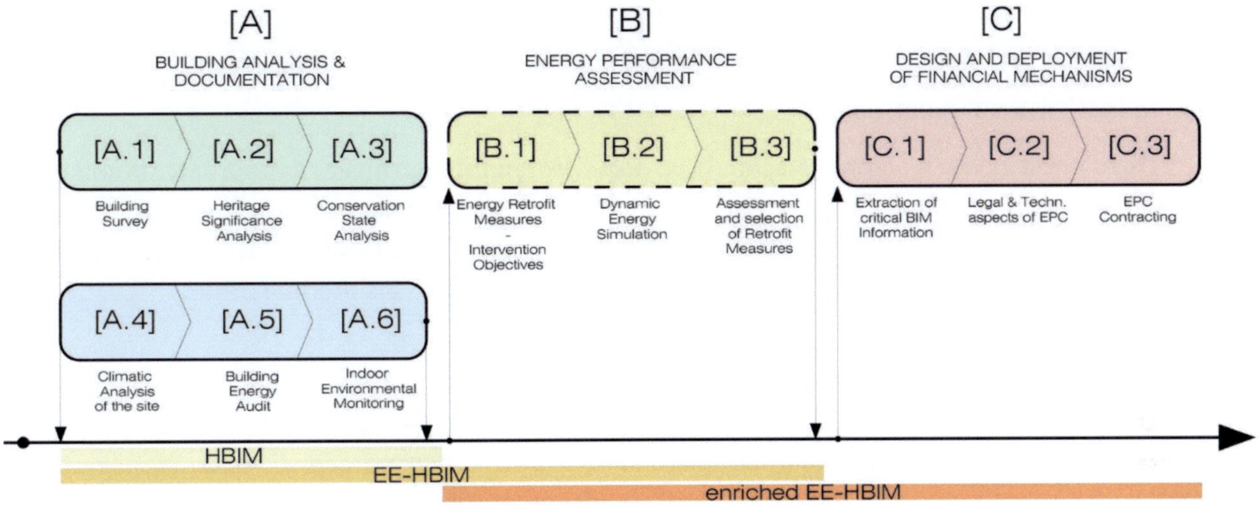


**Figure 2. Analytical EE-HBIM workflow**

## 2.1 BUILDING ANALYSIS AND DOCUMENTATION (A)

- Building survey:

  The careful documentation of the building involves the following: identification of legislative information, national zoning plans and other regulations, designation of heritage, technical documentation of the building geometry based on traditional and innovative techniques.

- Heritage significance analysis:

  The analysis entails archival research regarding the history and development of the building and its elements. This includes the study of historic layers (present and previous uses), the assessment of the heritage significance and vulnerability to change, as well as the conservation priorities or constraints on behalf of the heritage authorities.

- Conservation state Analysis:

  Documenting the building's structure and condition entails brief reporting regarding the construction materials, finishes, hygrothermal properties, decay phenomena and crack pattern analysis, identification of air leakage and moisture presence. These data are integrated in the HBIM model in the form of spreadsheets and general report sheets.

- Climatic analysis of the site:

  This set of data regard information about the local environment, climatic and topographic conditions of the area, physical interaction with adjacent objects (e.g. building, trees etc.) and assessment of the inherent passive strategies with regard to the local microclimate and the prevailing outdoor climatic conditions.

- Energy audit:

  The energy audit provides written documentation of the condition of the building and its energy-source systems. This involves the record of the technical characteristics of the existing mechanical systems, the lighting and plug loads, the water service systems, the setpoint and setback temperatures, as well as the operation and occupancy schedules.

- Indoor environmental monitoring:

  The current indoor environment should be documented through in-situ measurements and user surveys. The recommended monitoring period is one full calendar year. Indoor environmental data may involve the use of thermographic techniques, the installation of temperature, relative humidity, air movement, lighting and CO2 sensors in characteristic thermal zones.

### 2.2 HBIM MODELLING

BEEP methodology employs a dual building survey technique for creating a HBIM model at BIM Level 2. This is achieved by combining both traditional and innovative surveying techniques, i.e., topography, technical documentation and terrestrial laser scanning or photogrammetry. This collaborative documenting process can supply an accurate 3D point cloud with critical representation of its graphical rendering (points carrying RGB colour information) and a validating two-dimensional drawing file; together these two data types comprise the resources for the HBIM modelling initiation. This approach ensures the accuracy of the primary building structure orientation and scale, while maintaining important building details, such as artefacts or other decorative elements. In order to avoid time-consuming modelling processes, standardised modelling tools are used in the native BIM software. Mesh or irregular model geometry is not integrated. Modelling of building components characterising historic buildings, following conventional modelling tools of native BIM software could form the basis for establishing a rich database of intrinsic parametric BIM components contributing to international efforts on built heritage retrofitting.

### 2.3 ENERGY PERFORMANCE ASSESSMENT (B)

- Energy retrofit measures – Intervention objectives:

  Defining the objectives of intervention is of prime importance. Energy improvement measures can refer to alterations of the building envelope, energy supply and control, as well as, user's indoor occupancy and behaviour. In the framework of BEEP pilot actions, three scenarios will be assessed (short, middle and long term) with variable level of the scale of intervention, energy consumption reduction and financial requirements. Passive and active technologies are promoted, as long as they are compatible with international restoration policies.

- Dynamic energy simulation

  A whole-building energy model is used to estimate the energy performance and consumption of the proposed retrofit measures. Dynamic numerical simulations are performed, calibrated on energy bills, occupation patterns, and environmental monitoring.

- Assessment and selection of retrofit measures

  The assessment criteria for the selection of the retrofit measures is based on a risk-benefit scheme expanding to the categories of technical compatibility (e.g. hygrothermal and structural risks, reversibility), heritage significance, economic viability and energy savings potential.

## 2.4 EE-HBIM DEVELOPMENT

The data from the analysis of the existing building already in the HBIM platform, contain a series of input parameters for the dynamic energy performance simulation; i.e. properties of building materials, occupancy schedules, technical characteristics of systems and equipment etc. The rightful semantic organization of data and metadata in the EE-HBIM platform constitutes a major asset of the process, as this can become accessible to all involved professionals and building stakeholders at intra-business level. The Level of Detail (LoD) used in BEEP is L400. LoD 200 is applied in objects that have been severely damaged or removed, and their geometry is approximated based on historical documentation, i.e. historical photos.

Environmental and building performance evaluation through an interoperable workflow between the HBIM model and static or dynamic simulation tools is still at experimental stage. Energy Performance simulations enable the examination and optimisation of a historic building's performance, through the creation of behavioural models. These models however should only carry reduced building data and a building geometry simplified to a certain level of abstraction necessary to perform the simulation. In the pilot action presented here the exporting techniques that are being tested rely on IFC or gbXML data exchange schemas, which are also constantly under development. The BEEP project focuses on providing a functioning, semi-automatic data exchange process between the two steps, based on best practices and experimental feasibility tests.

## 2.5 ENRICHED EE-HBIM DEVELOPMENT (4D & 5D BIM IMPLEMENTATION)

At this stage, the operations execution schedules (4D) and the cost estimation (5D) of the three energy retrofit intervention scenarios are added to the BIM model. The cost of materials and products together with the work force and time estimation needed to retrofit the heritage building will be introduced to the BIM information model. These will secure the production of a trustful planning of the intervention.

## 2.6 DESIGN AND DEPLOYMENT OF FINANCING MECHANISMS (C)

- Extraction of critical BIM information to be used by financial institutes for EPC contracting.

  The aim at this stage is to find the information necessary to be exported from the BIM datasets to be used for EPC contracting. The information should be exported in Open format and should carry only the information which is crucial for financial institutes in evaluating the feasibility on signing an EPC for the building's energy refurbishment. Using this information, a WBS and a GANNT chart will be generated to further support the BIM preparation for EPC contracting.

- Technical Documentation of legal and technical aspects for EPC implementation.

  At this stage, the overall results will be analyses in order to define a common base for the evaluation of the ROI. This includes the analysis of the legal, economic and technical aspects in order to facilitate the process of approaching Energy Services Companies (ESCOS).

- EPC Contracting.

  The desired outcome of the proposed methodology is the financing of the refurbishment. In the framework of BEEP, a series of guidelines for both Public Administration and Financial Institute will be drafted. Two different guidelines, Strategic and Technical offers, are necessary as the decision makers (e.g. owners of the building) do not always have technical background to understand the EE-HBIM approach, while they need to be informed in a clear way about the great opportunities that this method brings to refurbishment mortgage market.

## 3. METHODOLOGY IMPLEMENTATION

The data collection and analysis regarding step *A: Building analysis and Documentation* of the presented methodology is fully implemented on the pilot case-study in Nicosia, Cyprus, while the activities of step *B: Energy Performance Assessment,* and the creation of EE-HBIM model are ongoing. Critical considerations and challenges faced so far regarding their implementation are presented below.

### 3.1 THE PILOT CASE STUDY BUILDING

The Cyprus pilot building is located just outside the walled city of Nicosia. The building hosted the British Cavalry club and was later used as the barracks of the Danish Canadian extract in Cyprus. It has been abandoned since the 1960s and is currently in dilapidated condition. It is a listed building and unique example of Cypriot architectural heritage, as it combines features of colonial architecture (1878-1929) and local rural architecture. It is rich in architectural features of the period, including fireplaces, arched openings, stone ornaments, courtyard and many details with significant historical value, such as the tall, angular stone turret that dominates the facade.

**Figure 3. The British Cavalry Club in Nicosia. Photographic archive from 1964 (left) and 2013 (right).**

### 3.2 MODELLING AND SIMULATION PROCESS

- A: Building survey & Documentation

  The H-BIM model development (Figure 4) included both 2D documentation of the complete georeferenced topographical and architectural survey and the generated 3D point cloud of the terrestrial photogrammetric survey, which have been linked to the BIM model. Although the level of accuracy of a Lidar system survey is higher compared to photogrammetric methods, the later was found as a more suitable solution, since it can capture narrow spaces and halls (common building attribute in heritage structures), while simultaneously supports colour registration to post-processing process for point cloud generation, and direct processing of image textures. The technique of photogrammetry was principally used to capture the entire building geometry, achieving an acceptable level of accuracy of the building's details, e.g. stone pediments, decorative stone profiles and artefacts, false ceiling decorative details, etc. The creation of a 3D point cloud model with reality capture tools accelerated the construction of various thematic maps, necessary for the conservation documentation and study of the building. All hosted model objects (families), such as doors, windows, stone ornaments and artefacts were parametrically designed to avoid mass modelling of similar objects. Severely damaged or removed objects were modelled as individual objects based on information collected during the step of the *Historical Significance Analysis*. The integration of the non-geometric information involved two separate categories of data: the information linked to the entire project model, i.e. conservation reports, and the information linked and assigned to particular project families/model objects, i.e. historical photos of damaged stone artefacts. For both data categories, open file formats were adopted.

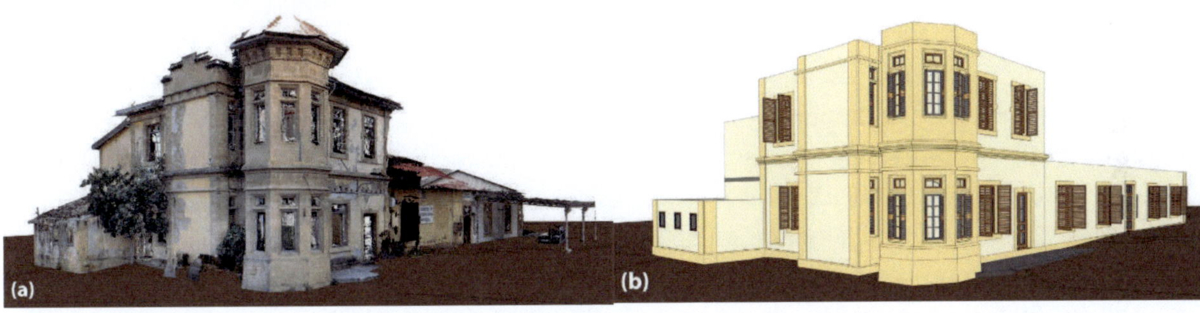


**Figure 4. EE-HBIM model development: a) reality capture point cloud model; b) BIM model.**

- B: Energy Performance Assessment

  As a number of questions regarding the compatibility of software interfaces, remain open for the scientific community, the authors' efforts focused on implementing the most efficient semi-automatic workflow, in order to streamline the process of BIM to BPS and avoid repeating modelling processes, or user errors during the exchange of data between the two software tools. The gbXML data exchange schema was adopted, while feasibility studies are currently conducted between the project partners to test the applicability of the approach on the pilot buildings selected. The complete implementation of the BEEP integrated EE-HBIM approach is estimated by the end of 2022. Custom tasks and processes are still under investigation and additional observations will be extracted.

- C: Design and Deployment of Financing Mechanisms

  The level of activity of energy auditors and ESPs in Cyprus is very low, despite the great potential for market development. The ESCO market penetration so far is considered to be at its initial stages. This may be due to a lack of confidence on the part of end users in the process and to a lack of know-how and experience on the part of ESPs and banks [9]. A preliminary assessment of the challenges in developing energy services in Cyprus has pointed to: the lack of information and awareness of the key actors and stakeholders, as well as of the public (property owners), institutional and legislative obstacles, financial obstacles and technical and administrative obstacles. In this environment, the implementation of BEEP is expected to benefit the market, as through its planned actions it will contribute to:

a) strengthening the current legal and institutional framework by removing obstacles to public procurement and making the recording of energy consumption in public buildings mandatory;
b) introducing new practices in the professional market, the benefits of which will be multiplied by the already secured engagement of local policy promoters, such as the Cyprus Energy Agency and the Scientific and Technical Chamber, as associate partners of BEEP;
c) promoting training and information by creating standard EPC forms and setting up an information platform for ESPs.

## 4. CONCLUSIONS

Building Information Modelling is increasingly recognised by the construction industry as a promising set of technologies. Built heritage is a great case to demonstrate the scalability and benefits of using BIM technology (in producing a H-BIM model) towards energy upgrading Europe's existing building stock. In this framework, the research project BEEP aims to create a comprehensive methodology for Energy Efficient HBIM, while supporting the financial decision-making through the enhancement of the Energy Performance Contracting process. The overall benefits focus on the promotion of digitalisation of energy refurbishments projects and the increasing of the volume, flow and access to financing mechanisms. The current implementation state of the integrated EE-HBIM approach in the Cypriot pilot building was presented, along with the encountered challenges. The central information model was used for incorporating data related to *A. Building Analysis and Documentation*, i.e. heritage significance information, building technical documentation (geometric survey) and environmental monitoring data. A semi-automatic iterative process regarding the incorporation of energy-related data was followed in stage *B. Energy Performance Assessment*. The activities regarding stage *C. Design and Deployment of Financing Mechanisms* are ongoing; along with the incorporation of time and cost related data, towards the creation of an enriched EE-HBIM model. Concluding, the paper presented how the BIM environment can become a valuable tool for the elaboration of all the geometrical and alphanumerical information for critical decision making during the retrofit scenario selection, and the financial evaluation of the intervention. Further investigation on the implementation of EE-HBIM is required in order to consolidate all the stages in the workflow.

## AKNOWLEDGEMENTS

BEEP: "Building Information Modelling for Energy Efficiency in the Public Sector" (2019-2021) is funded by ENI-CBC-MED Framework B.4.3. (2014-2020). http://www.enicbcmed.eu/projects/beep This paper has been produced with the financial assistance of the European Union under the ENI CBC Mediterranean Sea Basis Programme.

## 5. REFERENCES

[1] European Commission. (2019). The European Green Deal. Internet: https://eur-lex.europa.eu/legal-content/EN/TXT/?qid=1596443911913&uri=CELEX:52019DC0640#document2 [October 2020]
[2] B. Sodagar, "Sustainability Potentials of Housing Refurbishment," Buildings, vol. 3, (1), pp. 278-299, 2013.

[3] A. L. Webb, "Energy retrofits in historic and traditional buildings: A review of problems and methods," Renewable and Sustainable Energy Reviews, vol. 77, pp. 748-759, 2017. DOI: https://doi.org/10.1016/j.rser.2017.01.145.
[4] B. Succar, “Building information modelling framework: A research and delivery foundation for industry stakeholders,” Autom. Constr., vol. 18, no. 3, pp. 357–375, 2009, doi: 10.1016/j.autcon.2008.10.003.
[5] J. F. Tchouanguem Djuedja, M. H. Karray, B. K. Foguem, C. Magniont, and F. H. Abanda, Interoperability challenges in building information modelling (bim), vol. 9. Springer International Publishing, 2019.
[6] D. P. Pocobelli, J. Boehm, P. Bryan, J. Still, and J. Grau-Bové, “BIM for heritage science: a review,” Herit. Sci., vol. 6, no. 1, pp. 23–26, 2018, doi: 10.1186/s40494-018-0191-4.
[7] “BIM for Energy Efficiency in the Public sector (BEEP)" ENI CBC MED Research Project. [Online]. Available: http://www.enicbcmed.eu/projects/beep.
[8] Energy Efficient Mortgages Initiative. (2017). Internet: https://eemap.energyefficientmortgages.eu/ [November 2020]
[9] M. Economidou, P. Zangheri and D. Paci, “Long-term strategy for mobilizing investments for renovating cyprus national building stock.” Tech. Rep. D1.8, 2017.

# Reality capture technologies as a support for efficient energy diagnosis and simulation of heritage buildings: the case of the 'Alte Schäfflerei' in Benediktbeuern

S. Dubois, J. Desarnaud, M. de Bouw

Retrofitting and Heritage Laboratory, Belgian Building Research Institute, Limelette, Belgium.

**Abstract** – This paper proposes to study multiple 3D surveying techniques and evaluate their capacity to support the energy diagnosis of heritage buildings, with clear data processing workflows. The 'Alte Schäfflerei' of the Cloister of Benediktbeuern was chosen as a case study to support the analyses and was captured using a variety of airborne and terrestrial equipment.

**Keywords** – Photogrammetry, Structure From Motion, Building Energy Simulation, Laser scanning, Drones.

## 1. INTRODUCTION

Under no circumstances should the conservation of built heritage be synonymous with 'freezing' its condition. There is a strong risk that this would make it unsuitable for today's needs in terms of comfort and performance [1]. For buildings which were neglected during a long period, a clever transformation is sometimes a chance to regain interest. On the other hand, such actualization process cannot be done blindly. Standard energy optimization solutions can be detrimental to old materials [2], [3]. In Europe, many research efforts are ongoing on the energy retrofitting of old buildings with historical values. Here, the identification of adequate intervention strategies is crucial. Therefore, computer simulations are often used to explore widely differing combinations of retrofit measures. Calibrating whole building energy simulation tools remains challenging, given the amount and complexity of inputs [4]. The rapid and robust capture of data is thus here crucial to support modellers and scientific teams.

In previous research [5], multi-view photogrammetry (MVP) was proposed as a 'multi-purpose' tool for supporting the energy diagnosis of heritage buildings, not only to quickly collect meaningful data related to geometry but also to provide insights on materials and heritage values thanks to the detailed picture datasets created. And all of that can now be done at a relatively low investment cost. The method is non-destructive and disturbances when working with occupied buildings are also kept to a minimum. Depending on the type of photographic lens used and the typical capture distance, various scales of study can be considered, from material level up to site and even district level. Lastly, because the method is UAV-compatible, inaccessible or dangerous areas are also becoming easily diagnoseable. However, MVP also has some noticeable pitfalls. Many factors can affect the quality of the final 3D reconstruction. Among others, the protocol followed to capture the scene of interest has a major impact on the results. So do the textural properties of the object of interest. Nonetheless, historical buildings that have undergone little transformation seem good candidates for successful MVP surveys, whereas this technique appears less appropriate for modern, typically less-textured, buildings.

The accuracy of MVP for reconstructing Heritage buildings has been central in many researches [6]–[8]. Evaluating its performance in terms of geometric reconstruction fidelity often involves a comparison with the more robust terrestrial laser scanning approach. The strength of photogrammetry regarding textural reproduction is well documented but actual valorisation workflows for specific research domains, and for a large public, are still rare [9]. And yet, this technique may be a crucial ally when it comes to elaborate accessible capture strategies for the energy diagnosis of heritage buildings.

In the end, reality capture strategies are defined by the type of data to be extracted, so-called 'deliverables', and their foreseen final use. Hence, two main aspects should be considered when studying 3D surveying techniques: the creation of 3D assets (and other side datasets) and the processing of these assets. In this paper, a global view on the MVP potential for energy diagnosis and simulation applications is provided, through a practical case study where both photogrammetric and lidar surveying techniques were used. It is shown how the data stemming from different sources compares, and can be combined under various valorisation strategies without the necessary implementation of complex *heritage-BIM (HBIM)* models. A clear overview of the data process is provided and key deliverables linked to energy modelling tasks are identified. Because the data process can become complex and tortuous, some insights on automated data extraction schemes are also shown. Finally, reality capture strategies for heritage energy diagnosis are drafted. Those integrate contextual parameters that can influence the choice of one or several acquisition techniques.

## 2. CASE STUDY

### 2.1 DATA CAPTURE

The specific building chosen as case study here is shown in Fig. 1. The 'Alte Schäfflerei' is part of the former craftsmen's district in the Benediktbeuern monastery. This listed building dates to the second half of the 18th century and was originally used as storage space for barrels from the adjacent monastery brewery. The Fraunhofer IBP is putting the building to a new use by establishing the *Fraunhofer Center for Energy Efficient Building Renovation and Monument Preservation*.

The whole building was captured from the inside and from the outside, producing several datasets. A *DJI M210 V2 RTK* drone was used for aerial photography (Fig. 2), a *Leica RTC 360* laser scanner for terrestrial lidar acquisition and a *Sony a7r III* with various lenses for terrestrial photography. A *DJI X7* RGB camera was mounted on the drone, with a 24mm lens and a mechanical shutter. The drone was also equipped with a high accuracy RTK positioning system. A mobile GNSS base was used to provide the differential position data. Here is how the capture missions summarizes:

- Aerial imagery. The drone flights allowed to capture 311 usable pictures.
  - 59 were taken with the camera in 'top-down' position, following a grid pattern (**dataset A1**)
  - 252 were taken with the camera forming an angle between 30° and 60° from a horizontal position, following a 'perimeter' scheme around the building (**dataset A2**)
- Terrestrial photography.
  - Outside, 320 pictures were taken from the ground, following a 'perimeter-mission' pattern around the building. A 20mm lens was used for those wide-angle shots (**dataset T1**).
  - Inside the building, 2088 pictures were required to cover the entirety of accessible spaces. The 20mm lens was used for most pictures. A 12mm lens was useful for confined spaces. (**Dataset T2**).

- Terrestrial laser scanning.
  - Outside, 13 coloured scans were made around the building to provide sufficient overlap (**dataset S1**).
  - 82 scan positions were required on the inside (**dataset S2**). Dedicated registration targets were used to improve accuracy.
  - On all scanning positions, 360 panoramic pictures are generated to provide colour information to the point clouds (**dataset P1**).

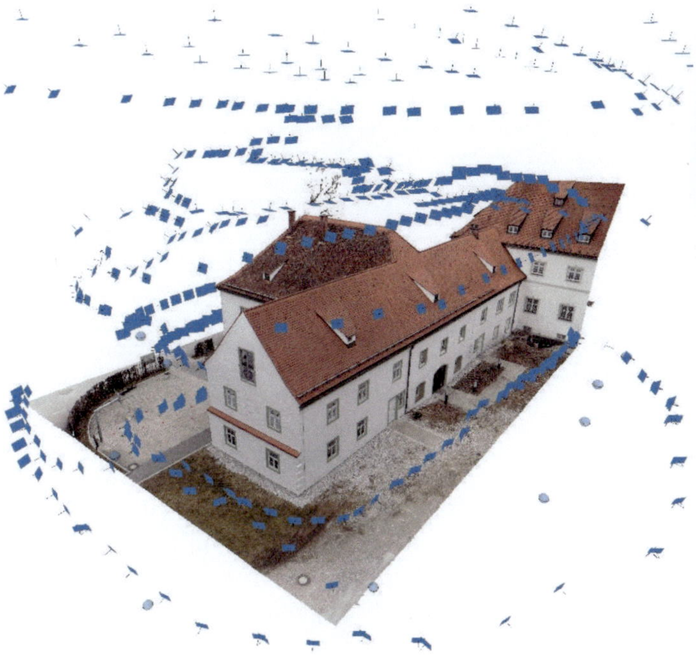

**Figure 1. Scanning the 'Alte Schäfflerei' of the Monastery of Benedikbeuern**

**Figure 2. The drone used for aerial photography**

## 2.2 DATA PROCESSING

Fig. 3 illustrates the data processing scheme followed for this holistic study and all the generated deliverables. It is a clear example on how captured data can be transformed according to many 'routes'. Each node of the diagram represent one specific type of data, which can be classified under a column that represents its nature (bidimensional, tridimensional, ...). A datatype node can have several inputs and several outputs, with some types being more 'transformable' than others. Naturally, this scheme constitutes only a part of what is possible – the diagram, despite its apparent complexity, is very simplified. Each data type could be further divided according to subtypes or according to the surveyed element (interior spaces, exterior, ...), for example. The actual processing stages from the 'Alte Schäfflerei' captured data are detailed in the following sections. Within this broad data transformation scheme, the focus is put on three main processing tools: the MVP software, the point cloud processing software and the image processing software.

### *2.2.1 3D reconstruction*

From the raw collected data (i.e. images, laser scans and mission metadata) the first processing stage consisted in creating high resolution 3D assets in the form of point clouds or meshes. Those are referred to as 'Level 1' deliverables. Such files are generally particularly heavy and their manipulation requires not only adequate hardware, but also specific technical knowledge.

Obtaining 3D assets is relatively straightforward when terrestrial laser scanning is used on site. The main task for the surveyor is to register the data. The 95 scans generated here were preregistered on-site using a SLAM technology embedded in the scanner. Later, the created links were optimized based on ground control points (solid targets). The possibilities in terms of 3D reconstruction with photogrammetric software were broader, especially given the variety and quantity of the collected data. Indeed, modern photogrammetry software solutions allow to automatically register laser scans and photos. Here, photo datasets were processed both with and without lidar datasets to assess the impact on reconstruction quality. *Agisoft Metashape*[1] and *Reality Capture*[2] *were used* using highest dense reconstruction quality settings. Both coloured point clouds and high resolution meshes were produced. When no laser dataset was used for photo alignment, ground control points were used to register the 3D reconstructions produced from images.

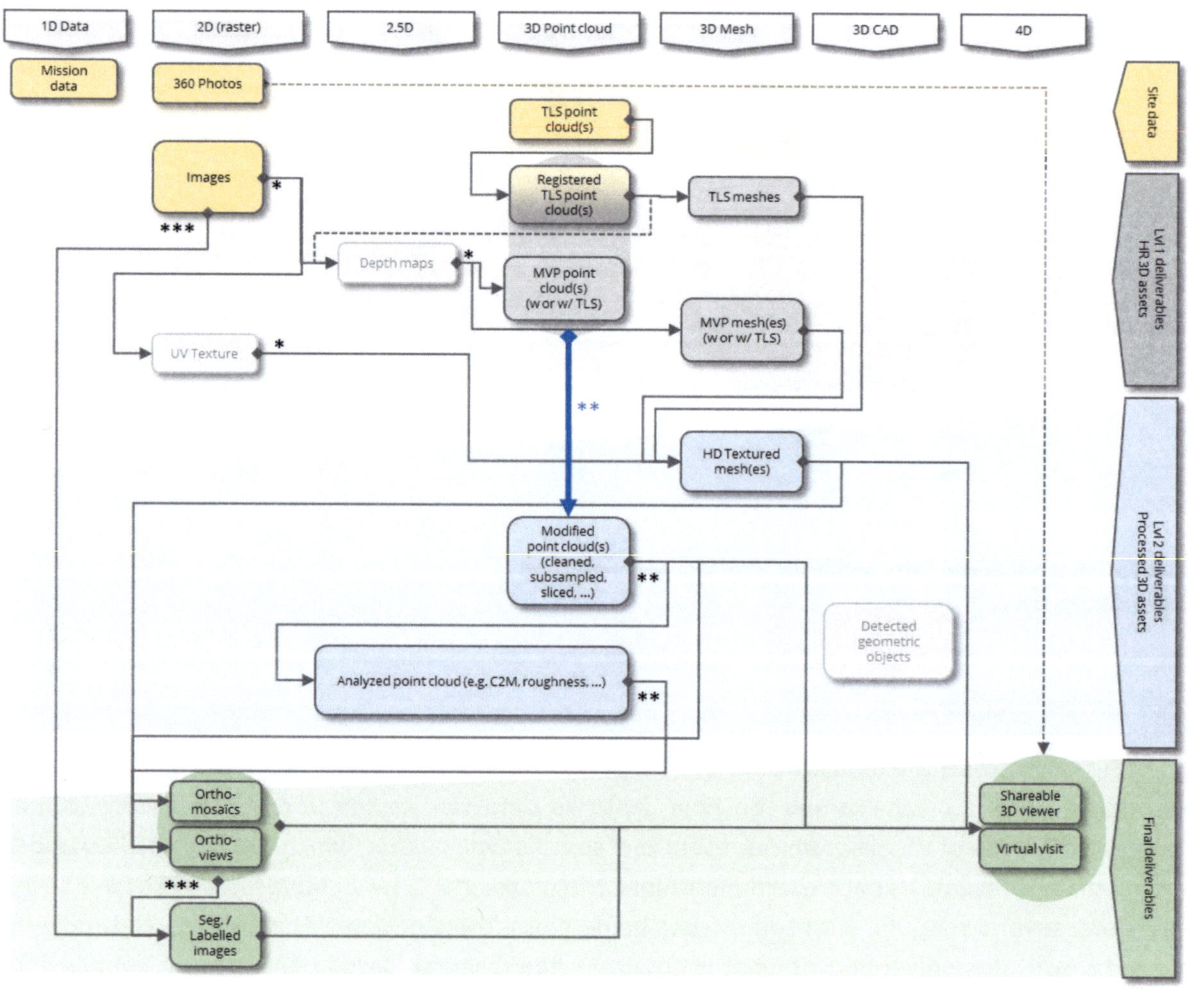


**Figure 3. Transformation of data types from one type to another: simplified view on how to valorise reality capture for energy diagnosis and simulation. The processes indicated with a * are generally performed in the photogrammetry software. Processes marked with ** are linked to the point cloud processing software. Processes marked with *** involve the use of image processing software.**

**Abbreviations: C2M = Cloud to Mesh, HD = High Definition, MVP = Multi-View Photogrammetry, TLS = Terrestrial Laser Scanning**

---

[1] Version 1.6.5

[2] Version 1.1

#### 2.2.2 *Processing 3D assets: making point clouds and meshes talk*

Processed 3D assets are referred here to as 'Level 2' deliverables. Such files would already be useful for energy diagnosis but are aimed towards people familiar with 3D technologies. For point clouds, the most basic processing steps consisted in cleaning, subsampling/resampling or slicing the datasets. Those actions do not add any information to the existing datasets. They rather serve the purpose of making the 3D files easier to manipulate or focusing on zones of interest. More complex processing actions involved computing additional scalar fields, i.e. computation. The simple and more advanced point cloud processing operations were all carried out in the open source software *CloudCompare*.

In their whole resolution state, the 3D meshes are particularly difficult to manage for the majority of visualization softwares. Decimation steps were required to make them more broadly exploitable. The number of polygons was reduced according to identified target softwares. For example, 3D PDFs required meshes with an extremely low amount of polygons. A particularly useful aspect of using meshes is here the possibility of reprojecting the high resolution colour information on low polygon density meshes, as shown in the results section on Fig. 5. The files are kept reasonably light while retaining interesting visual information. The decimation steps were performed in *Agisoft Metashape* or *Reality Capture*.

#### 2.2.3 *Formatting useful data from the prepared 3D files*

At a later stage, so-called 'final deliverables' were produced from the optimized 3D assets. Emphasis was placed on producing files that are usable for a larger public, and easily transferable. In order to evaluate the scanning technologies, many files were produced here for describing the 'Alte Schäfflerei' extensively. Table 1 provides a selection of the most important ones, with their potential use for energy diagnosis/simulation. Above, we have shown how data types can be transformed into each other. Yet, here, we show which particular dataset could be valued up to a specific use, in a common format.

Several remarks can be made. First, the useful information for energy diagnosis and simulation can be split into three main aspects: (1) the assessment of the geometry of the building, its environment and its subparts, (2) the identification/mapping of materials, components and systems and (3) the evaluation/mapping of the condition of the identified entities. Textural information is thus far from being neglectable, even crucial for aspects 2 and 3. Comparing 3D surveying technologies solely based on their geometrical accuracy would not cover the totality of relevant requirements. Secondly, images play a key role within the chosen end-user files. Indeed, thanks to the universality they retain, such deliverables ensure an effective communication between surveying teams and energy/heritage specialists – for geometric as well as for textural information (e.g. materials and pathologies). Images also constitute a widespread medium for performing advanced analyses like segmentation and labelling, within an image processing software. Especially with orthoviews, which add powerful quantification possibilities.

Naturally, choosing only images as communication medium will necessarily cause a loss of information. To avoid this, final image datasets were complemented by some 'immersive' deliverables, which allow anyone to manipulate 3D information. Firstly, *Potree* viewers were created to allow anyone to access the point cloud information, only using a web browser. Such WebGL solutions are simplifying the sharing of complex data. Later, a virtual visit application based on 360 photos navigation was implemented in *3DVista.* This appeared as a very satisfying solution to centralize, organize and contextualize all the generated deliverables.

**Table 1. Some of the chosen key final deliverables for energy diagnosis / simulation. Grey cells are further illustrated on Fig. 4.**

| Collected data | | Level 1 deliverable | | Level 2 deliverable | | Final deliverable | Main use |
|---|---|---|---|---|---|---|---|
| *TLS, interior (**S2**)* | → | *TLS, Point cloud of indoor spaces* | → | Distance maps of selected wall/floor/ceiling elements compared to reference planes | → | Orthoviews of the distance maps | Evaluating the **condition** of the wall/floor/ceiling element |
| *TLS, exterior (**S1**)* | → | *TLS, Point cloud of the envelope* | → | Distance maps of each façade compared to reference planes | → | Orthoviews of the distance maps | Evaluating the **condition** of fabric |
| *Photos, exterior (**A1 + T1**)*<br>*TLS, exterior (**S1**)* | → | *High poly mesh of the envelope* | → | High poly mesh of the envelope with RGB texture | → | Orthomosaïc photos of all façades and roof elements | Materials/pathologies **identification and mapping** (through image analysis and machine learning) |
| *Photos, exterior and interior (**A1 + A2 + T1 + T2**)*<br>*TLS, exterior and interior (**S1 + S2**)* | → | *MVP, Merged point cloud of the envelope and the interior spaces* | → | Horizontal sections of the point cloud at different levels | → | Plan of each storey | **Mapping** the internal organization of the building rooms |
| | | | → | 0.1m thick cross sections every 0.5m along the main building axes | → | Orthoviews of the sections | Encoding the building **geometry** and the thickness of envelope elements in whole-building energy models |
| *Photos, exterior (**A1 + T1**)* | → | *High poly mesh of the envelope* | → | Low poly mesh with RGB and normal map texture | → | 3D PDF of the mesh model | Materials/pathologies **identification** (direct observation) |
| Etc... | | | | | | | |

## 3. RESULTS

### 3.1 THE 'ALTE SCHÄFFLEREI', DIGITIZED IN DETAIL

As stated above, the quantity of generated data was extremely significant here. Only a general overview of produced files and performed analysis can be provided, along with some significant findings.

In *Agisoft*, A1, A2, T1 and T2 photos could be aligned without requiring any addition of manual tie points. The dense cloud reconstruction results show a high level of noise, which can be adequately filtered out using the 'confidence' level computation offered by the software. In Reality Capture, T1, T2, S1 and S2 datasets were aligned successfully. The sparse cloud created from aerial photosets (A1 & A2) could be aligned to this first dataset with the use of some manual tie points. Ultimately, the dense reconstruction from all those datasets resulted in a point cloud of 1.2 billion points. From this particular point cloud, Fig. 4 illustrates one of the specific and more relevant data workflow. It corresponds to the workflow which is greyed in Table 1. The final deliverables are here regularly spaces slices along all three axis and floor plans. Those orthoviews can be imported in the geometry modeler of a Building Energy Modelling (BEM) or Building Heat Air and Moisture (BHAM) software, such as the *Sketchup* plug-in for *WUFI iPlus* simulation.

With the density of point clouds that can be achieved with modern MVP software, it is expected to reach a certain level of visual realism. Fig. 5 shows how the different 3D datasets compared regarding this aspect. It can be seen that incorporating images into the 3D reconstruction process not only greatly improved the visual rendition of envelope elements but also reduced the missing data areas. Combining photographic and lidar acquisitions has made it possible to get the best of both worlds, namely high accuracy for geometrical reconstruction and fidelity in colour rendition.

The richness of visual data offered by MVP could be useful for many subsequent studies. However, the question of sharing of data remains critical, especially for huge point clouds. Working with meshes proved here to be really satisfactory to that end, as shown in Fig. 6.

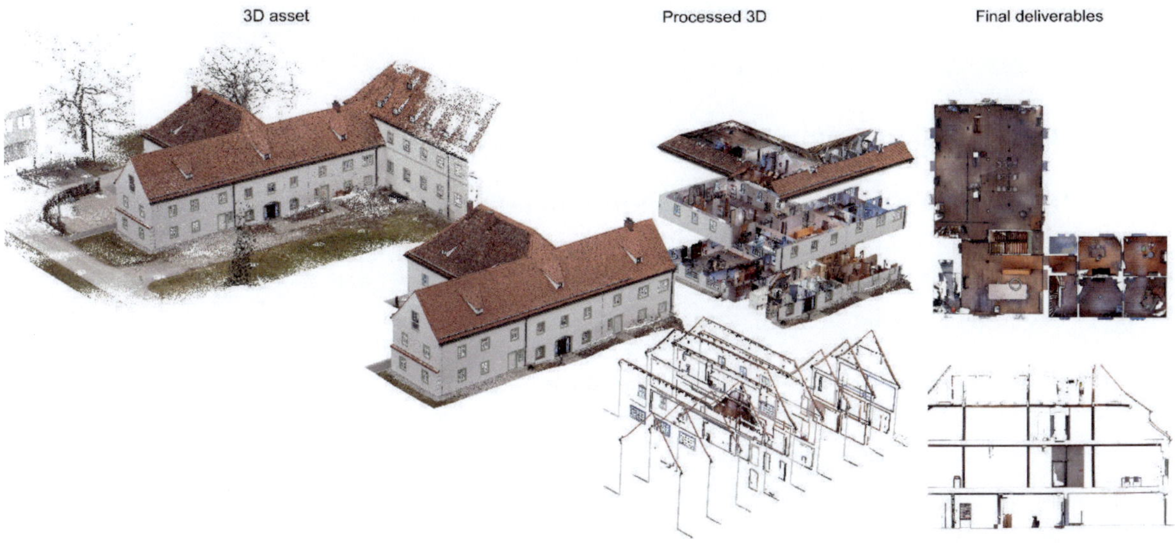


**Figure 4. An illustration of data processing: Merged point cloud of the envelope and the interior spaces processed into slices and then key orthoviews (images)**

**Figure 5. 3D reconstruction compared on a small interest zone. A: Laser scanning; B: Photogrammetry (only interior photos) from *Agisoft*; C: Photogrammetry (laser scans and interior photos registered together) from *Reality Capture***

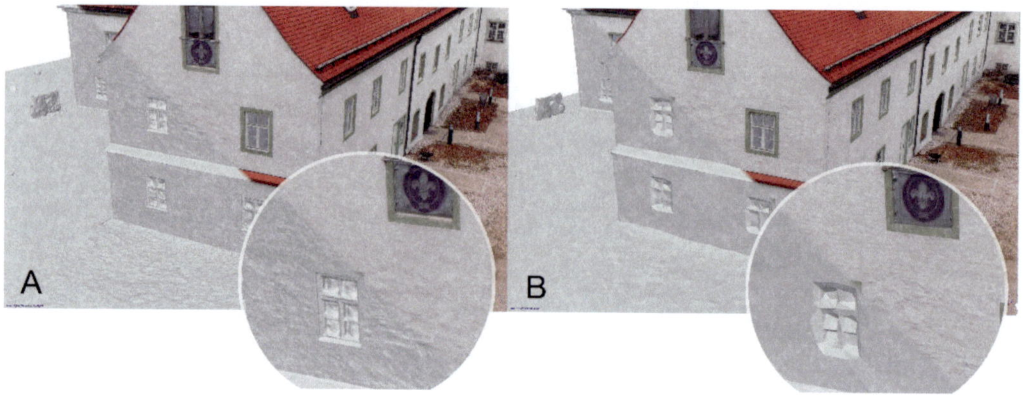


**Figure 6. Reprojection of colour information on low poly meshes**

### 3.2 REFLEXIONS ON 'REALITY CAPTURE' STRATEGIES FOR ENERGY DIAGNOSIS

Facing the many possibilities in terms of both data acquisition and data processing can become challenging, especially for the ones less familiar with modern scanning equipment, the opportunities they offer but also their intrinsic limits. To cope with that risk, we need to provide guidelines that would define adequate 'reality capture' strategies.

The actual final use, or uses, should always be the starting point when defining a 3D surveying mission. In turn, the needs will point towards adequate deliverables. The latter should be defined with clear specifications in terms of quality, such as accuracy or completeness of measurements. Depending on those specifications, a surveying technique, or a combination of techniques will be chosen. If only the building geometry matters and high measurement accuracy is sought after, then modern laser scanners might offer the ideal solution. If photorealism is a key aspect of the capture mission, then photogrammetry is an unavoidable step (see Fig. 7). Aerial surveying with drones offers a unique perspective for digitizing roofs or elevated surfaces (Fig. 8). However, using such technology may prove expensive or more cumbersome from an administrative point of view. Because beyond the technical considerations each energy diagnosis mission is also characterized by a specific socio-economic context: budget, building accessibility, time frame, or the locally available expertise are some of the aspects that will define the 3D survey specifications. A compromise might be necessary to define how to cope with operational or budgetary limitations. If so, priority deliverables must be put forward.

To summarize, the term 'reality capture strategy' covers multiple aspects: the definition of on-site surveying equipment, the elaboration of the acquisition plan, the processing of deliverables and the sharing/updating of data. A balance has to be found between the applicant expectations and the surveyor means. Table 2 provides an example of diagram that would allow evaluating scanning approaches for a specific mission. Providing such clear decision tools will be crucial in the future to encourage better retrofits thanks to better diagnosis and simulation campaigns.

**Table 2. Evaluating reality capture strategies according to quality and contextual parameters. '- -' = extremely poor, '-' = poor, '+' = satisfactory, '+ +' = excellent**

| Reality capture approach ↓ | Possibilities in terms of analyses and quality evaluation (completeness of data/relevance of data) | | | | | | | | | | | |
|---|---|---|---|---|---|---|---|---|---|---|---|---|
| | *Site* | | *Building* | | | *Envelope element* | | | | | | |
| | *Description* | *Dimensions of obstacles/ topography* | *Exterior dimensions* | *Typology* | *General diagnosis* | *Description of openings* | *Thickness* | *Surface materials identification* | *Damages and pathologies* | **Efforts for sharing of data** | **On site efforts** | **Off site efforts** |
| *Terrestrial images (low overlap), outdoor only* | -/+ | --/-- | -/- | + / ++ | - /+ | -/- | | - /+ | - /+ | ++ | ++ | ++ |
| *Terrestrial Panoramic images, indoor and outdoor* | +/+ | +/-- | +/-- | ++ / ++ | +/+ | +/- | | +/- | +/- | ++ | ++ | ++ |
| (...) | | | | | | | | | | | | |
| *TLS (no colour), indoor and outdoor* | +/- | +/+ | ++/++ | +/- | +/- | +/- | +/++ | +/-- | +/- | - | + | + |
| *TLS (with colour), indoor and outdoor* | +/+ | +/+ | ++/++ | +/++ | +/+ | +/+ | ++/++ | +/+ | +/+ | -- | - | + |
| (...) | | | | | | | | | | | | |
| *RGB TLS + indoor and outdoor photogrammetric photoset, including drone shots* | ++/++ | ++/++ | ++/++ | ++/++ | ++/++ | ++/++ | ++/++ | ++/++ | ++/++ | -- | -- | -- |

**Figure 7. Combining TLS and photos to maximize both geometrical accuracy and textural quality (illustration: mesh from combined photos and laser scans)**

**Figure 8. Drones allow detailed reproductions of roofs**

### 3.3 AUTOMATION OF DATA PROCESSING AND INFORMATION EXTRACTION

Valorising reality capture data is a time-consuming task. Although undoubtfully relevant, the many deliverables presented here are extremely demanding in terms of manual processing. It is thus natural to seek automation solutions. In this research, the importance of textural data was stressed. Indeed, MVP allows to produce detailed visual maps of many building components. Creating a large and organized set of orthoimages, which would provide a comprehensive visual dictionary of the building, appears as the ultimate goal. However, that seems hardly achievable using manual approaches.

Whereas state-of-the-art scan-to-CAD or scan-to-BIM undoubtedly have a bright future, the automatic transformation of complex building elements into geometric (or even semantic) objects is still in its initial stage [10]. At their stage of development, using them results in too much uncertainty. There is also a risk of oversimplifying or over-complexifying data for the energy modeller. Using simplified and robust algorithms wisely, on the other hand, could lead to significant improvements for the fast creation of orthoviews datasets.

In this exploratory research, the automatic detection of shapes was limited to using the RANSAC algorithm to detect geometric planes in the point clouds. These planes are relevant both as support for various analyses and for the segmentation of the building. CAD object were used only to process the 3D information but in the end images remain the main output of our automation efforts. A prototype app was created to perform five main operations (Fig. 9):

1) Performing a global RANSAC analysis and extracting the main constructive planes of the building as well as the points supporting those planes;
2) Analysing and labelling planes and their support points according to geometric (e.g. orientation, size, density) and textural (e.g. average colour of the support points, variation of colours) information;
3) Using the characteristics of the detected planes to provide semantic information to the point cloud, while segmenting it in large classes (points belonging to a specific storey or a specific façade, for example)
4) Generating key orthoviews (as listed in Table 1) automatically, using semantic information as support.
5) Proposing modern image segmentation/labelling approaches to extract information from the orthoviews

The application was developed in Python and relies intensively on *CloudCompare* command line interface and the *Open3D* library. The results are encouraging and show ways other than HBIM to enhance surveying data. In the future, it is planned to have each generated orthoview (step four) projected on its reference planar mesh object, which would be registered in space and exportable as a CAD file. By doing so, implementing the BEM or BHAM geometry would be made significantly easier. While particularly interesting, step five is still under development.

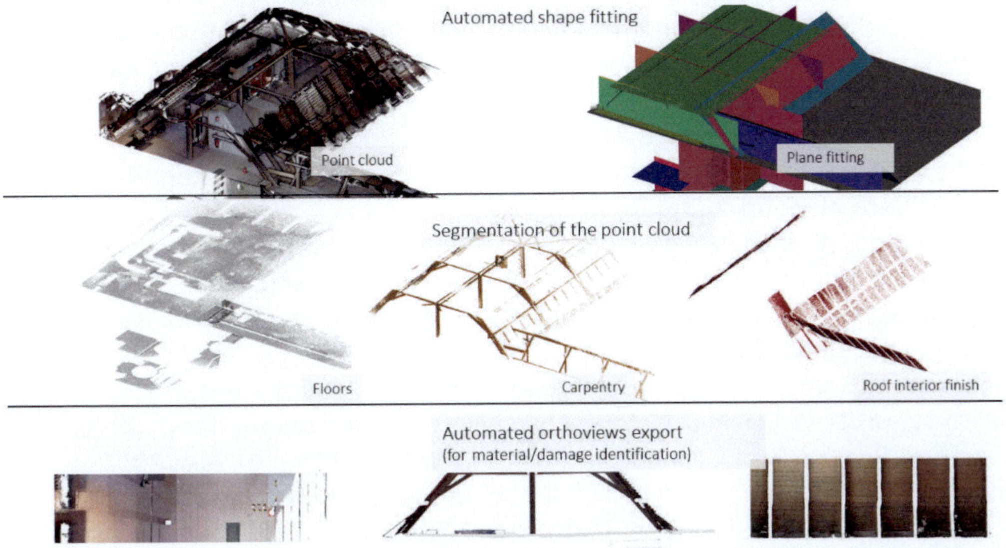


**Figure 9. Various automation processes implemented using CloudCompare CLI and Open3D library.**

## 4. CONCLUSIONS

The modern 'reality capture' technologies allow a better understanding of buildings before retrofits. For heritage buildings, the produced 3D data can improve and accelerate the implementation of dynamic simulation models. Not only do high definition surveys allow the collection of many details about the geometry of the building concerned, but also offer the possibility of linking accurate colour and radiometric information. Subsequently, these data can be processed in the form of files directly usable for energy simulation specialists. It includes image files, filtered point clouds, textured meshes or even virtual visit interfaces. Each type of file can be used at various stages of the energy modelling process, from initial geometrical domain creation to simulation results valorisation. An original automated data processing approach was proposed, with a focus on orthoimage datasets creation.

## AKNOWLEDGEMENTS

The authors would like to address special thanks to the *Fraunhofer Centre Benediktbeuern for Conservation and Energy Performance of Historic Buildings* of the Fraunhofer Institute for Building Physics IBP for its warm welcome and the fruitful collaboration offered within the scope of the EEHB2020 Conference.

## REFERENCES

[1] A. Martínez-Molina, I. Tort-Ausina, S. Cho, and J.-L. Vivancos. "Energy efficiency and thermal comfort in historic buildings: A review." Renew. Sustain. Energy Rev., vol. 61, pp. 70–85, 2016.

[2] Hisotric England (2011). "Energy Efficiency and Historic Buildings, Application of Part L of the Building Regulations to historic and traditionally constructed buildings." Internet: https://historicengland.org.uk/images-books/publications/energy-efficiency-historic-buildings-ptl/. [Dec. 11, 2020]

[3] A. Troi. Energy Efficiency Solutions for Historic Buildings – A Handbook. CH, Bâle: Birkhäuser, 2015.

[4] G. G. Akkurt et al., "Dynamic thermal and hygrometric simulation of historical buildings: Critical factors and possible solutions," Renew. Sustain. Energy Rev., vol. 118, p. 109509, 2020.

[5] S. Dubois et al. Contribution of photogrammetry and sensor networks to the energy diagnosis of occupied historical buildings. In Preventive Conservation - From Climate and Damage Monitoring to a Systemic and Integrated Approach, 1st ed., UK, London: Taylor & Francis, 2020.

[6] M. Bolognesi, A. Furini, V. Russo, A. Pellegrinelli, and P. Russo, "Accuracy of cultural heritage 3D models by RPAS and terrestrial photogrammetry," ISPRS - Int. Arch. Photogramm. Remote Sens. Spat. Inf. Sci., vol. XL–5, pp. 113–119, Jun. 2014.

[7] K. Themistocleous, C. Mettas, E. Evagorou, and D. G. Hadjimitsis, "The use of UAVs and photogrammetry for the documentation of cultural heritage monuments: the case study of the churches in Cyprus," in Earth Resources and Environmental Remote Sensing/GIS Applications X, Strasbourg, France, Oct. 2019, p. 18.

[8] A. Murtiyoso, M. Koehl, P. Grussenmeyer, and T. Freville, "Acquisition and processing protocols for UAV images: 3D Modeling of historical buildings using photogrammetry". ISPRS Ann. Photogramm. Remote Sens. Spat. Inf. Sci., vol. IV-2/W2, pp. 163–170, Aug. 2017

[9] H. Rahaman, E. Champion. "To 3D or Not 3D: Choosing a Photogrammetry Workflow for Cultural Heritage Groups." Heritage, vol. 2, p. 1835-1851, 2020.

[10] R. Volk, J. Stengel, and F. Schultmann, "Building Information Modeling (BIM) for existing buildings — Literature review and future needs," *Autom. Constr.*, vol. 38, pp. 109–127, 2014.

# Seismic Upgrade as an Opportunity for Energy Retrofitting Historic Unreinforced Masonry Buildings in Aotearoa New Zealand

P. Besen[1], P. Boarin[2]

[1] Huri te Ao School of Future Environments, Auckland University of Technology, & Future Cities Research Hub, University of Auckland, New Zealand, priscila.besen@aut.ac.nz.

[2] Future Cities Research Hub, School of Architecture and Planning, Faculty of Creative Arts and Industries, Waipapa Taumata Rau University of Auckland, New Zealand, p.boarin@auckland.ac.nz.

**Abstract** – Energy retrofitting historic buildings can help improve their indoor environmental quality, protect them from decay and obsolescence, reduce their energy use and related GHG emissions. However, in New Zealand, there are currently no policies to regulate energy retrofit in historic buildings and no substantial examples of this practice. On the other hand, there are significant regulations and practical examples of seismic retrofit, especially of unreinforced masonry (URM) buildings. As several seismic upgrades are taking place, this study explores the potential of applying energy retrofit concurrently with seismic strengthening. The research investigated three case studies, which are listed heritage URM buildings located in different climates in New Zealand. Their current performance was investigated, and retrofit scenarios were analysed through energy simulation. The potential energy savings from each intervention were balanced against their heritage impact. The study highlights the benefits of encouraging energy retrofit concurrently with seismic strengthening, so that historic buildings are more resilient not only to seismic threats, but also to a changing climate.

**Keywords** – Energy Retrofit; Unreinforced Masonry Buildings; Integrated Retrofit; Seismic Strengthening; New Zealand Heritage.

## 1. INTRODUCTION

Built heritage plays an important role in making history visible in Aotearoa New Zealand's cities and in creating vibrant and sustainable urban environments. The adaptation of historic buildings to current and future needs is of high importance, as the country has had countless examples of lost heritage [1] due to earthquakes, fire, lack of maintenance and decay related to inadequate indoor environmental conditions. New regulations came into force in 2017 for earthquake-prone structures, which set up timeframes for all buildings to achieve minimum structural standards [2]. As a result, there are many seismic strengthening projects taking place, especially in Unreinforced Masonry (URM) buildings – a historic type of construction identified as one of the most vulnerable in the country. However, the other future challenges for historic buildings will be to keep good levels of indoor comfort in the climate crisis and to minimise energy consumption [3,4]; these buildings shall be included in adaptation and mitigation programmes. So far, energy considerations have not been extensively included as parameters in retrofit projects in existing national policies or in practice, except for the mandatory upgrade of selected residential rental properties [5]. There are very limited data on their current energy performance – comprehensive studies have investigated energy use in existing buildings [6], but no in-depth studies focussed specifically on historic buildings. Few research projects have analysed deep energy retrofit of existing building fabric in Aotearoa [7] and there is a lack of information on retrofit options for heritage-listed buildings. Internationally, a few studies have explored the integration between energy and seismic retrofit in built heritage [8–11] and revealed the potential benefits of combining both interventions. To fill these gaps in knowledge, this research explored the opportunities of integrating energy retrofit and seismic strengthening as a way to safeguard historic URM buildings for future generations. The study analysed the current performance of selected buildings and possible scenarios integrating seismic and energy upgrades.

## 2. METHODOLOGY

To explore the possibilities of integrating energy retrofit with mandatory seismic upgrades in URM buildings, the research utilised case studies where quantitative and qualitative investigations were conducted. Three case study buildings were selected in different cities, one in each of New Zealand's climate zones (Table 1). All of them are University buildings, are built of URM construction and are listed in the national Heritage NZ Pouhere Taonga list [12]. Buildings A and B had already been seismically strengthened, while building C was in the design process for seismic upgrading. All buildings have single glazing and uninsulated masonry walls, uninsulated floors and partly insulated roofs.

**Table 1 – Overview of selected case study buildings**

| | **Case Study A** | **Case Study B** | **Case Study C** |
|---|---|---|---|
| |  |  |  |
| **Location and Latitude** | Auckland – 37° | Wellington – 41° | Dunedin – 46° |
| **Climate Zone** [13] | Zone 1 | Zone 2 | Zone 3 |
| **Heating Degree Hours** | 20 kKh/a | 42 kKh/a | 57 kKh/a |
| **Seismic Risk Area** [2] | Low Risk | High Risk | Low Risk |
| **Year of Construction** | 1904 | 1903-1904 | 1919-1920 |
| **Main architectural styles** | Italianate, Arts & Crafts | Gothic Revival, Edwardian | Gothic Revival |
| **Heritage NZ Listing** | Category 2 | Category 1 | Category 1 |
| **Seismic resistance capacity (before strengthening)** | 30% NBS (Earthquake-prone) | Earthquake-prone[1] | 10-15% NBS (Earthquake-prone) |
| **Seismic Strengthening Status** | Retrofitted in 2014-2016 | Retrofitted in 1990-1993 | To be retrofitted (currently at design stage) |
| **Main seismic strengthening systems** | Plywood diaphragms with tie rods | Sprayed concrete, steel portal frames | Post-tensioning systems |
| **Treated Floor Area** | 273$m^2$ | 5078$m^2$ | 1161$m^2$ |
| **Current Use** | Offices and meeting rooms | Offices, meeting rooms, lecture theatres | Offices, lecture theatres, laboratories |

[1] *Percentage of New Building Standard not shown because ratings were subject to different standards and classifications at the time of strengthening.*

Due to the limited knowledge about deep energy retrofit in New Zealand, the case studies were of exploratory and illustrative character. The use of three case studies allowed the exploration of the possibilities in energy retrofit and the evaluation of potential risks and benefits in different climatic contexts in New Zealand to a level of detail that provided reasonable accuracy for the study. The case studies were also of illustrative character, as they demonstrated how energy retrofit projects should be proposed and evaluated for URM buildings, taking into account heritage impact and seismic resilience considerations. The assessment of case study buildings and proposal of energy retrofit scenarios were guided by EN 16883 [14], which provides guidelines for improving the energy performance of historic buildings. As this standard does not include specific technical parameters and targets for energy retrofit, the study utilised the EnerPHit standard developed by the Passive House Institute [15] as a reference for specific technical requirements and methodologies. EnerPHit was selected because it provides a reliable, performance-based methodology to improve the energy performance and thermal comfort in existing buildings [16–18]. The study was structured into the following five main phases:

(A) Literature review investigating the URM building stock in New Zealand, current challenges and opportunities in integrating energy retrofit to seismic upgrade projects, presented by the authors in past publications [19,20];

(B) Selection and analysis of case study buildings, including an assessment of energy performance, technical, historical and indoor environmental factors, guided by EN 16883 [14];

(C) Hygrothermal simulation of the interventions that presented high interstitial condensation risks to determine suitable materials to be utilised in the proposed energy retrofit. Simulation was developed in WUFI Pro, a software which has been validated by detailed comparison with measurements obtained in laboratory and outdoor testing fields [21].

(D) Development of retrofit scenarios through energy simulation in the Passive House Planning Package (PHPP), assessing the possibility of achieving the EnerPHit standard through the energy demand method [15];

(E) Assessment of impact of retrofit scenarios on heritage conservation, based on EN16883, to understand how interventions would affect the building physically and its heritage significance [14].

This publication focuses on stages B-E. The energy retrofit scenarios proposed for each building (phase D) aimed to investigate the potential savings and benefits from upgrading the building envelope in conjunction with seismic retrofit works. Six progressive retrofit scenarios (Figure 1), ranging from the least invasive works to the most comprehensive upgrades, were identified based on extant literature and best practices [4,22–24].

**Figure 1 – Retrofit scenarios investigated for each case study**

Each scenario adds additional components in relation to the previous one. Scenario 1 represents the baseline, consisting of the original building before seismic strengthening or energy retrofit take place. Scenario 2 includes only seismic strengthening works, aiming to analyse how structural elements impact on the building performance, especially trough thermal bridging. Scenario 3 includes roof and floor insulation, which are some of the most common retrofit interventions in New Zealand, and the least invasive in terms of visual impact. Scenario 4 includes upgrades to windows through the addition of secondary glazing, and improvements to airtightness by sealing air leaks in windows and junctions. Scenario 6 includes the addition of a heat recovery mechanical ventilation system to the buildings. Scenario 6 includes the addition of wall insulation to the inside face of walls, while keeping the facades intact. The investigation aimed to test how the selected New Zealand case study buildings could achieve the EnerPHit standard through compliance with the criteria of the energy demand method.

Structural elements were considered as fixed factors in the analysis – information on the proposed seismic retrofit systems was collected from structural design drawings for each case study. Cost investigations were outside the scope of the study. Other limitations of the study were the inability to perform blower door tests and in-situ measurements of U-values – these factors were estimated based on the literature review and on the building plans and details.

## 3. RESULTS AND DISCUSSION

The energy audit revealed the main issues in energy performance and indoor environmental quality in the selected buildings. The energy consumption in case study C was 40% higher than new buildings with a similar use in the same institution. Thermal imaging revealed gaps in ceiling insulation leading to significant heat losses. An IEQ questionnaire with occupants showed that, in all locations, there was dissatisfaction with indoor temperatures in summer as well as in winter. One occupant in one of the buildings commented that "when the radiators haven't been on, the office is frigid. Mondays can be terrible, as it can take almost all day for the temperature to go up after being off over the weekend." Other comments were that "people bring in bar heaters (from home), both officially and unofficially." Personal heaters and fans were seen in the visual inspections of the buildings, indicating that the building fabric and the mechanical systems were not sufficient to ensure indoor comfort. Single-glazed windows, draughts and lack of insulation were indicated as some of the sources of discomfort for occupants. Even in the cold climate of Dunedin, over half of occupants were dissatisfied with summer indoor conditions.

Considering these demands, retrofit scenarios were proposed and tested based on EN16883 and EnerPHit. Figure 2 illustrates the results for case study C, and similar patterns were found in the other case studies. Significant reductions of heating demand were identified through the energy models when comparing the most comprehensive retrofit scenario with the baseline: 92% reduction in Dunedin, 91.6% in Wellington and 89.9% in Auckland. The only retrofit package to achieve the EnerPHit standard in the simulation was scenario 6, the most comprehensive one – it included roof, floor and wall insulation, as well as upgraded windows through secondary glazing, airtightness and heat recovery ventilation. It is also worth mentioning that scenario 5, a relatively less invasive package without wall insulation, also achieved significant savings in heating demand. Frequency of overheating was very low in case study C, under 1%, in all scenarios. In case study A, located in the warmer Auckland climate, frequency of overheating reduced from 9.6% in retrofit scenario 1 to 6.3% in retrofit scenario 6.

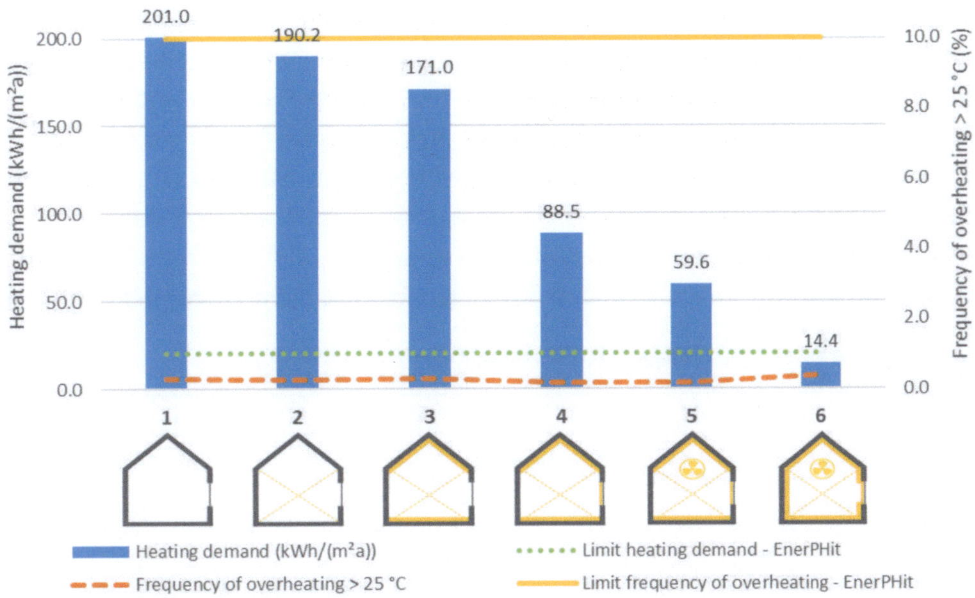


**Figure 2 – Retrofit scenarios: heating demand and frequency of overheating results for case study C**

Seismic strengthening can impact energy performance positively or negatively, depending on the systems utilised. It was found that, in case study A, the use of plywood diaphragms in the ceiling can be useful to improve airtightness, if appropriately taped in junctions. On the other hand, systems utilising steel beams need to be carefully designed with thermal breaks to minimise thermal bridging. However, in many cases, steel elements were located on the outer layers of the building envelope, not in direct contact with insulation layers, thus the potential thermal bridging was not very significant in these circumstances. External post-tensioning systems, which were considered as an option by the design team for case study C, would be a beneficial technique to avoid the contact between highly conductive materials with internal insulation layers. The use of sprayed concrete in case study B can negatively impact the possibility of adding internal insulation, as internal spaces are already reduced with concrete by itself.

Considering the risks of introducing internal insulation, a range of different materials were investigated for the study. Overall, perlite boards and calcium silicate boards achieved the best results in hygrothermal performance, due to their capillary-active properties. In case study C, it was found that mineral wool, wood fibre and cellulose would create significant moisture issues, due to the wall assembly configuration with stone cladding and its reduced drying capacity.

The principle of minimal intervention [25] was applied for building elements with high historic significance. This was the case in the Council Chamber room in case study B, where the presence of heritage significant stained glass windows, decorated ceilings and walls led to the proposal of excluding them from energy retrofit interventions. EnerPHit already provides exemptions for heat transfer coefficients of the exterior envelope components to be exceeded if required by heritage authorities [15]. However, given the heritage significance and the less frequent use of this room, a better solution could be to exclude this space completely from the thermal envelope considered for EnerPHit certification. Although this could bring challenges in separating the room from the treated building envelope, it was found as an appropriate solution in this case of high heritage impact. For an appropriate application of the EnerPHit standard in historic buildings, this type of exemption needs to be considered for further development of the standard in order to allow more flexibility for heritage buildings.

To assess the impact of retrofit scenarios on heritage conservation, the study utilised the tabular risk-benefit scheme by EN16883 to identify the most effective measures and eliminate inappropriate interventions. Two categories from the standard – economic viability and impact on the outdoor environment – were not included in this study as they were outside of its scope. As the focus was on the relationship between energy and seismic upgrades, a new category was proposed: integration of retrofit solutions with seismic strengthening, to assess the applicability of each scenario concurrently with seismic strengthening. Table 2 illustrates the assessment for each retrofit scenario in Case Study C: high risk (red), low risk (yellow), neutral (grey), low benefit (light green), high benefits (dark green).

**Table 2 – Assessment of impacts based on EN 16883 (2017) – case study C**

| Assessment categories | Assessment criteria | Retrofit Scenarios | | | | |
|---|---|---|---|---|---|---|
| | | 2 | 3 | 4 | 5 | 6 |
| Technical compatibility | Hygrothermal risks | | | | | |
| | Structural risks | | | | | |
| | Corrosion risks | | | | | |
| | Salt reaction risks | | | | | |
| | Biological risks | | | | | |
| | Reversibility | | | | | |
| Heritage significance of the building and its settings | Risk of material impact | | | | | |
| | Risk of visual impact | | | | | |
| | Risk of spatial impact | | | | | |
| Energy | Energy performance and operational energy demand | | | | | |
| Indoor Environmental Quality | Indoor environmental conditions suitable for building fabric preservation | | | | | |
| | Indoor environmental conditions suitable for achieving good occupant comfort levels | | | | | |
| Aspects of use | Influence on the use and the users | | | | | |
| | Ability of building users to manage and operate control systems | | | | | |
| Integration of retrofit solutions with seismic strengthening[1] | Compatibility with proposed seismic strengthening systems | 2 | | | | |
| | Access allowed by seismic strengthening (i.e. fabric already affected by interventions) | 2 | | | | |

[1]Proposed new category of assessment, specific to buildings subject to combined energy and seismic retrofit.
[2]Not applicable, as retrofit scenario 2 considers seismic strengthening only.

All retrofit strategies were assessed in terms of their impact on heritage fabric, according to the scale proposed by EN 16883, which includes criteria such as potential reversibility and compatibility. The intervention with most benefits in this scale was the upgrade of windows with secondary glazing, and the intervention with highest risks was wall insulation. Wall insulation was especially problematic in case study B, where seismic retrofit with sprayed concrete had already created a visual impact in internal walls and depth of window reveals, and further increasing wall thickness can be problematic, leading to further reductions in internal space. This analysis proved to be a useful tool to ensure a holistic approach in the retrofit decision-making process.

## 4. CONCLUSIONS

URM buildings are valuable pieces of New Zealand's historic heritage and provide multiple benefits to their surroundings and communities. Their seismic performance issues are well-known in Aotearoa, and this study revealed their energy performance challenges through an energy audit and IEQ assessment of case studies. Although the selected buildings had high energy consumption, their indoor environments were not comfortable: occupants demonstrated significant dissatisfaction with temperatures both in winter and in summer. Without retrofitting these buildings, these issues are likely to be accentuated with more frequent temperature extremes due to climate change.

Overall, the study demonstrated that it is possible to improve the building envelope in URM construction and achieve significant energy savings by utilising the EnerPHit standard in a sensible way, if applied in conjunction with the methodology proposed by EN 16883. The energy models demonstrated a reduction of up to 92% in heating demand when comparing the most comprehensive retrofit scenario (package 6) with the baseline in the coldest climate studied. However, the assessment based on EN 16883 showed that this retrofit package would also lead to significant impacts on heritage fabric. Hygrothermal simulation was performed to assess the addition of wall insulation in this scenario. It showed that materials such as perlite boards and calcium silicate boards achieved the best results in hygrothermal performance, due to their capillary-active properties. Retrofit scenario 5, without wall insulation, provided a good balance between energy performance and heritage impact, although it would not achieve the EnerPHit standard. The frequency of overheating was also significantly reduced in the warmer climate of Auckland in retrofit scenarios 4, 5 and 6.

The analysis highlighted how the retrofit of URM buildings should be guided by a holistic approach that encompasses seismic, energy efficiency and heritage conservation considerations, among many other disciplines. This balancing act can be complex, especially in projects with very limited budgets, Therefore, the use of a thorough assessment framework can be helpful to guide project teams make informed decisions for each unique building. The presented method can be replicated to other URM buildings in New Zealand and other countries with similar building stocks and similar demands in terms of seismic risks, energy efficiency standards and indoor environmental quality demand. Overall, the research demonstrated that current seismic upgrade projects can be an opportunity to integrate energy improvements to historic URM buildings through sensitive interventions to the heritage fabric. This integrated approach can help improve the resilience of these valuable buildings and ensure they can continue to serve a useful purpose in a post-carbon future.

## 5. ACKNOWLEDGEMENTS

We would like to thank PhD co-supervisor Prof. Errol Haarhoff, external advisor Dr. Kara Rosemeier, the staff at universities who manage the case study buildings, the design teams who shared information about the projects, and all the building occupants who participated in our surveys.

## 6. REFERENCES

[1] R. Wolfe, New Zealand's Lost Heritage. Auckland, New Zealand: New Holland Publishers, 2013.

[2] Ministry of Business Innovation & Employment. Building (Earthquake-prone Buildings) Amendment Act. Wellington, New Zealand: New Zealand Government, 2016.
[3] S. Roaf, D. Crichton, F. Nicol. Adapting Buildings and Cities for Climate Change: a 21st Century Survival Guide. Oxford, UK: Elsevier Architectural Press, 2009.
[4] A. Troi, Z. Bastian. Energy Efficiency Solutions for Historic Buildings. A Handbook. Basel, Switzerland: Birkhäuser, 2015.
[5] New Zealand Government. Residential Tenancies (Healthy Homes Standards) Regulations 2019. Wellington, New Zealand: New Zealand Government, 2019.
[6] L. Amitrano, N. Isaacs, K. Saville-Smith, M. Donn, M. Camilleri, A. Pollard, M. Babylon, R. Bishop, J. Roberti, L. Burrough, P. Au, L. Bint, J. Jowett, A. Hills, S. Cory. BEES Part 1: Final Report Building Energy End-use Study SR 297/1. Judgeford, New Zealnd: BRANZ, 2014.
[7] P. Leardini, M. Manfredini, M. Callau. "Energy upgrade to Passive House standard for historic public housing in New Zealand". Energy and Buildings, vol. 95, pp. 211–218, 2015.
[8] G.M. Calvi, L. Sousa, C. Ruggeri. "Energy Efficiency and Seismic Resilience: A Common Approach". Multi-Hazard Approaches to Civil Infrastructure Engineering, pp. 165–208, 2016.
[9] I. Nardi, T. de Rubeis, M. Taddei, D. Ambrosini, S. Sfarra. "The energy efficiency challenge for a historical building undergone to seismic and energy refurbishment" Energy Procedia, vol. 133, pp. 231–242, 2017.
[10] E. Negro, M. D'Amato, N. Cardinale. "Non-invasive Methods for Energy and Seismic Retrofit in Historical Building in Italy" Frontiers in Built Environment, vol. 5 (125), 2019.
[11] P. la Greca, G. Margani. "Seismic and energy renovation measures for sustainable cities: A critical analysis of the Italian scenario". Sustainability (Switzerland), vol. 10 (254), 2018.
[12] Heritage New Zealand Pouhere Taonga. "The Heritage New Zealand Pouhere Taonga List". Available: https://www.heritage.org.nz/the-list [March 9, 2021].
[13] Standards New Zealand. NZS 4218: 2009 Thermal Insulation – Housing and Small Buildings. Wellington, New Zealand: NZS, 2009.
[14] European Committee for Standardisation. EN 16883:2017 Conservation of cultural heritage — Guidelines for improving the energy performance of historic buildings, 2017.
[15] Passive House Institute. Criteria for the Passive House, EnerPHit and PHI Low Energy Building Standard (2016). Available: https://passipedia.org/_media/picopen/9f_160815_phi_building_criteria_en.pdf [April 10, 2021].
[16] C. Jimenez-Bescos, R. Prewett. "Monitoring IAQ and thermal comfort in a conservation area low energy retrofit". Energy Procedia, vol. 147, pp. 195–201, 2018.
[17] F. Moran, T. Blight, S. Natarajan, A. Shea. "The use of Passive House Planning Package to reduce energy use and CO2 emissions in historic dwellings". Energy & Buildings, vol. 75 pp. 216–227, 2014.
[18] F. Rodrigues, M. Parada, R. Vicente, R. Oliveira, A. Alves. "High energy efficiency retrofits in Portugal". Energy Procedia, vol. 83 pp. 187–196, 2015.
[19] P. Besen, P. Boarin. "Energy retrofit of historic buildings in Aotearoa New Zealand: current practice, challenges and opportunities". Imaginable Futures: Design Thinking, and the Scientific Method, 54th International Conference of the Architectural Science Association 2020. Auckland, New Zealand: Auckland University of Technology, pp. 471–480, 2020.
[20] P. Besen, P. Boarin, E. Haarhoff. "Energy and Seismic Retrofit of Historic Buildings in New Zealand: Reflections on Current Policies and Practice". Historic Environment: Policy and Practice, vol. 11, pp. 91–117, 2020.
[21] Fraunhofer Institute for Building Physics IBP, What is WUFI®?, WUFI. (2020). Available: https://wufi.de/en/software/what-is-wufi/ [May 20, 2020].
[22] V. Belpoliti, G. Bizzarri, P. Boarin, M. Calzolari, P. Davoli. "A parametric method to assess the energy performance of historical urban settlements. Evaluation of the current energy performance and simulation of retrofit strategies for an Italian case study". Journal of Cultural Heritage, vol. 30, pp. 155–167, 2018.
[23] P. de Berardinis, M. Rotilio, C. Marchionni, A. Friedman. "Improving the energy-efficiency of historic masonry buildings. A case study: A minor centre in the Abruzzo region, Italy". Energy and Buildings, vol. 80, pp. 415–423, 2014.
[24] K. Fabbri, L. Tronchin, V. Tarabusi. "Energy Retrofit and Economic Evaluation Priorities Applied at an Italian Case Study". Energy Procedia, vol. 45, pp. 379–384, 2014.
[25] ICOMOS New Zealand. "ICOMOS New Zealand Charter for the Conservation of Places of Cultural Heritage Value", 2010. Available: https://icomos.org.nz/wp-content/uploads/2020/12/NZ_Charter.pdf [March 9, 2021].

# Legal and Social Challenges to a Fossil-Free Historic Built Environment: A Comparative Review of Laws and Policies from Sweden and India

H.V. Bhati[1], M. Geijer[2], T. Broström,[2,] Y. Epstein[1]

[1] Dept. of Law, Uppsala University, Uppsala and Gotland, Sweden, harsh.bhati@jur.uu.se; yaffa.epstein@jur.uu.se.

[2] Dept. of Art History (Conservation), Uppsala University, Uppsala and Gotland, Sweden, mia.geijer@konstvet.uu.se; tor.brostom@konstvet.uu.se.

**Abstract** – This conference paper presents an ongoing project that uses the Hanseatic town of Visby, Gotland and the Jaipur Walled city as case studies to examine how the political goals of protecting cultural heritage and combating climate change are implemented in law and in practice. It provides a brief review of laws and policies concerning sustainable energy use in historic or culturally valuable buildings in Sweden and India.

**Keywords** – European Union Law; Sweden-India; Sustainable energy use; Historic preservation; World Heritage; cultural heritage.

## 1. INTRODUCTION

In this conference paper, we briefly review some of the most significant heritage protection and sustainable energy laws and policies in Sweden (including laws of the European Union (EU) that impact Swedish heritage protection) and India to present an overview of the current legal situation pertaining to measures for sustainable energy use in culturally valuable buildings. By 'sustainable energy laws', we mean both measures to replace the use of fossil fuels with renewable energy, and measures intended to reduce energy use. By 'culturally valuable' or 'historic' buildings, we mean buildings that have been protected by law in some way as historically or culturally valuable. This paper is part of the ongoing project 'Väga Rätt?' [1], in which we investigate whether and how legal systems balance the goals of sustainable energy use and historic preservation in Sweden and India in order to understand the role of law in achieving greenhouse gas (GHG) emission and energy targets in historic districts. In this paper's final section, we present some preliminary conclusions and planned future research.

### 1.1 BACKGROUND

Carl Elefante [2], former president of the American Institute of Architects, said 'we cannot build our way to sustainability; we must conserve our way to it'. There is a common misconception that historic preservation and environmental sustainability cannot go hand in hand [3]. In many cases, effective solutions are available to reduce energy use or expand the use of renewable energy in culturally valuable buildings [4-10].

Without energy retrofits however, old and heritage buildings generally consume more energy than modern buildings. The average heating energy demand of a building in Europe built after 1990 is around 40 kWh/m2 while buildings built before 1930 in average demand 170 kWh/m2 [11]. Therefore, improvement in energy efficiency, or transitioning to renewable energy sources, provides an opportunity to reduce GHG emissions from existing building stock. According to 2019 report by the World Green Building Council, buildings account for 39 percent of energy related global carbon dioxide emissions: 28 percent from operation emissions (energy needed to cool, heat and power them) and 11 percent from material and construction *i.e.*, embodied carbon [12]. Existing buildings account for 41 percent of energy consumption and 36 percent of carbon dioxide emissions in the EU [13]. At present, nearly 35 percent of the EU's buildings are over 50 years old, and almost 75 percent of these are energy inefficient [14]. Old buildings, therefore, represent a substantial part of the EU building stock and improving their energy sustainability can play a vital role in advancing energy sustainability in cities and while paving a path towards sustainable living heritage [15].

Culturally valuable buildings are a non-renewable resource which not only contain embodied energy and carbon but also transmit the spirit and identity of a country from one generation to the next [11]. Retrofitted, these buildings can not only be more comfortable for those who use the buildings, but also reduce energy demand [11]. Moreover, preserving existing structures avoids the extensive use of energy that would go into demolishing and rebuilding, sends less demolition waste to landfills and 'capitalizes on traditional energy efficient building materials and techniques' [16, pp. 290]. Energy retrofits of historic buildings therefore contribute both to reducing energy use and GHG emissions, and to preserving their heritage values for the future generations while enabling the current generations to more comfortably enjoy them [17].

In the next section of this paper, we describe the laws in place that impact measures for sustainable energy use in culturally valuable buildings in Sweden and India. Because EU law is essential to understanding Swedish law, we start with a brief overview of EU laws in this area.

## 2. REGIONAL STUDIES

### 2.1 THE EUROPEAN UNION

Although the EU surpassed its 2020 emission reduction targets of 20 percent, it still requires resilient and realistic pathways to achieve a carbon neutral society by 2050 [18]. The climate and energy framework goals for 2030 set a target of a 40 percent reduction in GHG emissions compared with 1990 as well as 32 percent share of renewable energy and at least 32.5 percent improvement in energy efficiency targets by 2030 [19]. The EU and its Member States have ratified the Paris Agreement [20]. The 2030 framework in part aims to meet the EU's responsibilities according to the Paris Agreement and is consistent with the longer-term objective of the '2050 low-carbon economy roadmap', which sets the EU ambition to reduce its GHG emissions by 80 percent compared with 1990 level [21].

The EU is committed to developing a sustainable, competitive, secure and decarbonised energy system. The Energy Performance of Building Directive (EPBD) plays an essential role. The first EPBD directive came in 2002, followed by amendments in 2010 and 2018 [22-24]. The objective of the

EPBD directive is to 'promote the improvement of energy performance of buildings within the Union, taking into account outdoor climatic and local conditions, as well as indoor climate requirements and cost-effectiveness' [24].

Article 4(1) of EPBD Directive sets a requirement for Member States to take the necessary measures to ensure that minimum energy performance requirements are set for existing and new building stock with a view to achieving cost-optimal levels. However, Article 4(2) provides discretion to Member States, which may or may not decide to apply Article 4(1) on officially protected buildings or buildings with special architectural or historical merit, to the extent that compliance with minimum energy performance requirements would 'unacceptably' alter their character or appearance [24]. Article 8 requires Member States to apply system requirements to existing buildings for the purpose of optimising the energy use of technical building systems [24-25]. Furthermore, Article 4(1) requires that Member States set minimum energy performance requirements which are cost-effective over the estimated economic lifecycle [24]. Article 10 and 11 also requires Member States to establish a system of certification of the energy performance of buildings and provide appropriate financial incentives to catalyse the energy performance of buildings and the transition to nearly-zero emission buildings [24]. The EPBD Recital 18 further encourages research and testing for 'improving the energy performance in historical buildings and sites, while also safeguarding and preserving cultural heritage' [24].

In December 2021, the European Commission adopted a legislative proposal to revise the EPBD, as part of a broader overhaul of EU climate and energy legislation referred to as the 'Fit for 55' package [26]. This revision aims to accelerate use of renewable energy in building renovations, reduce GHG emissions and energy consumption and support better air quality and digitalization of energy systems to make them more resilient and accessible [26].

Although EU has a huge number of historic buildings, the majority are still regarded as energy inefficient. While historic buildings and cultural heritage are protected under international and regional treaties in EU [27], the problem is that that EU has no specific directives when it comes to preservation of cultural heritage and historic built environment [9]. However, it should also be noted that although there is EPBD and Energy performance directives within EU, there is no specific international treaty regarding sustainable energy use, energy efficiency and minimum energy performance requirements in historic buildings or for existing buildings. A commonly held position in many EU nations is that heritage buildings, particularly those under listed buildings, should be exempt from having to be equipped with new energy retrofit measures [28]. Many old heritage buildings in EU have however already gone through changes to allow 'running water, centralised space heating, cooling, ventilation, electricity and telecommunication networks' [28]. Without these changes, many of these old heritage buildings would already be considered unusable today [28]. Similarly, energy sustainability measures can contribute to the conservation of culturally valuable buildings in the long term. With the proper consideration and involvement of relevant stakeholders, it is in many cases possible to introduce energy efficient measures, as well as to transition to the use of renewable energy sources, in historic buildings while keeping their characteristics and values intact.

The impact of legal barriers faced by owners of historic buildings while communicating and applying for a permit for alteration/refurbishment/retrofit in Member-states of EU is not yet well-defined and documented. In the next section, we review how these EU laws, and other international agreements, are implemented in Sweden.

## 2.2 SWEDEN

Heritage preservation and conservation laws in Sweden have roots in the late 17th century. The royal Placat of 1666 issued by the governing council under the minority of King Charles XI of Sweden is often hailed as the first antiquities legislation in Europe [29, pp. 1]. In Sweden, the current legal paradigm for heritage preservation was partly influenced by the widespread demolition of city centres between 1945 and the early 1970s to make way for modern buildings and urban renewal [30-31]. The concept of *sanering* [32] was extensively used as an argument for demolishing 'outdated' housing units and to promote urban renewal [33]. However, international influences also increased the attention paid to how people related to their living environment, for instance the Council of Europe's declaration of 1975 as the European Architectural Year [33], and the ratification of the Convention concerning the Protection of the World Cultural and Natural Heritage in 1972 [34] and the Amsterdam Declaration of 1975 [34]. Sweden ratified the 1972 World Heritage convention in January 1985 and fifteen sites are inscribed on the World Heritage list as of 2022 [36].

The oil crisis of the mid 1970s pushed Sweden to seek alternative energy sources, and a program was launched to reduce energy demand in the building sector by 25-30 percent over the course of a decade [37]. District heating was promoted by the official authorities and the government subsidies were provided to replace oil with other energy sources [38]. As electricity was relatively cheap in Sweden, electric heating became a popular alternative where the district heating nets did not reach. As a result, the shift away from oil heating took place already during the end of the 20th century [31]. Grants were also introduced for energy retrofit measures, such as insulation and change to modern windows.

There are three main laws which apply to the use and preservation of the built heritage environment in Sweden. The Swedish Environment Code (1998:808) (EC) aims at promoting sustainable development to assure a healthy and good environment for present and future generations [39]. Chapter 1 §1 requires that EC shall be applied in such a way as to ensure valuable natural and cultural environments are protected and preserved [39]. Chapter 3 §6 focuses on management of land and water areas. In particular, land and water areas, the physical environment in general and areas of national and public interest having natural or cultural values shall be protected against measures that damage the natural or cultural environment [39]. Chapter 6 §2 clarifies that when evaluating environmental impacts, not just natural but also cultural environments must be considered [39].

The Swedish Historic Environment Act (1988:950) (HEA) protects cultural heritage of national concern [40]. Chapter 3 §§1-2 of the HEA states that County Administrative Board (CAB) can list a building that is of particularly high cultural and historical value and prescribe the way in which a heritage building shall be conserved and maintained [40]. Chapter 3 also contains provisions about alteration and refurbishment of listed buildings under §§14-15 [40]. Section 14 raises the possibility to obtain permission to alter the building from the CABs. If the owner of a building would like to integrate sustainable energy measures or make changes for other reasons, such permission can be granted according to Chapter 3 §14 [40]. It is the CABs who evaluate if the reasons given for the alteration is valid and whether the alteration will lead to distortion of the cultural and historical values that the listing is supposed to protect. The CABs may impose conditions on the permit such as how the alteration shall be executed to ensure the long-term sustainability of the building. Moreover, there are guidelines given by the Swedish National Heritage Board for better clarification of content of the provisions of HEA [41]. There is also close cooperation between the CABs where such issues are discussed.

The Swedish Planning and Building Act (2010:900) (PBA) aims at promoting 'a societal progress with equal and proper living conditions and a clean and sustainable habitat, for people in today's society and for future generations', according to its Chapter 1 §1 [42]. Chapter 8 §13 prohibits the distortion of culturally valuable building, and §§14 and 17 requires that they are properly maintained and that caution is used in making any alterations, so that their characteristics and values are protected [42]. The local municipalities of Sweden are responsible for the implementation of the PBA, where Planning and Building Boards and Building Committees deal with spatial land use planning and building permit processes [9]. Buildings and built areas of cultural value can be protected through detailed development plans or designated through various forms of preservation plans. These regulations may affect the building permit process if a property owner applies for energy retrofits for installation of devices [9].

All in all, these three laws, namely EC, HEA and PBL refer to each other and must all be considered in the management and protection of culturally or historically valuable buildings.

The Energy Performance Certification Act (2006:985) for buildings and an Act on Energy Measurements in buildings (2014:267) have been introduced to implement the requirements prescribed in the EU Energy Efficiency Directive 2012/27/EU and EU Energy Performance of Buildings Directive 2010/31/EU [9]. The Swedish Energy Agency [43] and Boverket- the Swedish National Board of Housing and Building [44] also play key roles in the implementation of renewable energy measures in heritage buildings. The latter is responsible for the implementation of PBL on a national scale. More significantly, Boverket has overall responsibility for the policy area 'gestaltad livsmiljö (designed living environment)' [45]. Cultural and historical values are highlighted as an important part of the designed living environment [46].

In 2017, Sweden's Riksdag (the Swedish Parliament) introduced a new climate policy framework [47]. The framework consists of climate goals, a climate act and a climate policy council. Sweden's overarching goal is to have net zero GHG emissions into the atmosphere by 2045. The Swedish energy target for 2020 was that renewable energy would be at least 50 per cent of the total energy consumption [48]. In fact, Sweden surpassed its 2020 target by reaching 60 percent of production level from renewable energy by 2020 [49].

The Swedish National Heritage Board has a national registry holding information about its listed and designated buildings. The Bebyggelseregistret (Data Base of Built Heritage) has around 13,000 buildings listed as national monuments, historical buildings and Church monuments until 2011 [50]. The total number of listed buildings and areas is outdated and unknown due to lack of unified accounting. However, the National Antiquities Office is now producing a new register which will eventually contain information and knowledge base on all buildings with identified cultural values in Sweden by 2023-24 [51].

The European Committee for Standardization (CEN) produced the 'Conservation of cultural heritage – Guidelines for improving the energy performance of historic buildings' [52]. This standard was developed with participation of Uppsala University and the Cultural Conservation Department at the Swedish National Heritage Board [53]. In 2021, the results of an evaluative study conducted on usability of this standard showed low motivation for adoption by potential users due to uncertain benefits, lack of practical knowledge to carry out steps suggested in the standard and lack of external pressures [54]. Based on the results, the authors suggested to incorporate further support towards easily accessible information, more guidance on practical aspects to implement the standard, training for professionals in the field and demand of use of standard by authorities and stakeholders [54].

## 2.3 INDIA

Historic conservation has a long history in India. One of the main reasons is that India is one of the world's oldest civilizations, and endowed with a long history of tangible and intangible heritage and cultural wealth. In the 14th century AD, Firoz Shah Tughlaq decreed that ancient buildings must be protected [55, pp. 3]. During the British East India Company rule in India, the Bengal Regulation (XIX) and the Madras Regulation (VII) were passed in 1810 and 1817 respectively [55, pp. 3]. These rules and regulations entrusted the government with the duty to act if public buildings were under threat of misuse [55, pp. 3].

Cultural heritage protection entered a new era when the Ancient Monuments Preservation Act, 1904 (the 1904 Act) was promulgated. The 1904 Act provided effective preservation of public and privately owned ancient monuments and facilitated government acquisition of ancient monuments and objects of archaeological, historical or artistic interest. In 1951 following independence, the Ancient and Historical Monuments and Archaeological Sites and Remains Act (1951 Act) was enacted by the Parliament of India. Consequently, all the historical monuments and archaeological sites and remains protected under 1904 Act were re-declared as of national importance under the 1951 Act. The 1951 Act was repealed and replaced by the Ancient Monuments and Archaeological Sites and Remains Act, 1958, which provided 'for the preservation of ancient and historical monuments and archaeological sites and remains of national importance' [56]. India ratified 1972 World Heritage Convention in November 1977 and forty sites inscribed on the World Heritage list as of 2022 [57].

Early in the first decade of the 21st century, energy efficiency and energy conservation related laws were introduced in India. In 2001, the Government of India enacted the Energy Conservation Act to 'provide efficient use of energy and its conservation' [58]. Following this, Bureau of Energy Efficiency (BEE) was constituted [59], and the Electricity Act 2003 and several policies have been launched to promote energy conservation and to better coordinate development of the power sector in India.

Buildings are accountable for around 35 percent of India's total energy consumption, and their energy use is growing at 8 percent yearly [60]. The building sector projected to emit seven times more CO2 by 2050 compared with 2005 levels [61]. This scenario makes it crucial to ensure that building sector consumes energy in a sustainable and efficient manner. The energy Conservation Building Code (ECBC) was promulgated for commercial buildings in 2007 [62]. Later in 2018, the Ministry of Power has announced the ECO Niwas Samhita 2018, which is the Energy Conservation Building Code for residential buildings (ECBC-R) [63]. However, there has not been a specific development with respect to Energy Conservation Building Code for heritage and culturally-historically-environmentally valuable buildings of India.

The ECBC has been included in unified building byelaws of some states (Karnataka, Telangana, Rajasthan, etc.). However, the nationwide implementation at large has been slow [64]. In August 2022, the Union Cabinet [65] approved the India's updated nationally determined contribution (NDC) as required by Paris Agreement. India now pledges to reduce the emissions intensity of its gross domestic product (GDP), that is emissions as a unit in relation to GDP, by 45 percent by 2030 from 2005 level and to increase the non-fossil fuel-based electricity to 50 percent by 2030 and a long-term goal of reaching net-zero by 2070 [66]. Keeping the ever-growing energy demand of India's buildings in mind, it is essential to implement, practice and encourage energy sustainability in existing buildings.

India is a huge reservoir of historic buildings. With only 40 of these listed as World Heritage sites by UNESCO [67], about 3,650 to be protected by national agencies, several thousand to be protected by state-level agencies [68], the task of studying energy efficiency use in heritage buildings in India is onerous. For India, to put more emphasis on use of energy retrofit measures in historic buildings, there is a need for a new ECBC particularly focused on heritage buildings, policies for energy audits of historic buildings, implementation of energy efficiency projects through energy service companies and BEE, measurements and verification of existing heritage building stock. Given that the Indian environment is much warmer than Sweden, emphasis should be placed on cooling systems, thermal comfort and proper ventilation of old and heritage buildings. Without careful consideration, the gadgets and appliances used for cooling can distort and alter the characteristics and appearance of heritage buildings.

## 3. FUTURE RESEARCH

The brief review of laws and policies of EU, Sweden and India reflects that the need for and interest in sustainable energy use in buildings and built areas of cultural value will continue to grow as countries look for new ways to meet goals and obligations for reducing emissions. These laws and policies show potential for sustainable future of built heritage, but efficient enforcement is not without challenges. Further research is needed to support both the enforcement of existing laws and proposals for legal reforms. Furthermore, guidelines for the implementation of laws and regulations within the legal systems and the training of government officials and consultants at all levels should reflect the goals of sustainable energy use and sustainable preservation.

In our ongoing project [1], we focus on selected World Heritage Sites and analyse the application of international and national law at local levels. We will examine the legal systems of Sweden and India with special attention towards preservation of historic building stock and use of energy retrofit measures, and analyse the potential of these laws to achieve emissions and energy targets, particularly in the historic districts of Visby in Gotland, Sweden and the Jaipur Walled city in Rajasthan, India.

In our cases studies, we will select culturally valuable buildings in three types of legal situation: a) buildings which have been denied permits for energy retrofitting; b) buildings which have been retrofitted to reduce energy use or the use of fossil fuels; and c) buildings that are either currently under energy renovation or where the owners do not opt for energy retrofitting at all. We will review permit applications and decisions relating to these buildings. We also plan to conduct semi-structured interviews with public officials, building owners, policy makers and other experts. We will use this information to help us answer three questions. First, why do many culturally valuable buildings continue rely on fossil-fuel based energy in 21st century or not undergo sustainable energy retrofits? Second, what are the challenges or limitations faced in order to adopt sustainable energy measures in culturally valuable buildings? Third, what kind of enforcement mechanisms are there, and how does the post implementation review help in achieving the sustainable energy use and heritage preservation? Answering these questions will contribute to understanding the extent to which the legal system supports energy sustainability in built heritage, and what improvements are possible.

This project will contribute to filling the research gap on legal approaches to sustainable energy use in culturally valuable buildings. This project aims to support mutually beneficial learning amongst these regions to support sustainable energy use in historic districts and culturally valuable buildings. Overall, we hope to contribute further towards preserving the past for the future.

## 4. REFERENCES AND NOTES


[1] Getting the Balance Right? (Väga Rätt?) is the larger ongoing project between Department of Law and Department of Art History and Conservation at Uppsala University, where both the faculties are hosting two doctoral student projects together, which makes it interdisciplinary in nature. Both the researchers are also part of the Graduate School in Sustainability Studies (GRASS), Campus Gotland and taking part in various research activities and in discussions with local municipal authority *i.e.*, Region Gotland. Available: https://www.campusgotland.uu.se/sustainability/graduate-school/getting-the-balance-right-/ [accessed 25 August 2022].

[2] C. Elefante. (2012). "The Greenest Building Is... One That Is Already Built." Forum Journal. [Online]. 27(1), pp. 62-72. Available: https://www.muse.jhu.edu/article/494514 [accessed 25 August 2022].

[3] District of Columbia Office of Planning. "Sustainability Guide for Existing and Historic Properties" Internet: https://planning.dc.gov/sites/default/files/dc/sites/op/publication/attachments/Sustainability%20Guide%20for%20Existing%20and%20Historic%20Properties.pdf 2019 [accessed 25 August 2022].

[4] T. Broström and K. Svahnström. "Solar energy and cultural-heritage values." World Renewable Energy Conference Linköping, Sweden, 2011, pp. 2034-2040.

[5] T. Broström and G. Leijonhufvud. "Heat pumps for conservation heating." In: Proceedings of the 8th Symposium on Building Physics in the Nordic Countries. Dept. of Civil Engineering, Technical University of Denmark, 2008, pp. 1143-1150.

[6] F. Berg, A. C. Flyen, Å. L. Godbolt, T. Boström. (2017). "User-driven energy efficiency in historic buildings: review." Journal of Cultural Heritage. [Online]. 28, pp. 188-195. Available: https://doi.org/10.1016/j.culher.2017.05.009 [accessed 25 August 2022].

[7] T. Broström, A. Donarelli and F. Berg. (2017). "For the categorisation of historic buildings to determine energy saving." AGATHÓN | International Journal of Architecture, Art and Design. [Online]. 1, pp. 135-142. Available: https://doi.org/10.19229/2464-9309/1212017 [accessed 25 August 2022].

[8] P. Eriksson, V. Milić and T. Brostrom. (2020). "Balancing preservation and energy efficiency in building stocks." International Journal of Building Pathology and Adaptation. [Online]. 38(2), pp. 356-373. Available: https://doi.org/10.1108/IJBPA-02-2019-0025 [accessed 25 August 2022].

[9] A. Christiernsson, M. Geijer and M. Malafry. (2021). "Legal Aspects on Cultural Values and Energy Efficiency in the Built Environment—A Sustainable Balance of Public Interests?." Heritage. [Online]. 4(4), pp. 3507–3522. Available: https://doi.org/10.3390/heritage4040194 [accessed 25 August 2022].

[10] Petra Eriksson. "Balancing Building Conservation with Energy Conservation- Towards differentiated energy renovation strategies in historic building stocks." Ph.D dissertation, Department of Conservation, Göteborgs universitet, Sweden, 2021. Available: http://hdl.handle.net/2077/68356 [accessed 25 August 2022].

[11] J. R. Carreón. "Review on techniques, tools and best practices for energy efficient retrofitting of heritage buildings." Utrecht University, 2015. Available: https://refomo.eu/wp-content/uploads/sites/105/2015/11/Review-on-techniques-for-energy-efficient-retrofitting-of-heritage-buildings_Final.pdf [accessed 25 August 2022].

[12] World Green Building Council. "Bringing Embodied Carbon Upfront Report." 2019. Available: https://www.worldgbc.org/sites/default/files/WorldGBC_Bringing_Embodied_Carbon_Upfront.pdf [accessed 25 August 2022].

[13] T. Parejo-Navajas. (2015). "A Legal Approach to the Improvement of Energy Efficiency Measures for the Existing Building Stock in the United States Based on the European Experience." Seattle Journal of Environmental Law. [Online]. 5(1), pp. 341-413. Available: https://scholarship.law.columbia.edu/sabin_climate_change/66 [accessed 25 August 2022].

[14] European Commission. "New rules for greener and smarter buildings will increase quality of life for all Europeans." Internet: https://ec.europa.eu/info/news/new-rules-greener-and-smarter-buildings-will-increase-quality-life-all-europeans-2019-apr-15_en 15 April 2019 [accessed 25 August 2022].

[15] J. O. Lewis, S. Ni. Hógáin and A. Borghi. "Building energy efficiency in European cities." URBACT II Capitalisation, 2013. Available: https://urbact.eu/sites/default/files/import/general_library/19765_Urbact_WS6_ENERGY_low_FINAL.pdf [accessed 25 August 2022].

[16] D. A. Lewis. (2015). "Identifying and Avoiding Conflicts Between Historic Preservation and the Development of Renewable Energy." N.Y.U. Environmental Law Journal. [Online]. 22(3), pp. 274-360. Available: https://www.nyuelj.org/issues/volume-22/ [accessed 25 August 2022].

[17] UN 2030 Agenda for Sustainable Development. Available: https://sdgs.un.org/2030agenda [accessed 25 August 2022]. Specifically SDG 7: Ensure access to affordable, reliable, sustainable and modern energy for all and SDG 11: Make cities and human settlements inclusive, safe, resilient and sustainable and 11.4: Strengthen efforts to protect and safeguard the world's cultural and natural heritage.

[18] European Environmental Agency. "Total greenhouse gas emission trends and projections in Europe." Internet: https://www.eea.europa.eu/ims/total-greenhouse-gas-emission-trends 18 November 2021 [accessed 25 August 2022].

[19] European Commission. "2030 climate & energy framework." Available: https://ec.europa.eu/clima/eu action/climate-strategies-targets/2030-climate-energy-framework_en [accessed 25 August 2022].

[20] Council of the European Union. "Paris Agreement on climate change." Available: https://www.consilium.europa.eu/en/policies/climate-change/paris-agreement/ [accessed 25 August 2022].

[21] European Commission. "2050 Long-term Strategy" Available: https://ec.europa.eu/clima/eu-action/climate-strategies-targets/2050-long-term-strategy_en [accessed 25 August 2022].

[22] Council Directive 2002/91/EC of the European Parliament and of the Council of 16 December 2002 on the energy performance of buildings [2003] OJ L 1. Available: http://data.europa.eu/eli/dir/2002/91/oj [accessed 25 August 2022].

[23] Directive 2010/31/EU of the European Parliament and of the Council of 19 May 2010 on the energy performance of buildings (recast) [2010] OJ L 153. Available: http://data.europa.eu/eli/dir/2010/31/oj [accessed 25 August 2022].

[24] Council Directive (EU) 2018/844 of the European Parliament and of the Council of 30 May 2018 amending Directive 2010/31/EU on the energy performance of buildings and Directive 2012/27/EU on energy efficiency [2018] OJ L 156. Available: http://data.europa.eu/eli/dir/2018/844/oj [accessed 25 August 2022].

[25] According to Article 2(3) of the Directive 2018/844/EU, 'technical building system' means technical equipment for space heating, space cooling, ventilation, domestic hot water, built-in lighting, building automation and control, on-site electricity generation, or a combination thereof, including those systems using energy from renewable sources, of a building or building unit.

[26] A. Wilson. (February 2022). "Revision of the Energy Performance of Buildings Directive: Fit for 55 package [EU Legislation in Progress]." European Parliamentary Research Service. Available: https://epthinktank.eu/2022/02/04/revision-of-the-energy-performance-of-buildings-directive-fit-for-55-package-eu-legislation-in-progress/ [accessed 25 August 2022]. The Fit for 55 package is part of the European Green Deal and European Climate Law, where EU has set a binding target of achieving 55 percent reduction in emission by 2030 and climate neutrality by 2050.

[27] M. Pasikowska-Schnass. (June 2018). "Cultural heritage in EU policies' (2018)." European Parliament . Available: https://www.europarl.europa.eu/RegData/etudes/BRIE/2018/621876/EPRS_BRI(2018)621876_EN.pdf [accessed 25 August 2022].

[28] A. G.Gonzalez, P. Bouillard, C. A. A. Román and M. Stevanovic. "KEEPING THE HISTORICAL HERITAGE ALIVE: Methodology for the energy renovation of the historic residential stock of the east extension in Brussels." PLEA2013- 29th Conference, Sustainable Architecture for a Renewable Future, Munich, Germany, 2013. Avialable: http://web5.arch.cuhk.edu.hk/server1/staff1/edward/www/plea2018/plea/2013/ [accessed 25 August 2022].

[29] James K. Reap. (2022). "Introduction: Heritage legislation and management." Built Heritage. [Online]. 6(9), pp. 1-3. Available: https://doi.org/10.1186/s43238-022-00059-9 [accessed 25 August 2022].

[30] M. Geijer. "Ett nationellt kulturarv – Utveckling av en professionell vård och förvaltning av statliga byggnadsminnen." licentiatavhandling, KTH, Stockholm 2004 Stockholm 2004.

[31] M. Geijer Energipolitik. (January 2021). "bebyggelse och kulturvärden i 1900-talets offentliga utredningar." REKO Report, Uppsala University. Available: https://img1.wsimg.com/blobby/go/9ac771ea-d620-467f-b048-917d5f8806fe/downloads/Energipolitik%20Rapport%20%2020210128.pdf?ver=1617002422861 [accessed 25 August 2022].

[32] Sanering means in construction technology a comprehensive renewal of one or more buildings or entire urban areas which can also include demolition.

[33] M. Legnér & G. Leijonhufvud. (2019). "A Legacy of Energy Saving: The Discussion on Heritage Values in the First Programme on Energy Efficiency in Buildings in Sweden, c. 1974–1984." The Historic Environment: Policy & Practice. [Online]. 10(1), pp. 40-57. Available: https://doi.org/10.1080/17567505.2018.1531646 [accessed 25 August 2022].

[34] UN Educational, Scientific and Cultural Organisation, Convention Concerning the Protection of the World Cultural and Natural Heritage (adopted 16 November 1972, entered into force 17 December 1975). Available: https://whc.unesco.org/en/conventiontext/ [accessed 25 August 2022].

[35] International Council on Monuments and sites, The Declaration of Amsterdam-1975. Available: https://www.icomos.org/en/and/169-thedeclaration-of-amsterdam [accessed 25 August 2022].

[36] UNESCO World Heritage Convention, Sweden- State Party.
Available: https://whc.unesco.org/en/statesparties/se [accessed 25 August 2022].

[37] M. Claesson and T. Broström. "Eight years of energy efficiency in historic buildings." *EECHB-2016 Energy Efficiency and Comfort of Historic Buildings*, eds. Michael de Bouw, Samuel Dubois, Liesbeth Dekeyser and Yves Vanhellemont, Flanders Heritage Agency, 2016, pp. 2-9.

[38] K. Lygnerud. (2018). "Challenges for business change in district heating." Energy, Sustainability and Society. [Online]. 8, pp. 1-13. Available: https://doi.org/10.1186/s13705-018-0161-4 [accessed 25 August 2022].

[39] The Swedish Environment Code (1998:808).

[40] The Swedish Historic Environment Act (1988:950).

[41] Swedish National Heritage Board. "Guidance Ch. 3 KML." 23 June 2020. Available: https://www.raa.se/lagar-och-stod/kulturmiljolagen-kml/byggnadsminnen-3-kap/vagledning-3-kap-kml/ [accessed 25 August 2022].

[42] The Swedish Planning and Building Act (PBA) (2010:900).

[43] The Swedish Energy Agency. Available: http://www.energimyndigheten.se/en/.

[44] Boverket – the Swedish National Board of Housing, Building and Planning. Available: https://www.boverket.se/en/start/.

[45] Boverket, About designed living environment. Available: https://www.boverket.se/sv/samhallsplanering/arkitektur-och-gestaltad-livsmiljo/gestaltad-livsmiljo/ 6 November 2020 [accessed 25 August 2022].

[46] Boverket, Policy for designed living environment. Available: https://www.boverket.se/sv/samhallsplanering/arkitektur-och-gestaltad-livsmiljo/gestaltad-livsmiljo/politik/ 10 February 2021 [accessed 25 August 2022].

[47] The Swedish Climate Policy Framework (2017), Ministry of the Environment and Energy, Government Office of Sweden. Available: https://www.government.se/495f60/contentassets/883ae8e123bc4e42aa8d59296ebe0478/the-swedish-climate-policy-framework.pdf [accessed 25 August 2022].

[48] The Swedish Energy Agency, Energy in Sweden 2021 - an overview. Available: http://www.energimyndigheten.se/en/news/2021/an-overview-of-energy-in-sweden-2021-now-available/> 14 July 2021 [accessed 25 August 2022].

[49] Smart City Sweden. "Sweden has exceeded 50 % renewable energy goal." Available: https://smartcitysweden.com/sweden-has-exceeded-50-renewable-energy-goal/ 17 January 2022 [accessed 25 August 2022].

[50] The Swedish National Heritage Board. Available: https://www.raa.se/in-english/digital-services/about-bebr/.

[51] The Swedish National Heritage Board. "New register for digital information about the built cultural heritage." Available: https://www.raa.se/2022/04/nytt-register-for-digital-information-om-det-byggda-kulturarvet/> 3 June 2022 [accessed 25 August 2022].

[52] SS-EN 16883: Conservation of cultural heritage - Guidelines for improving the energy performance of historic buildings. Available: https://www.raa.se/kulturarv/byggnader/byggnadsvard/energieffektivisering-i-kulturhistoriskt-vardefulla-byggnader/standardisering-inom-energieffektivisering/ 22 September 2020 [accessed 25 August 2022].

[53] From Sweden, Tor Broström, Professor at Uppsala University, and Camilla Altahr-Cederberg and Therese Sonehag from the Cultural Conservation Department at the National Antiquities Authority participated in the development of the Standard SS-EN 16883: 2017.

[54] G. Leijonhufvud, T. Broström, A. Buda. (October 2021). "An Evaluation of the Usability of EN 16883:2017- Suggestions for enhancing the European guidelines for improving the energy performance of historic buildings." SHC Task 59 | EBC Annex 76 | Report D.B2. Available: DOI: 10.18777/ieashc-task59-2021-0002.

[55] Directorate General: Central Public Works Department India. Handbook of Conservation of Heritage Buildings. 2013, pp. 1-89. Available: https://smartnet.niua.org/content/2e7124ac-31b4-4b32-8286-43a1308ad7d2 [accessed 25 August 2022].

[56] Ancient Monuments and Archaeological Sites and Remains Act, 1958, Ministry of Culture, Government of India. Available: https://www.nma.gov.in/rules-and-acts [accessed 25 August 2022].

[57] UNESCO World Heritage Convention, India- State Party. Available: https://whc.unesco.org/en/statesparties/in [accessed 25 August 2022].

[58] The Energy Conservation Act, 2001 (Act No. 52 of 2001), Government of India.

[59] Under the provisions of Chapter II of the Energy Conservation Act 2001, the Bureau of Energy Efficiency was established in 2002.

[60] R. Khosla and K. B. Janda. (2019). "India's building stock: towards energy and climate change solutions." Building Research & Information. [Online]. 47(1), 1-7.
Available: https://doi.org/10.1080/09613218.2019.1522482 [accessed 25 August 2022].

[61] RMI India. (June 2021). "India's Buildings Sector Moonshot: Corporate Climate Commitments Can Forge the Path." Available: https://rmi.org/indias-buildings-sector-moonshot-corporate-climate-commitments-can-forge-the-path/ [accessed 25 August 2022].

[62] Energy Efficiency, Ministry of Power, Government of India.
Available: https://powermin.gov.in/en/content/energy-efficiency [accessed 25 August 2022].

[63] ECBC Residential, Bureau of Energy Efficiency, India. Available: https://beeindia.gov.in/content/ecbc-residential [accessed 25 August 2022].

[64] N. Shandilya and A. R. Ghorpade, "ECBC Compliance in Indian Cities- Handbook (Supporting Sustainable built environment under Smart City Mission)." ICLEI- local governments for Sustainability, South Asia, 2019. Available: https://shaktifoundation.in/wp-content/uploads/2019/07/Handbook-ECBC-compliance-in-Indian-Cities.pdf [accessed 25 August 2022].

[65] The Union Cabinet is a smaller executive body that is a subset of the Union Council of Ministers, is India's supreme decision-making body. It is headed by the Prime Minister of India and include other Ministers of Cabinet rank appointed under article 75 of the Constitution of India.

[66] Press Information Bureau. "Cabinet approves India's Updated Nationally Determined Contribution to be communicated to the United Nations Framework Convention on Climate Change." Government of India. Available: https://pib.gov.in/PressReleaseIframePage.aspx?PRID=1847812 3 August 2022 [accessed 25 August 2022].

[67] Out of these, 32 are cultural, 7 are natural and one is mixed sites. Indian properties inscribed on the World Heritage Convention, UNESCO (2022). Available: https://whc.unesco.org/en/statesparties/in [accessed 25 August 2022].

[68] Built Heritage, Ministry of Culture, Government of India. Available: https://www.indiaculture.nic.in/built-heritage.

# Damage-free energetic window renovation in old and listed buildings

## Hygrothermal aspects of box-type windows

S. Bichlmair[1], M. Krus[1], D. Stiegler[2], Ch. Milch[1]

[1] Dept. Hygrothermics, Fraunhofer Institute for Building Physics Holzkirchen Branch, 83626 Valley, Germany, stefan.bichlmair@ibp.fraunhofer.de; martin.krus@ibp.fraunhofer.de; christine.milch@ibp.fraunhofer.de

[2] Thieltges Sachverständigenbüro, 83022 Rosenheim, Germany, ds@thieltges.com

**Keywords –** hygrothermal simulation, thermal bridging, historic buildings, Tracergas measurement, box-type windows

The maintenance and renovation of existing windows depends above all on the durability and usability of the renovation measure and the energy standard achieved. The aim of the research project is to make the energetic refurbishment of sustainable existing windows safer and thus preserve historic windows, prevent damage, promote energetic refurbishment and increase the market share of specialized window manufacturers. This is achieved by the targeted research of the essential parameters U-values of window pane combinations, air exchange conditions and the hygrothermal conditions resulting from this in the course of the year. Detailed measurements of the refurbished windows in the historic building Old Cooperage, housing the Fraunhofer-Center Benediktbeuern, generated the necessary data on energetic and hygrothermal behavior of the refurbished windows solution. With additional measurement campaigns of airtightness with tracer gas method the real air exchange behavior of the cavity of the box-type window could be determined. Hygrothermal calculations with software WUFI® Plus are carried out to investigate the effects of different refurbishment variants on a box-type window. Based on the measured values, a box-type window model is created to calculate the local climate in the cavity as a function of the flow conditions, U-values and the indoor and outdoor climate. The general effect of changed tightness on the humidity conditions in the cavity can thus be confirmed. The simulation of the effects of different U-values of the window levels on the humidity conditions in the cavity result in better understanding how to refurbish and preserve historic windows into damage free low energy box-type windows.

## 1. INTRODUCTION

The box-type window consists of two casement windows, which are connected to each other in the reveal with a wooden cladding. The space between the two casement levels or pane levels is also called cavity. In the case of energy-saving the original casement window is refurbished to a box-type window, or in case of box-type windows in general, moisture-related problems can (rather rarely) occur in the space of the box-type windows (cavity). Known phenomena are condensation on the glazing of the external casement, mould growth as a result of increased humidity in the cavity, and wood-destroying fungi in the case of persistent high humidity. The evaluation of window constructions when free from damage (surface temperature, risk of mould, etc.) is carried out on the basis of known regulations [1-3] on the inner window surface. However, there are no specifications or regulations for the evaluation of the cavity. Only traditional construction recommendation regarding the air tightness of "inner casement tight" and "external casement leaky" are intended to ensure durability. This leads to the main question for the investigations: How do different tightness level of the window planes affect the moisture in the cavity and is it possible to quantify?

## 2. METHOD AND MEASUREMENT CONCEPT

By measuring the relative humidity and air temperature indoors, in the cavity and the outside air, the moisture content of the air (absolute humidity AH) is determined. In the cavity sets a certain moisture content (moisture concentration), depending on the airing ratio of outside and room air. The mixture of two streams of humid air is applied to the system of the box-type window. The air mass balance, moisture balance and energy balance of humid air [4] result in a mixing ratio of the moisture concentration when applied to the box-type window system. The mixing ratio MR is calculated on the basis of the moisture differences of the external (xe) and internal absolute humidity (xi) to the absolute humidity in the cavity (xcav) using the simplified approach (1). Figure 1 shows the horizontal section through a box-type window with outside casement (historic window), added new casement inside and the assignment of the designations.

$$MR = \frac{m_e}{m_i} = \frac{x_i - x_{cav}}{x_{cav} - x_e} \quad [-] \tag{1}$$

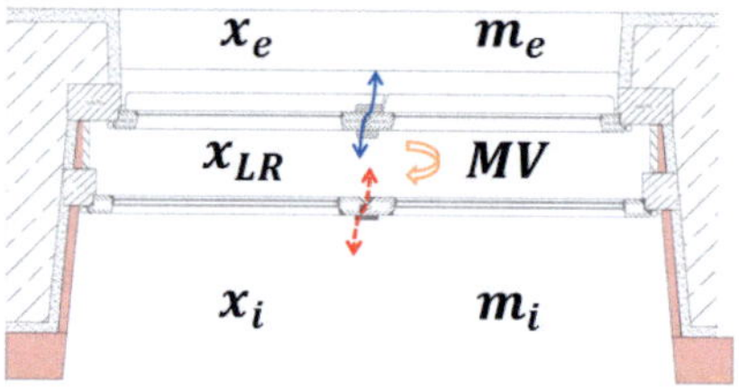


**Figure 1. Illustration of the box-type window F 1.25 in horizontal cross section with designation and assignment of the physical quantities determining the mixing ratio (MR) of the air humidity or air masses. The applied interior and reveal insulation is shown with a red coloured background.**

The ratio of the air volume flows (air mass flow $m_e$ and $m_i$) results also in the mixing ratio MR. The approach and the result of MR calculated with the absolute humidity can thus be verified. The air volume flows are measured with the tracer gas dilution method. With the help of a tracer gas it is possible to determine the air exchange rate without additional applied pressure difference. To determine the air exchange rates, the tracer gas method with "decreasing concentration" is used [5]. For this purpose, two holes are made at the upper and lower wood frame of the casement through which the injection of the gas and the gas concentration measurement are carried out in a circulation procedure. After injecting a defined quantity of the tracer gas (SF6), the gas concentration is measured with the gas analyser in an interval of about half a minute. The air exchange rate or the mass flow can be determined by analysing the decreasing gas concentration.

With the knowledge of the stationary thermal behaviour of the box-type window, the unsteady hygrothermal calculation of the box-type window is carried out with a simplified 3D model with WUFI® Plus [8] using real climate data. Furthermore, the moisture conditions in the cavity are calculated with different tightness and different $U_W$-values of the window planes.

## 3. EXAMINED WINDOW CONSTRUCTION

As part of the energy-saving refurbishment measures in the Old Cooperage, a total of seven windows on the upper floor and six windows on the ground floor were refurbished in various ways. The windows 1.25, 1.26 and 1.27 relevant for these investigations are located on the north side of the upper floor (Figure 2 at the top).

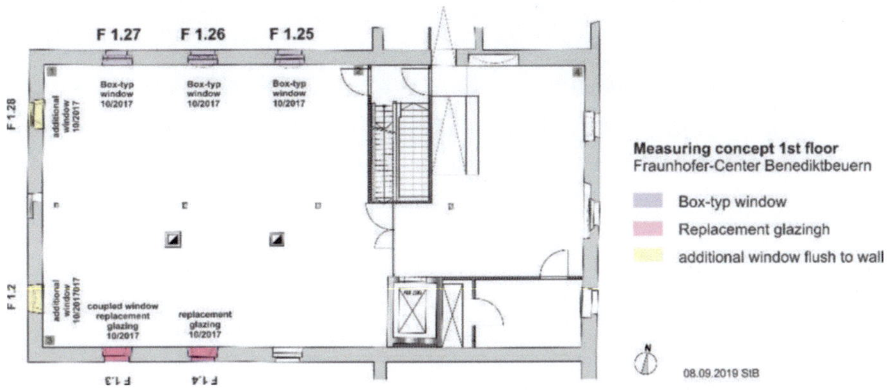


**Figure 2. Arrangement of the renovation variants of the windows on the upper floor of the Old Cooperage with the three examined windows on the north side F 1.25, F 1.26 and F 1.27 (top of the picture).**

The windows 1.26 and 1.27 have been extended to form a box-type window with single-leaf casement with a double pane insulating glass (Ug 1.1 W/m²K) on the inner window level. The window 1.25 is refurbished with double casement on the inner window level with also double pane insulating glass (Ug 1.3 W/m²K). The original historic window, now on the outer window plane, was left in place and hand repaired in the interest of the preservation. The profile geometry of the new casement window was deliberately made as slim as possible, in order to achieve a similar appearance to that of the historic casement windows. Figure 3 shows the original state in position 1.26 with interior and exterior views.

**Figure 3. Old Cooperage upper floor, window 1.26 on the north side with exterior and interior views, historic casement window in its original state. Frame dimensions 1.14 m x 1.40 m (1.60 m²).**

Figure 4 shows the interior views at window position 1.27 with closed and open inner window casement. The outer walls including the window reveals are cladded with internal insulation. The inner window sill (plastered brickwork) is made without thermal insulation for reasons of monument protection.

**Figure 4. Interior view of window 1.27 with closed (left) and open inner window casement. A heat flow sensor can be seen in the middle of the lower right glazing.**

## 4. EXAMINATIONS AND RESULTS

### 4.1 MIXING RATION IN THE CAVITY

Figure 5 shows the absolute humidity for the cavity of windows 1.25, 1.26 and 1.27 in the period from 27 to 31 December 2017. In addition, the absolute humidity of the indoor air and outside air is also included in the diagrams. The room air is conditioned to approx. 50 % RH and 20 °C and is recorded directly in front of the window (1.26 indoors) and additionally in the middle of the room. The measurement results of the two sensors differ only marginally. This allows us to conclude that the absolute humidity is homogeneously distributed in the room. The moisture content indoors is significantly higher than in the outside air due to humidification. As expected, the absolute humidity in the cavity is between the humidity content indoors and the outside air. It is noticeable that the absolute humidity of window F 1.27 is slightly lower than that of the other two windows, although windows F 1.27 and F 1.26 are identical in construction.

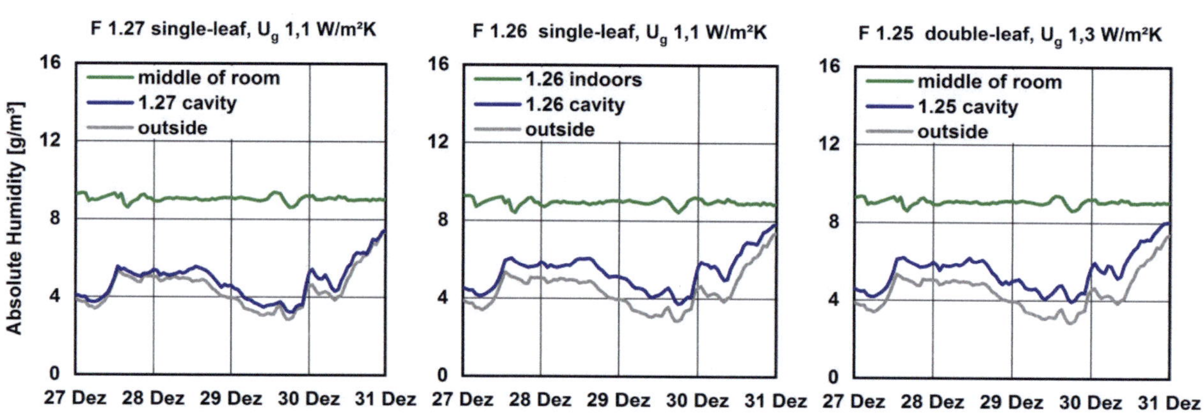


**Figure 5. Measurement of the absolute humidity in the cavity at windows 1.25, 1.26 and 1.27 with absolute humidity of indoor and outside air. Period 27th to 31st December 2017.**

Figure 6 shows the calculated mixing ratios for the box windows 1.25, 1.26 and 1.27 for the period from 23.12.2017 to 1.5.2018 with hourly values and moving weekly average. The course of the windows 1.25 and 1.26 is approximately the same. In contrast, window 1.27 (identical in construction to F 1.26) shows considerably higher values. With increasing temperature of the outdoor climate in spring, the values of the mixing ratios decrease for all windows.

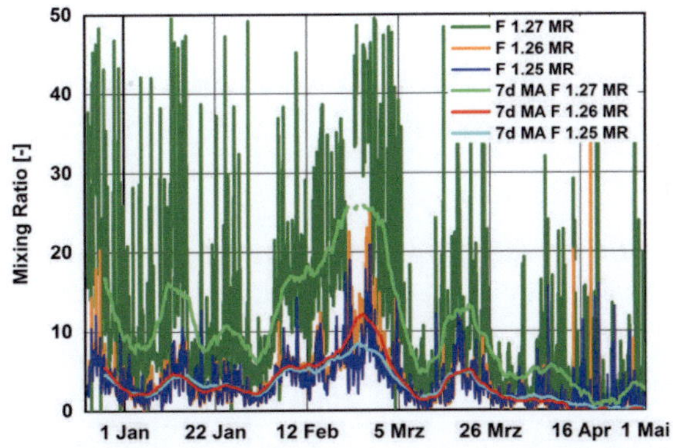


**Figure 6. Mixing ratio (MR) in the cavity of windows 1.25, 1.26 and 1.27 in the period from 23.12.2017 to 1.5.2018 with hourly values and moving weekly average (7d MA)**

The measurement of the air exchange rate with tracer gas is carried out on the three box-type windows. The original window at the outer window level has only a simple rebate without sealing strips. The inner new window level is designed with a rubber sealing strip. The air exchange rate determined in the unchanged installation condition takes into account the infiltration or exfiltration at both window levels. In order to obtain more detailed information on the infiltration behaviour of the inner and outer window levels, the individual window levels are additionally sealed in separate measurements. For this purpose, the window casements are sealed with sealing tape, in order to completely exclude air exchange to the inner or outer window levels. Based on the measured air exchange rates with additionally sealed window levels, the resulting mixing ratio me / mi is determined, see Table 1.

**Table 1. Air exchange rates of the cavity of windows 1.25, 1.26 and 1.27 determined from the tracer gas measurement and from this calculated mixing ratio (MV).**

| Var. | unchanged situation | | inner casement tight | | outside casement tight | | MR |
|---|---|---|---|---|---|---|---|
| | $m_{cav}$ [m³/h] | ACH [h⁻¹] | $m_e$ [m³/h] | LW [h⁻¹] | $m_i$ [m³/h] | ACH [h⁻¹] | $m_e/m_i$ [-] |
| F.1.25 | 0,49 | 2,54 | 0,39 | 2,02 | 0,07 | 0,39 | 5,6 |
| F.1.26 | 0,66 | 4,09 | 0,63 | 3,89 | 0,14 | 0,82 | 4,5 |
| F.1.27 | 0,91 | 5,63 | 1,1 | 6,83 | 0,05 | 0,30 | 22,8 |

For windows 1.25 and 1.26, the total air exchange rates with the window plane sealed on the inside and outside correspond approximately to the air exchange rates in the normal installation condition. With window 1.27, however, it is noticeable that the measurement with the internally sealed window plane shows a higher air exchange rate compared to the unchanged situation. This is due to air pressure fluctuations during the tracer gas measurement, which have an unfavourable effect on the measurement. The mixing ratios calculated from the mass flows correspond approximately to the mixing ratios determined from the absolute humidity.

## 4.2 HYGROTHERMAL SIMULATION WITH WUFI® PLUS

The measured indoor and outdoor climate from 1 January 2018 to 1 January 2019 is used for the simulation. First, a basic variant (V10) is simulated, which forms the starting point for the parameter study. The aim is to examine how accurately the real situation can be depicted over the course of the year using the measured climate data and which irregularities must be taken into account in the evaluation. Different air exchange rates are varied due to changed tightness at the inner and outer window levels (V50 and V90) and different $U_W$-values of the inner and outer window levels (V120 and V130), see Table 2.

**Table 2. Listing of the simulated variants with the parameters changed in each case (in bold). The specified U-values were calculated using the detailed method described in [3].**

| Var. | parameter | U-value window [W/m²K] | | |
|---|---|---|---|---|
| | | inner $U_{W,i}$ | outside $U_{W,e}$ | $U_{W,tot}$ |
| V10 | Basic model | 1,48 | 4,36 | 1,09 |
| V50 | **outside casement tight** (ACR 0 $h^{-1}$) | 1,48 | 4,36 | 1,09 |
| V90 | **inner casement tight** (ACR 0 $h^{-1}$) | 1,48 | 4,36 | 1,09 |
| V120 | inner glazing $U_g$ = 0,6 W/m²K | **1,08** | 4,36 | 0,86 |
| V130 | outside glazing $U_g$ = 1,3 W/m²K | 1,48 | **1,78** | 0,80 |

The simulation result of the basic variant is compared with the measured climate in the cavity. The relative humidity calculated by simulation shows a higher deviation from reality. On average over the year, the simulation deviates by 7.2 % RH from the real measured condition. Here, the simulation shows a deviation of 8 % RH in autumn to a mean value of approx. 20 % RH of too high relative humidity during the transition period in spring (Figure 7 left). The short-term fluctuations in the diurnal cycle are significantly lower in the simulation. On average, however, there is good agreement except for the transition period in spring.

Temperatures show good agreement in the transition periods from February to June and from September to November, with slight deviations in February. Only in summer the measured temperature is slightly higher. On average, the temperature over the year in the simulation is 0.72 K lower compared to the measured values. In the case of absolute humidity, the spring period from February to mid-March shows an increased humidity content for the simulated climate (Figure 7 right), whereas the measured values fluctuate more clearly in the short term during the course of the day.

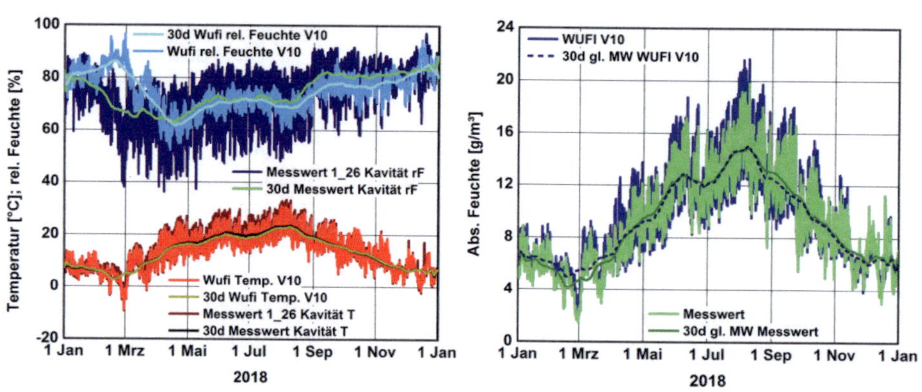


**Figure 7. Relative humidity, temperature and absolute humidity of the measured data and simulated climate data of the cavity with hourly values and sliding monthly averages (30d MW).**

A comparison of the other variants in Table 2 with the basic variant follows. The temperature curves of the two variants V50 (outside casement tight) and V90 (inner casement tight) change only marginally compared to the basic variant V10. The temperature curve for variant V50 is about 0.1 K lower and that for V90 about 0.1 K higher compared to V10 (Figure 8 left). With an additional improvement in the glazing of the inner window plane of variant V120 from $U_g$ 1.1 W/m²K to $U_g$ 0.6 W/m²K, the air temperature drops slightly, especially in the cold season. In contrast, the air temperature in the cavity of variant V130 increases considerably, as expected, with an improvement in the glazing of the outer window plane from $U_g$ 5.1 W/m²K to $U_g$ 1.3 W/m²K. The temperature increase in February is up to 3.5 K.

The absolute humidity at V50 (outside casement tight) increases enormously compared to the basic V10 variant and leads to an unfavourable humidity level in the cavity. Conversely, with the window level at V90 (inside casement tight), the absolute humidity decreases as expected in the cold season. This confirms the generally known design rule of "tight on the inside and open on the outside" as the right approach. The absolute humidity in V120 with a better $U_g$-value of the inner glazing increases slightly compared to the basic variant V10. With V130 with better outer glazing, the absolute humidity falls noticeably below the level of the basic variant V10.

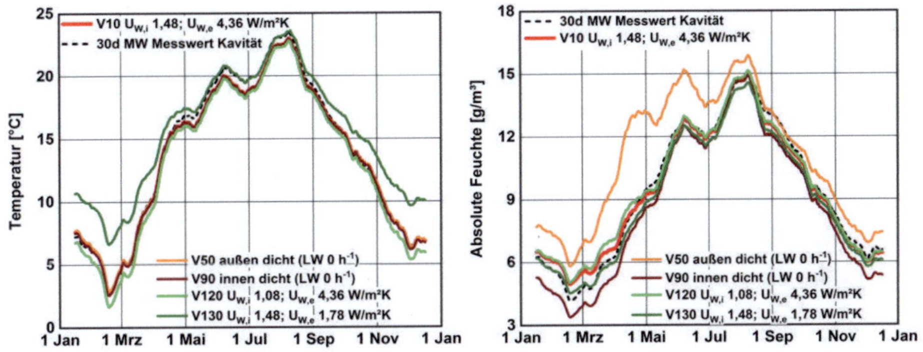


**Figure 8. Air temperature in the cavity (left diagram) and absolute humidity (right diagram) as moving monthly averages (30d MW) over the course of the year for 2018. The graph of variant V10 is covered by the two graphs V50 and V90 in the left diagram.**

The relative humidity in the cavity of variant V50 (outside casement tight) is considerably higher compared to the original variant, Figure 9. In contrast, a completely sealed inner plane leads to a significant reduction in relative humidity, especially in the cold season. The variants V120 and V130 with improved $U_g$-value of the inner and outer pane respectively also show an opposite behaviour. With an improvement of the inner pane in V120, the relative humidity increases significantly. On the other hand, the relative humidity of the V130 variant with an improved outer pane even drops below the level of a completely sealed inner window casement.

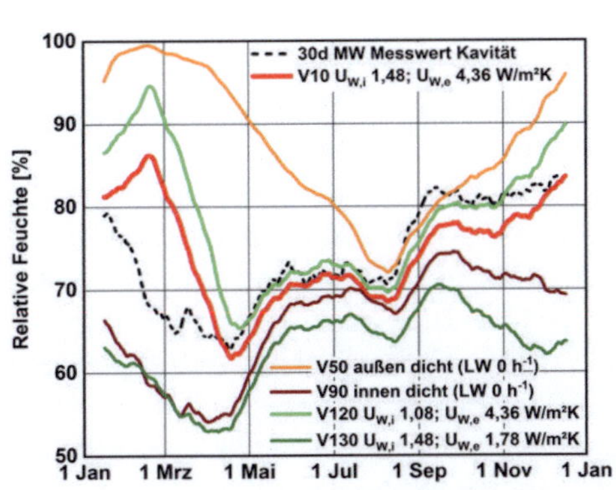


**Figure 9. Course of the relative humidity in the cavity of the variants to the basic variant V10 and the measured value as sliding monthly mean values (30d MW) for the simulation year 2018.**

## 5. CONCLUSION

For the humidity conditions in the space between the boxes (cavity), the airtightness of both window levels is of decisive importance in addition to the room and outside climate. To determine the air exchange conditions between room air and cavity as well as cavity and outside air, the moisture contents of the indoor and outdoor climate and the cavity are measured. By means of the moisture concentration, the mixing ratio between indoor and outdoor air can be determined by applying the mixing rule. The comparison of the three box-type windows in the Old Cooperage in Benediktbeuern, which were measured in detail, resulted in different mixing ratios depending on the tightness, which in turn resulted in different humidity levels in the cavity. Quantified exchanged air volumes from the tracer gas measurement are used to confirm the mixing ratios determined from the moisture measurement. In WUFI® Plus, a box-type window model is created in which climate of the cavity is calculated as a function of the flow conditions and the indoor and outdoor climate. The investigations are based on the general effect of humidity conditions in the cavity. The effect of different window tightness on the humidity conditions in the cavity can thus be confirmed. The U-values of the glazing also have an effect on the air change rate and mixing ratio due to the temperature-dependent pressure conditions via the temperature change in the cavity. With improvement of the $U_W$ of the inner window plane, the requirement for tightness increases in order to avoid critical humidity conditions in the cavity. With an improvement of the $U_W$ of the outer window level, the climatic situation in the cavity relaxes. Due to the higher air temperature in the cavity, the relative humidity decreases and the mixing ratio changes, which leads to a lower absolute humidity.


## ACKNOWLEDGMENT

The investigation was funded by the Research Initiative on Future Construction (BBSR) (reference number: SWD-10.08.18.7-17.42). The production and assembly of the renovation variants in accordance with the requirements of preservation of listed buildings were carried out in an excellent manner by the window manufactory companies RSP GmbH, Zoller-Prantl Gesellschaft mbH, Holzmanufaktur Rottweil GmbH and Glas Zange GmbH & Co.KG.


## REFERENCES


[1] DIN EN ISO 13788: Hygrothermal performance of building components and building elements - Internal surface temperature to avoid critical surface humidity and interstitial condensation - Calculation methods. German version. Beuth-Verlag. 2013.

[2] DIN EN ISO 10077-1: Thermal performance of windows, doors and shutters - Calculation of thermal transmittance - Part 1: General. German version. Beuth-Verlag 2006.

[3] DIN EN ISO 10077-2: Thermal performance of windows, doors and shutters - Calculation of thermal transmittance - Part 2: Numerical method for frames. German version. Beuth-Verlag. 2008.

[4] Rietschel, H.; Esdorn, H. (Hrsg.): Raumklimatechnik. Band 1: Grundlagen. Springer Verlag 2008.

[5] DIN EN ISO 12569: Thermal performance of buildings and materials - Determination of specific airflow rate in buildings - Tracer gas dilution method. German version. Beuth-Verlag. 2010.

[6] Flixo. Software. Flixo-Handbuch. Das thermische Analyse- und Reporting-Programm Infomind GmbH 2018. https://www.flixo.de/media/1350/flixo-handbuch.pdf

[7] DIN EN ISO 10211: Thermal bridges in building construction - Heat flows and surface temperatures - Detailed calculations. German version. Beuth-Verlag. 2018.

[8] WUFI® Plus. Software. Fraunhofer-Institut für Bauphysik IBP Holzkirchen. https://wufi.de/de/software/wufi-plus/.

[9] Hermes, Markus: Ein U-Wert kommt selten allein. Forum Holz/Bau/ Energie Köln 2009.

# Transparent internal wall insulation

S. Bichlmair, M. Krus, R. Kilian

Dept. Hygrothermics and Cultural Heritage Reserach, Fraunhofer Institute for Building Physics Holzkirchen Branch, 83626 Valley, Germany (stefan.bichlmair@ibp.fraunhofer.de; martin.krus@ibp.fraunhofer.de; ralf.kilian@ibp.fraunhofer.de)

**Abstract –** For reasons of monument protection, interior insulation is often the only option for energy-efficient renovation of the exterior walls. In order not to cover and hide walls with interesting historic surface structures or paint layers, a transparent internal wall insulation based on insulating glazing has been developed and tested. Due to the uneven inner surface of the outer wall and the supporting structure of the glazing, always an air gap exists between the outer wall and the glazing. This cavity has to be airtight towards the warm and moist indoor air in order to avoid accumulation of moisture because the diffusion-tight glazing inhibits re-drying of this enclosure towards indoors. But air exchange with the room air can always happen via leaks. This can lead to mould or moisture damage on the original surfaces. A test setup with transparent internal wall insulation with double glazing was installed in the historic building of the Alte Schäfflerei (Old Cooperage) in order to observe the real behavior of the system. To avoid mould growth conditions in the cavity, an electric heating cable was installed to control the relative humidity of the cavity air. Via a control unit the heating cable warms the cavity air when a certain limit is exceeded so as to lower relative humidity. The main reason for installing a transparent internal wall insulation is to reduce the heat flux of the wall in order to save energy. Therefore surface temperatures, relative humidity of the cavity, airtightness, heat flux and energy consumption of the heating cable were measured. The measured data shows that the controlling of the relative humidity in the cavity is necessary and the overall energy balance of the system shows that efficient energy retrofit is possible in this innovative way.

**Keywords** – internal wall insulation; insulating glass; monument protection; listed buildings; historic wall paintings; preventive conservation

## 1. INTRODUCTION

### 1.1 BACKGROUND

For reasons of monument protection, interior insulation is often the only option for energetic refurbishment of exterior walls. But sometimes the inner surface also needs to be conserved at the same time. Therefore, reversible systems were developed in order to protect the inner plaster and paintings when covered with internal wall insulation [1]. Through this technological progress the inner surface can be protected but then the surface is hidden, which would lead either to a loss of visibility of historic surface or to a not executed energetic refurbishment. In order not to hide original internal wall surfaces of outside walls with conventional internal wall insulation, a transparent internal wall insulation made of heat-insulating glazing is developed.

The idea to use glazing in front of historic walls is not new. Typically glazing is used as a protection for conservational purpose. Sometimes climatization technics are additionally used to ensure an appropriate local conservational climate behind the glass. New instead is the approach of using glazing for energy saving primarily and benefit additionally from other characteristics of glazing, like transparency and protection. The structure of the construction with smooth glazing and uneven inner surface of the outer wall creates inevitably an air gap between the glazing and the outer wall. As it is well known, an air gap between internal wall insulation and the original wall must be avoided and so, most importantly, the air infiltration from indoor air to the air gap [2]. This is in order to prevent infiltration of mould spores and moisture into the air gap and in consequence possible mould growth on the original surface.

The glazing is truly tight for air infiltration but there are necessarily construction joints between glazing, load bearing metal sheets and the original wall, so that we assume that the airtightness cannot be ensured completely. To guarantee satisfying air conditions in the air gap and on the original surface respectively we choose an innovative climatization technique with heating cable and control unit to sustain a sufficient local climate.

### 1.2 DESCRIPTION OF THE SYSTEM AND CONTINIOUS MONITORING

The transparent internal insulation with the glazing (U-value 1.1 W/m²K) is arranged on the west-facing exterior wall on the upper floor of the listed building of the Alte Schäfflerei [3] at the Benediktbeuern monastery.

The existing window (see middle of Figure 1) has some special features, on the left half it is partly covered with masonry, but only on the indoor side. This was apparently necessary for implementing temporary wall installations for house-in-house apartments housing refugees in the late 1940s. In order to refurbish this complex wall/window section energetically and to keep this interesting part of the building and its history still visible, thermally insulating glass was installed in front of the wall. Since this area is aired with outside air by the existing window no moisture problems are to be feared. Therefore an additional test field was installed in front of an undisturbed wall section in order to test thermally insulating glass as transparent internal wall insulation. The wall surfaces were examined and conserved by conservators in advance [4].

The tested area of the transparent internal wall insulation is shown in Figure 1 with red dashed rectangle. The size of the visible part of the glazing is 0.83 m x 3.07 m = 2.55 m² and the thickness of the air gap is ca. 0.10 m. The sealing of the holding construction was carried out very carefully with sealing tapes and an additional permanently elastic joint. The feed-through of the measuring cable and power line for the heating cable was also sealed.

Before installing the glass construction, sensors (temperature, relative humidity) were placed at three different heights (0.2 m, 1.53 m and 2.86 m). Additionally, at the main measurement position at 1.53 m (middle of the construction) temperature and heat flux meters were installed on the original wall surface, on the glazing as well as on the outside and inside (only temperature sensors). With the measurement setup it is possible to calculate the in situ measured R-value of the construction. The different measurement positions distributed over the height of the wall enables the observation of the local distribution of temperature and relative humidity. The heating cable is located at the bottom of the construction inside the air gap on the horizontal glass holding construction. The energy consumption of the heating cable is measured with an electricity meter. The measured data is recorded in a 1-minute interval.

**Figure 1. Transparent internal wall insulation on the west-side in the upper floor of the Alte Schäfflerei. The left part of the transparent internal wall insulation (dashed rectangle) is examined and evaluated.**

## 2. METHOD

Depending on the moisture content of the outer wall (protection against driving rain) and the amount of air that possibly flows in from the room air, different moisture conditions can arise in the air gap. In order to avoid high air humidity and thus the risk of mould, the air gap in the construction is heated by means of the heating cable. Because of the temperature dependency of the relative humidity, a temperature increase leads to a decrease in the relative humidity in the air gap. This method enables a sufficient control of the relative humidity in air gap and on the original surface respectively.

The air change rate is measured with the tracer gas method "decreasing concentration" according to the standard [5]. One major problem is to satisfy the boundary condition mixing of air with tracer gas because the air gap is not accessible to bring in e.g. ventilators for optimal mixing. Therefore, the heating cable is turned on during the measurement to support air mixing by convection.

Heating the air gap to raise the temperature, however, requires energy. For reasons of economy and energy efficiency, the amount of energy used for this purpose should remain well below the amount of energy saved by the transparent interior insulation. Based on the relative humidity and temperature measured in the air gap, the surface humidity can be determined as a control variable for the energy-saving operation of the heating cable, if the surface temperature is known. The control algorithm for this test is determined in such a way that the surface humidity is to be kept below 70 % RH. The real test is intended to demonstrate the suitability of the transparent interior insulation in principle.

For comparison, the course of the surface temperature and the heat flow were also recorded on a corresponding uninsulated wall area on the east side (reference wall). The amount of energy savings by transparent internal wall insulation is calculated with integral of heat flux measurement of the wall with transparent internal wall insulation minus integral of heat flux measurement of the reference wall plus the energy consumption by heating cable. Additionally, the U-values and R-vales of the walls and wall components are calculated according to the standard [6].

During the heating period, the indoor air of the upper floor is heated to the target value of 20 °C and humidified to 50 % RH, so that a comparatively high humidity load is present.

## 3. RESULTS

According to the tracer gas measurement, the air change rate is about 0.26 $h^{-1}$ when the heating cable is on. Due to poor satisfaction of boundary conditions according to the standard, this is a rough estimation. The poor $R^2$ value of 0.9 of the linear regression shows the unfavourable mixing of tracer gas with air, Figure 2. But the tracer gas measurement confirms that there is an air exchange with indoor air. The moist indoor air infiltrates the air gap, despite high effort to seal and tighten the joints and connections of the construction.

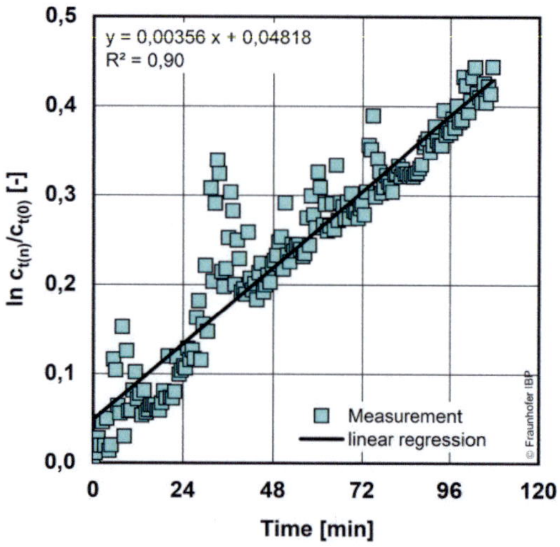


**Figure 2. Results of tracer gas measurement with linear regression of the data to obtain the parameters for calculation of the air change rate, which is at n = 0.26 $h^{-1}$.**

The transparent internal wall insulation was installed in September 2016. The course of the relative humidity in the air gap is well below the chosen limit of 70 % RH all the time up to now. But the relative humidity on the surface exceed in the first winter period the limit (2016/2017). This is due to adjustments of the control algorithm and control of heating cable (Figure 3). In the second winter period 2017/18 the heating system limits the relative humidity on the surface to 70 % RH. In the following years there is no need for heating within the air gap, except for the winter period 2020/21 and 2021/22, see Figure 3 (upper diagram, red circles). The electric heating power is recorded in Figure 3, bottom diagram. The heating power is at the highest in the winter period 2017/18. In the following years only single events can be observed for testing the system. In 2020/2021 the measurement system failed, but the effect of heating is clearly to see in a strict control of RH. The heating power in the period 2021/22 is much lower compared to the season 2017/18.

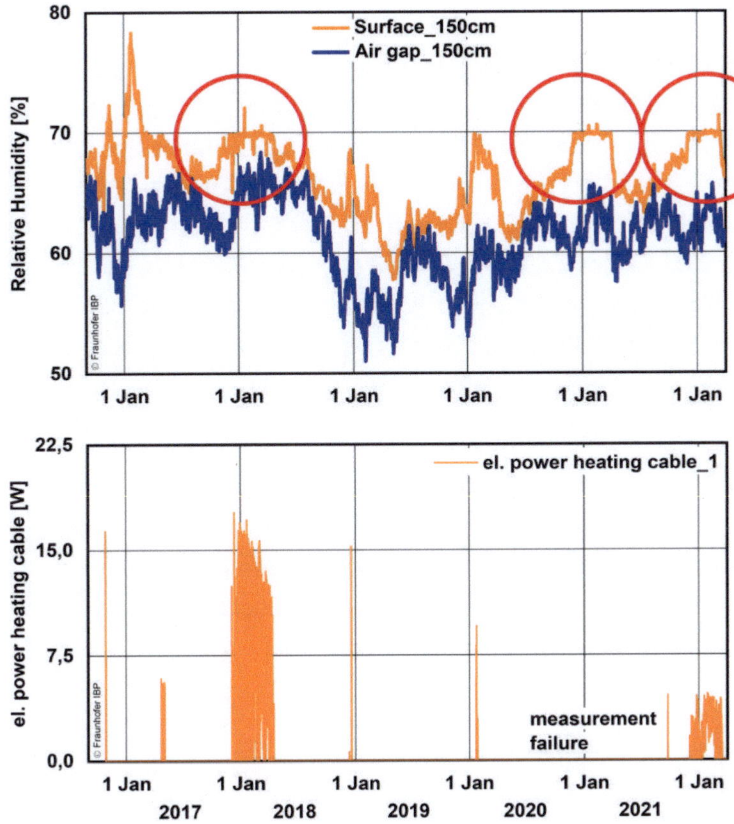


**Figure 3. Measurement results of relative humidity (upper diagram) and energy consumption of the heating cable from September 1st 2016 to April 1st 2022. Data is displayed as an hourly average of measured 1-Minute interval.**

The necessary heating power depends (beside the temperature level indoors and outdoors) on the moisture of the air gap, which is influenced by the unknown moisture content in the wall and the indoor air. A closer look at the relative humidity of the indoor air shows a high RH level in the season 2017/18 with values around 52 % RH and 48 % of the weekly average values (red circle on the left, Figure 4).

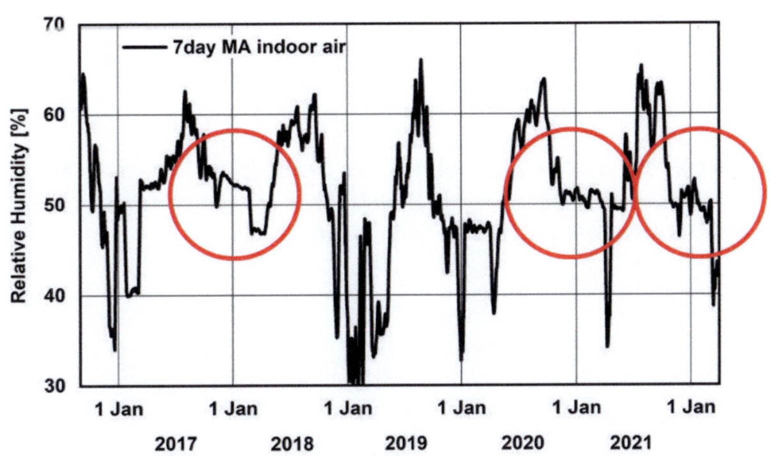


**Figure 4. Relative humidity of indoor air from September 1st 2016 to April 1st 2022. Data is displayed as a moving 7 day average (7d MA) of measured 1-Minute interval.**

In the following winter period 2107/2018 there is almost no humidification of indoor air, which results in low relative humidity for a longer time span and only in a higher RH level of indoor air for a short period. In that season a lower RH level in the air gap can be observed and no heating is necessary to limit the RH on the original wall surface. In the season 2019/20 the indoor air is humidified to ca. 48 % RH for the most time. The relative humidity rises on the original wall surface to the limit of 70 % RH but heating is almost not necessary. In the seasons 2020/21 and 2021/22 the indoor air is well humidified to a level of about 50 % RH (red circles on the right). In the same periods the relative humidity on the wall surface rises and heating is necessary to limit the RH level to 70 % RH. The moisture level in the air gap respectively on the wall surface seems to be clearly influenced by the moisture level of the indoor air. This confirms also the measured air infiltration by tracer gas measurement.

The thermal performance is expressed with thermal resistance R and thermal transmittance U. These values are necessary for dimensioning and calculation of performance certificates or heat load for heating systems of buildings. The transparent insulation consists of a thermally insulating glass pane with a U-value of 1.1 W/m²K according to manufacturer. The thermal resistance R and thermal transmittance coefficient U can be calculated with the in situ measurement of surface temperatures and heat flux according to [6], see Figure 5 and Table 1.

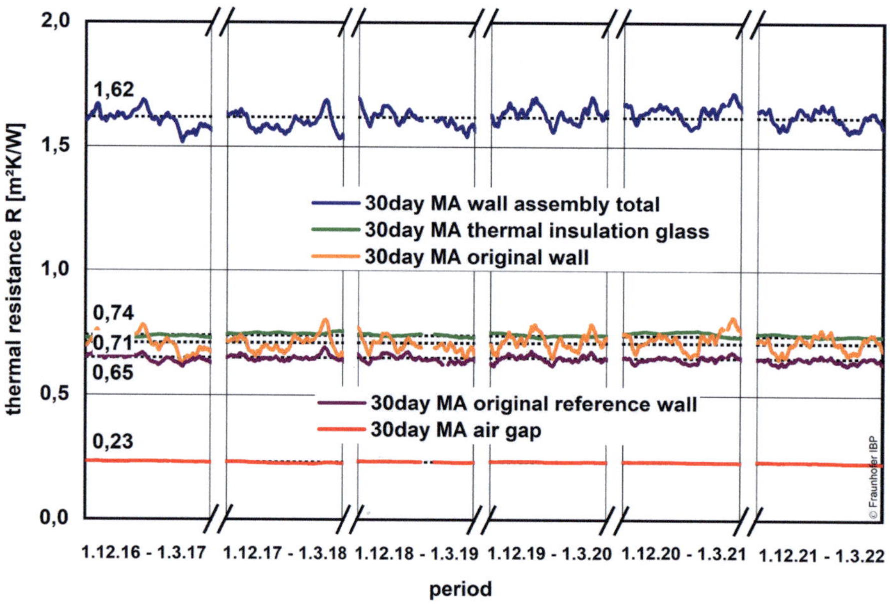


**Figure 5. R-values calculated from measured data for each winter period from December 1st to March 1st between 2016 and 2022. Data is displayed as 30 day moving average (30d MA) of measured 1-Minute interval. The numbers in the diagram give the average values for all winter periods.**

**Table 1. Calculated U-values (thermal transmittance) and R-values (thermal resistance) out of the measured data of the different wall components. The U-value is calculated out of the R-value with standard heat transfer coefficients.**

| | U-value [W/m²K] | R-value [m²K/W] |
|---|---|---|
| Original wall | **1.13** | 0.71 |
| Glass pane (acc. to manufacturer) | 1.1 | 0.74 |
| Wall total (with transparent glass insulation) | **0.56** | 1.62 |
| Reference wall | 1.22 | 0.65 |

The thermal insulating glazing plus the thermal resistance of the air layer is halving the U-value of the original wall of 1.13 $W/m^2K$ to 0.56 $W/m^2K$. The measured R-value of the glazing meets quite well the values provided by the manufacturer. Comparing the U-value of the original wall to the reference wall shows a slightly higher U-value for the reference wall. That implies an overestimation of the performance of the transparent internal wall insulation of about 7 % compared to the reference wall.

The difference between the measured heat flux of the reference wall and the heat flux of the transparent wall insulation gives the energy savings of the thermal insulating glazing. The calculated hourly difference of those heat fluxes and the hourly energy input with 1day moving average (1day MA) is displayed in Figure 6 (left diagram) for the period from December 1st 2017 to April 15th 2018. Figure 6 (diagram in the middle) shows the cumulative difference between the determined heat flow of the wall with transparent thermal insulation and the uninsulated reference wall as well as the cumulated energy consumption of the heating cable for this period. The cumulated difference to the reference wall is 29.5 $kWh/m^2$ compared to the cumulated energy input via heating cable of ca. 7.5 $kWh/m^2$. This results in energy savings of ca. 22 $kWh/m^2$ for the period 2017-2018. The second period from December 1st 2021 to March 30th 2022 results also in ca. 22 $kWh/m^2$ energy savings (Figure 6, right diagram).

Not considered here is the overestimation of the slightly higher U-value of the original reference wall compared to the U-value of the original wall with internal transparent insulation and the different orientation (west and east). Also the influence of thermal bridging of the framing of the transparent internal wall insulation is not considered in detail. For comparison of the cumulated heat fluxes, the heat flux meter is used, which is installed on the original surface. The additional heat fluxes via thermal bridging of the holding system of the glazing are therefore partly included in the heat flux measurement.

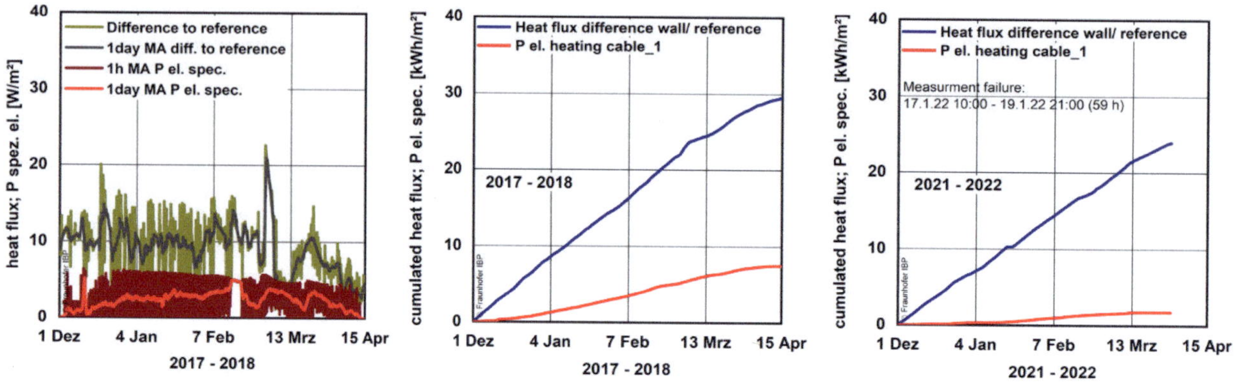


**Figure 6. Left: Difference of heat fluxes of uninsulated reference wall and transparent insulated wall as well as specific energy output of heating. Middle and right: Difference of heat fluxes cumulated and additional cumulative specific energy output of heating for the period 2017/18 and 2021/22.**

The areas to the right and left of the area intended for transparent interior insulation are covered with opaque interior insulation. The thermographic image from outside that was taken before the installation of the transparent insulation clearly shows the higher heat transmission through the still uninsulated central field (Figure 7 left image, red frame). The thermographic image made after the installation instead (Figure 7 right image, red frame), shows the effect of the transparent thermal insulation.

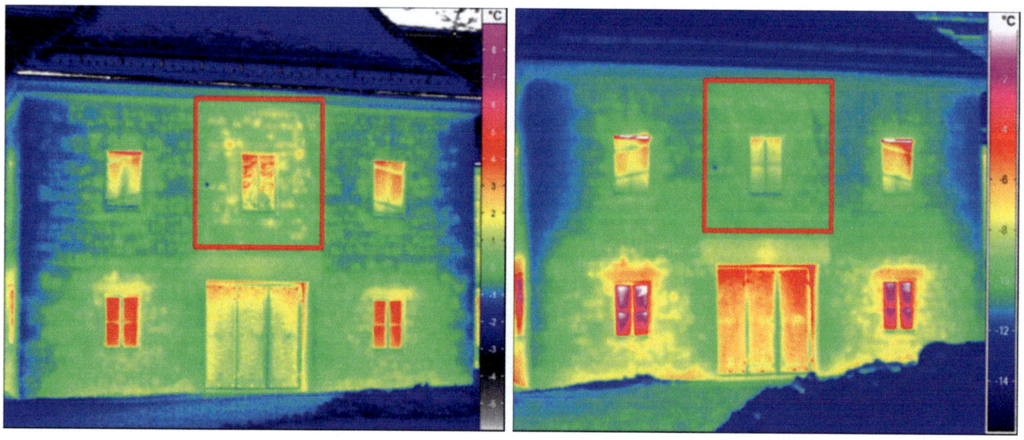

**Figure 7. IR-image of the west side of the Alte Schäfflerei before (left image) and after (right image) installation of the transparent internal wall insulation (red rectangle). After the insulation, the outside wall is colder, which indicates the reduction in heat losses.**

## 4. SUMMARY

The results show that interior insulation is possible even for valuable interior surfaces that are worth seeing. The transparent internal wall insulation with insulating glass presented here brings significant energy savings despite the intermittent heating of the air space between the glazing and the exterior wall, which is necessary to prevent moisture problems or mould growth. The humidity level of the indoor climate influences the necessary energy input via heating cable. Also, the outside temperature level influences the system as well as the characteristics of the original wall. Therefore detailed planning is necessary in forefront of application.

A very simple solution with heating cable and a controller is used. With this type of insulation, the inside face of the wall remains visible while saving energy but is also protected from moisture-related and mechanical damage. Appropriate control algorithms can also be used, either to achieve further energy savings, to mitigate the problem of salt-contaminated surfaces or to comply with certain conservational climate requirements.

## 5. ACKNOWLEDGMENT

The project was supported by Saint-Gobain Isover G+H AG.

## 6. REFERENCES

[1] Kilian, R. et al. Energetische Untersuchungen und Optimierung von Innendämmungen. Abschlussbericht zum Forschungsprojekt, IBP-Bericht HTB-003/2017, Fraunhofer Institut für Bauphysik. Holzkirchen, 2017.
[2] Bichlmair, S.; Krus, M.; Kilian, R. "Testing a new method for VIP interior insulation for heritage buildings" in: 10th Nordic Symposium on Building Physics NSB 2014. Full Papers, pp. 1349–1356. Lund, Sweden.
[3] www.denkmalpflege.fraunhofer.de
[4] Friedl, C.; Gietz, C.; Kimmerle, D. "Untersuchung und Konservierung eines Teilabschnitts der westlichen Innenwand in der Alten Schäfflerei des Klosters Benediktbeuern." Projektarbeit. TU München, 2015.
[5] EN ISO 12569 Thermal performance of buildings - Determination of air change in buildings – Tracer gas dilution method. German version 2001. Beuth-Verlag.
[6] ISO 9869 Thermal insulation - Building elements - In-situ measurement of thermal resistance and thermal transmittance. 2014.

# Improving the (energy) performance of (historic) buildings and communities: towards "neighbourhood Energy Performance Certification" (nEPC)

N.S. Brito [1,2]

[1] modular, arq:i+d, lda, Tv. de Montarroio, 2. 3000-288 Coimbra, Portugal. info@modular.pt
[2] University of Coimbra, ADAI, Department of Mechanical Engineering/MIT Portugal Program

**Abstract** – Energy Performance Certificates (EPCs), a "one size fits all" comparison initially designed to know more about European buildings, suddenly became a prescriptive energy efficiency strategy/funding metric and a "shadow" European Norm. To favour European buildings' comparability, centuries of local empiric knowledge are ignored towards "linear cycle" strategies (adding materials, trashing/replacing equipment, ...) that only deliver costs, letters, and virtual energy savings. As mandatory "minimum energy performance standards" are proposed, are EPCs the solution? Or the source for social unrest/disbelief fuelled by populist stakeholders? Acknowledging that every existing community has a history, specific needs and expectations, diverse buildings and people, this paper proposes that effective results require collective engagement. Neighbourhood scale EPCs—framed with/within the communities, scales (*micro, meso, macro*) and commitments they are part of—can be matched with 21st-century strategies/technologies to deliver what the EU needs: real decarbonization, energy security/poverty prevention, resilience, and a circular economy—with engaged citizens.

**Keywords** – Historic buildings; Neighbourhoods; Energy Efficiency; Decarbonization; Circular Economy

## 1. INTRODUCTION

*"At the current pace, the decarbonisation of the building sector would require centuries. (...) Numerous barriers stand in the way of higher renovation rates."* [1].

Collective problems cannot be solved individually, yet the European Union (EU) residential building decarbonization strategies favour individualism: from EU members' Directive transposition to home-by-home Energy Performance Certificates (EPCs), improvement measures, advice, financing, and local aid, almost all are processed and measured individually. This bureaucratic individualism imposes excessive costs, unneeded worries, reduced negotiation capacity and millions of European citizens' wasted hours while scattering the scale needed for robust circular business opportunities.

To illustrate 20 years of EPCs' failure "Introduction" exposes the oversimplified residential "science" that delivers fragmented practices/virtual results. "Arguments for neighbourhood scale approaches" identifies the advantages of tackling existing/historic buildings with(in) their background, constructive solutions and problems—well beyond energy— using 21st-century strategies. "Discussion" will focus on the challenges to include users and multidisciplinary teams—from programmers to experienced craftsmen—in the quest for contemporary ways to optimize the past, tackle neighbourhood scale issues, and mass customization to deliver more value with the same investment. "Conclusion" will propose "neighbourhood (Energy) Performance Certification" (nEPC) as local strategies to identify, validate and intertwine Traditional Knowledge with progressive decarbonization, towards the future we anticipate in the European 2050 long-term strategy [2].

## 1.1 ENERGY PERFORMANCE CERTIFICATION: 20 YEARS OF SELF-RECURRING SIMPLIFICATIONS

The Energy Performance Certificates (EPCs), designed back in 2002 "for consumers to compare and assess the energy performance of the building" [4, p. L1/68] opted for a division between large-scale office/service buildings (over 1000sqm) and residential buildings/small service buildings. This division assumed that only large stakeholders could afford the teams/ calculation capacity required to accompany the evolving science and legislation, dispatching residential buildings with a simplified version. In the Portuguese residential EPCs this residential certification process assumes, for all households—ancient, contemporary, or innovative experiments—many simplifications:

- Number of users defined by "number of rooms +1": if someone can sleep there (9sqm and a window), then this room, even if not used as such, defines the water consumption baseline.
- all these virtual users are always at home spending energy, and ubiquitously at work too: if one office/school building assumes their virtual presence, they are also there from "9 to 5".
- households are considered as fully heated/cooled 24h/365 days while assuming that users will not adequately act to improve their comfort by correctly operating windows/shades.
- Energy consumption beyond acclimatization and domestic hot water is averaged nationally.
- Overall values divided by floor area: a family living in a small historic house with lower energy consumption often gets worst EPCs letters than equivalent families in new suburban houses.
- "Energy Efficiency Measures" (EEMs) are calculated from these overestimated consumptions while excluding maintenance costs. A state-sanctioned document induces its "clients" in error by publicizing a "payback" that does not include significant annual maintenance costs.
- only these EEMs are considered for financial support, and only replacement (linear cycle) is supported, with maintenance or improvement (circularity) strictly excluded from financing.

Relying on assumptions instead of consumptions means that a better certification letter does not reflect lower carbon emissions. In two identical homes with 2 bedrooms the use patterns vary for a retired widow or a young family of 5, and so do the risks of technology. Nevertheless, similar prescriptive "improvement measures" are mandatorily proposed for similar home typologies.

## 1.2 EXPOSING THE IMPACTS OF THE ENERGY PERFORMANCE CERTIFICATION (EPC) LINEAR WORKFLOWS

The current EPC approach is a linear workflow, in process and results. Tackling each fraction of a building individually, requiring one EPC expert visit as it is sold/rented/rehabilitated induces excessive travels, hampers contextual approaches, and makes collective learning impossible. In almost 20 years EPCs evolved from comparing ancient buildings with new ones (comparing "apples to oranges") to a strategy proposing orange's peel and pith to better "protect" apples. The 2021 Revision of the EPBD [1] now proposes a *"phased introduction of mandatory minimum energy performance standards (...), and to extend progressively the requirements to other buildings"*, an alias for a "shadow"-standard proposing all apples to look like oranges. Are these prescriptive and mostly imported fossil-energy-based solutions the only way forward? Will investment in insulation plastics, leaky greenhouse gas emissions heat pumps, rare earth renewable energy sources, new suburban "green" Nearly Zero Energy buildings, high embodied energy electric cars, ... cut European emissions? Will displacing our emissions to less "climate-engaged" countries solve our worldwide problem?

Figure 1 illustrates a map of EPCs issued for 25 residential buildings around the Montarroio case study within a universe of 76 buildings in the image. Between 2012-22 only one-third (25 in 76) were certified by 13 different experts on different dates. Assuming an average EPC cost of 400€ per building with 3 floors (all taxes included), and around 10 km drive per visit implies that certifying 25 buildings twice (initial buy and after renovation) cost around 20.000€ and imposed 500km drive inside Coimbra. By regulatory imposition 13 experts, including the author, visited the area. Were they there, and aware of where they were? Searching for "Montarroio" in the EPCs portal [3] five decades old abandoned references to "Eastern"/"Western" streets of Montarroio, and an EPC for an inexistent "1st floor left" in the building this author owns. In short, an absolute absence of context.

The letters in Figure 1 represent the worst performing fraction of most buildings after renovation: an excessive cost for those results. The prescriptive improvement measures follow the European guidelines of exterior/interior insulation and double-glazing windows that would alter the image, thus mostly discarded in this UNESCO protection area. The proposed and/or applied measures are not available in the EPCs´ portal [3], yet Figure 1 shows no solar panels although regulations allow "5 % of total roof area (...) up to 3sqm", while forbidding heat-pump evaporators even if disguised [4].

**Figure 1: Energy Performance Certification (EPC) of Montarroio case study, boxed in red, and neighbourhood. In 10 years of EPC, only 33% were certified and none of the "rehabilitated" residential buildings includes solar panels, although allowed up to 3sqm. Source: author sketches over GIS [4] and Google maps.**

Although the Montarroio Case Study leaflet [5] demonstrates that most prescriptive EE measures are inadequate for historic areas, there is no incentive for these 13 EPCs "Qualified Experts" to enforce scale and reduce installation/maintenance costs, nor to use the municipal roof space (in blue rectangles) for safe/aesthetic renewable energy installations, making use of the great potential in neighbourhood scale approaches [6]. In short, individualist approaches residential buildings increase technological risks, operation and maintenance costs— as more time is spent in scattered travels than in specialized work—, and lower owners' negotiation capacity. And all this individualism postpones community-scale approaches, able to solve local issues, hampers attractive win-win circular economy business opportunities, maintenance habits and local green jobs.

## 2. ARGUMENTS FOR NEIGHBORHOOD SCALE APPROACHES

Many historic buildings and neighbourhoods (HB&N) managed centuries of versatility and resilience with low energy needs by applying Traditional Knowledge [7]. When energy was expensive the use, reuse, and recycling of local materials, together with an investment in good construction and shared practices ensured negligible energy needs, and lower construction and operational costs. Together with their users, historic buildings and neighbourhoods (HB&N) already portray centuries of alignment with the New European Bauhaus [8] and Circular Economy goals.

The higher availability/lower cost of (fossil) energy sources favoured new contexts when Climate Change was not an issue, and nuclear promised never-ending energy. New constructive processes, uses for energy and forms of living evolved from high-embodied energy materials and linear cycle approaches, promoting urban sprawl and globalization. At that time reducing the exterior wall thickness favoured more interior space, decreased the use of materials/transportation, lowered construction costs and promoted higher urban density. At that time, it was the obvious strategy.

The recent awareness of collective environmental, economic, social and cultural objections to the contemporary *status quo* make space for renewed approaches to residential buildings. A 5W approach will guide the arguments for neighbourhood-scale approaches.

### 2.1 WHY ARE NEIGHBORHOOD RESIDENTIAL EPCS INTERESTING AND/OR FEASIBLE?

Individual residential EPCs are often oversimplified views (see 1.1) of individual homes: complex events are averaged to deliver streamlined economic "one size fits all" solutions, as detailed characterization would render individual EPC costs prohibitive. Air renovation rates, assumed as constant, are a useful example: prescriptive "air engines"—natural as stack effect or mechanically induced—"promise" indoor air quality 24 hours per day for a defined number of users, present or not. This "steady state" does not identify, nor react to, absence, a family party or simple failure.

Large service buildings use software like Energyplus [9] and WUFI [10] to identify scenarios and sensors/actuators —meteorological stations, CO2 sensors, fans, filters, and heat recovery systems— to retrieve information and deliver dynamic comfort/safety while reducing emissions and bills. These would be too expensive in individual residential buildings, but at a neighbourhood scale a shared meteorological station, IoT sensors and actuators would deliver more comfort, safety and lower bills. Are there valid reasons to keep the current "individual residential EPCs" strategy?

Table 1 proposes a SWOT analysis based on information retrieved from the Montarroio area (1.2), a rising commercial-pressure historic area with renovation rates of 33%. In 20 years of EPBD (2002-2022) most certified buildings are those sold and renovated/under renovation, while the other two-thirds have no information. Can local governments act with no/inadequate information?

**Table 1: SWOT analysis of the current European RESIDENTIAL Energy Performance Certificates strategy (PT)**

| **Strengths** | **Opportunities** |
|---|---|
| EPCs being issued daily for each rented/sold fraction<br>20 years of data (of varying assumptions and quality)<br>Recognized (mandatory) document | EPCs require improvement and conceptual updates<br>Growth: 66% out-of-market fractions uncertified<br>To become more than a legal obligation/paper |
| **Weaknesses** | **Threats** |
| Low coverage rates: 33%, only when mandatory!<br>Expensive & inefficient linear cycle approaches (1.2)<br>EPCs are individual snapshots, not collective *loci* views<br>Residential emissions stable since 1990 [11]<br>Virtual EPC letters are useless for real investments | Virtual letters instead of real consumptions decouple investments from real savings/ability to react to risks<br>Excessive dependence on imported products/parts<br>*"lack of trust in the energy savings that renovation will achieve"* [1], and risk of social unrest if imposed; |

There is space for improving the EPBD with 21$^{st}$-century IoT low-cost approaches, to scale up lower carbon emissions neighbourhood by neighbourhood, and to extend value well beyond energy. Can it be done by making use of the already existing strategies and institutions?

## 2.2 "WHAT" & "WHEN": URBAN BUILDING ENERGY MODELLING (UBEM)

"Neighbourhood" EPCs are possible in EPBD [1] for repetitive multifamily residential fractions. Extrapolation to historic neighbourhoods can integrate constructive traditions, collective needs, and much more. Documenting complementarity, "pooling" and "bundling", is essential for "Alternative financing schemes for energy efficiency in buildings" [12]. UBEM uses representative buildings and their consumptions as archetypes for "bottom-up urban building energy models (UBEM)" [13], but the detail is needed to match low-exergy solutions with "the density and diversity of loads in urban systems" [14]. The legal frameworks, technologies and software are already available now, but only numeric parameters depict people and communities. Will humans (users, owners, communities, decision-makers, ...) mobilize to change only with "spreadsheet" approaches?

## 2.3 WHO? ONBOARDING, NEIGHBOURHOOD BY NEIGHBOURHOOD

Innovation is often confused with a *tabula rasa*, yet its Latin etymology, *innovare*, points to 'renewed, altered' [15], a "re-usage" approach familiar to those who understand historic buildings. Continuity with evolution, including humans, will render more probability of success. Owners/renters would appreciate an EPC for a lower/no cost as long as "Minimum Energy Performance Standards" (MEPS)/aka mandatory "improvements" do not become a risk. Local communities' participation can deliver more than decarbonization, as change must be attractive to many. Business decision-makers need real information and a generic direction to better plan their future, and validate potential investments, as economical sustainability matters. And (local) governments need the information to prioritize investments to reach the 2030/50 goals. A "win-win" for all stakeholders?

## 2.4 HOW? A HUMAN ROLE IN DIGITAL TECHNOLOGIES

One home cannot afford multidisciplinary teams, but a neighbourhood can. Yet participatory processes often forget that most people cannot read architectural plans/cuts or understand the sustainability value of "shaving peak loads" by authorizing demand-side load control. Collective debates require context to make conversations attractive, and here digital technologies have great potential. Matching visually attractive illustrations of building physics (Figure 2) with 3D printed models and the knowledge of those who use spaces enhances proposals and ensures awareness of their advantages and limitations, essential for better use and future optimization.

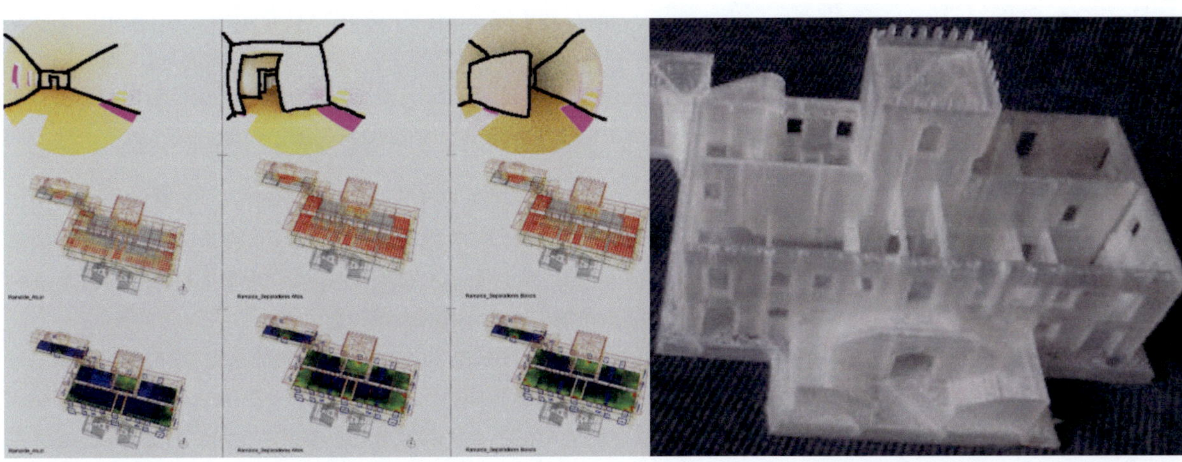

**Figure 2: "Hardcore" building physics made simple. The columns to the left depict a graphic comparison of lighting parameters in the original open space (left), and for partition alternatives: complete (middle) and partial (right). Aligned with a 3D printed model and some explanations, graphical comparisons facilitate collective debate/experimentation with building users (Source: author, from data retrieved at DRCN)**

Aligning graphic representations of the original with alternatives (Figure 2) allows untrained users to visualize change in parameters, whatever they are; while 3D prints facilitate a "birds-eye" view that triggers new thoughts and proposed actions with bits of paper: a "hands-on" debate.

In neighbourhoods, these "graphic approaches" can build upon EN16883 on "Guidelines for improving the energy performance of historic buildings" [16] to organise neighbourhood context. Using EN16883 for neighbourhood scale enforcement of good practices—from participatory diagnosis and digital approaches to optimization—can deliver new strategies to the *"decarbonisation of the building sector"* [1]. Yet this European Norm, a paid "paper" document, must become a free digital process, optimized to embrace contributions by all (local) stakeholders.

## 3. DISCUSSION: FROM EEHB TO DECARBONIZED HISTORIC NEIGHBORHOODS

Assuming buildings as an energy problem is a problem in itself, as "Buildings don't use energy: people do" [17]. Although the Energy Performance of Buildings Directive (EPBD) in revision [7] is a subsidiary of other higher-level European commitments, it is not clear how its goals, objectives, and performance indicators will align with global intents. As the "New European Bauhaus" [8] strives to reduce emissions by at least 55% by 2030 in desirable pathways (Energy Transition / Security, Circular Economy, ...) within a "beautiful|sustainable|together" [8] path, improvement is possible.

The "European Cultural Heritage Green Paper**"** [18] illustrates the potential, scope and diversity of inputs that cultural heritage has to contribute to European commitments. Can a sector that traditionally focused on heritage landmarks now help (historic) buildings users and the communities they define? Can digital technologies already common in new construction/deep renovation overcome the resistance in heritage professionals that advocate all interventions in historic buildings as handcraft? Although nothing can replace the knowledge of artisans and heritage experts, technology can reduce costs by providing more and better information, reducing unplanned travels, optimizing repetitive tasks, accelerating delivery, and fostering a carbon-efficient operation.

Historic buildings and neighbourhoods (HB&N) were originally designed mostly in the way the 2050 goals of decarbonization/circularity are now aiming for. Adjustments—to buildings and ways of use—are necessary to fit current/future energy sources, increased density and comfort standards. Opportunities exist at the core of New European Bauhaus policies to match the EPBD with the Green Deal, Circular Economy, and citizen engagement efforts. From start to optimization, neighbourhood-scale streamlined processes liberate time/resources while matching global/local expectations. Better results for more people with smaller costs—economic and environmental—are possible by:

- evolving EN16883 to address neighbourhoods and digital approaches. Upgrading this norm into a "digital template" for (historic) neighbourhoods would guide most European municipalities into cooperative approaches, facilitate the work of project teams and favour traceability and measurable results, key factors for learning curves and optimization.
- matching low and high-detail digitalization at the neighbourhood scale to ensure that the "levels-of detail" depicts the needs/expectations of neighbourhoods well beyond energy and decarbonization. Health, quality of life and community engagement really make the difference.
- training interdisciplinary teams for applied investigation/practice to streamline replicable solutions—from design to better use— to optimize the value of community scale.
- Tackling energy (efficiency, sufficiency, action,...) [19] as addictive: aiming for energy sobriety?

- making use of Artificial Intelligence to better match centuries of tangible and intangible Traditional Knowledge with the complexity of new technologies, and the people using them.
- engaging with pilot communities/municipalities/industry to solve the "last mile": transforming an uninteresting "Business-to-consumer" market into attractive decarbonization strategies.

Neighbourhoods will only join if solutions solve local problems, enhance their resilience—not their dependence—and foster new perspectives. Matching our common 2050 goals with local community needs while engaging younger generations can deliver the missing diversity and scale.

## 4. CONCLUSION

*"Europe needs results, not nice EPC letters. And Historic Buildings can help deliver."* [20]

In the name of "Energy Efficiency", existing European buildings (historic, traditional, post-war, contemporary and those yet to be built) were framed by "one solution fits all" comparisons enforced by the Energy Performance Certificates (EPCs). This instrument, initially designed to know more about European buildings, became a "shadow" European Norm aiming to replace centuries of empiric knowledge, and future innovation, with reminiscences of 20th century "linear cycle past" characterized by the addition of materials and efficient equipment; and assumedly failed [1].

Aiming for a "performance" oriented future, Energy Performance Certificates (EPCs) imposed unified solutions to diverse buildings/communities while dismissing decades of neighbourhood adaptations and centuries of empirical knowledge, also known as culture. By tackling one household at a time, by proposing high-embodied energy "solutions", and by sourcing materials and/or equipment from outside Europe, EPCs are displacing European carbon emissions (and investment capacity) to less "climate-engaged" countries, delivering virtual letters instead of measurable results.

The Energy Efficiency in Historic Buildings (and Neighbourhoods) (EEHBn) must evolve from a niche to a leading stream. To acknowledge that all existing neighbourhoods, historic or not, have a history that makes them like they are, and not something else, is to admit that the EEHB community can influence over 98% of European neighbourhoods: the other 2% are still under construction.

As "Energy Efficiency" cannot be a goal "in itself", neighbourhood Energy Performance Certification (nEPC) strategies offer the scale and diversity to match interdisciplinary teams with local communities towards attractive decarbonised neighbourhoods. Matching the respect for buildings' history and their communities with 21st-century digital approaches can deliver what the EU needs: 2050 decarbonization, energy security, resilience, and a circular economy—with engaged citizens.


## ACKNOWLEDGMENTS

This research evolved with the support of the "Sustainable Energy Systems" University of Coimbra/MIT Portugal Program (SFRH/BD/51017/2010 FCT grant) and engaged commitment of modular, arq:i+d, lda, Bosch Termotecnologia, S.A., DRCN (Heritage North) and highly engaged staff. Specific thanks to Snapmaker (https://snapmaker.com/) for supplying the "3-in-1"printer used for the model in Figure 2 and TUCAB (https://www.tucab.pt) for Portuguese-made filament for the great results only briefly presented. A special highlight to all of us organizing the workshop "Recording historic buildings using digital workflows – Designing the intersection from 3D model to building simulation" for a glimpse of a streamlined future (https://www.eehb2022.org/?page_id=88).

## REFERENCES


[1] "Revision of the EPBD (2021 Inception Impact Assessment)," *Have your say*, Mar. 22, 2021. https://ec.europa.eu/info/law/better-regulation/have-your-say/initiatives/12910-Energy-efficiency-Revision-of-the-Energy-Performance-of-Buildings-Directive (accessed Mar. 23, 2021).

[2] "2050 long-term strategy," *Climate Action - European Commission*, Nov. 23, 2016. https://ec.europa.eu/clima/policies/strategies/2050_en (accessed Apr. 02, 2021).

[3] "Pesquisa de Certificados," *Sistema de Certificação Energética dos Edifícios*. https://www.sce.pt/pesquisa-certificados/ (accessed Apr. 20, 2022).

[4] "Aviso n.º 2129/2012 | DRE," *Diário da República Eletrónico*. https://dre.pt/ (accessed Apr. 24, 2022).

[5] N. S. Brito. "IEA EBC Annex 56 Detailed Case Study 'Montarroio,'" in *IEA EBC Annex 56 on "Cost-Effective Energy and Carbon Emissions Optimization in Building Renovation,"* University of Minho., Guimarães, Portugal: University of Minho, 2016. [Online]. Available: DOI: 10.13140/RG.2.1.1029.9600

[6] University of Coimbra / EfS, "EfS student awarded with the 1st prize in the 'VINCI Innovation Awards 2015' contest," 2016. http://www.uc.pt/en/efs/destaques/2016/vinci (accessed Nov. 07, 2016).

[7] N. Brito, G. Brites, R. Castela, P. Fonseca, and M. G. da Silva. "Learning from Traditional Knowledge towards engaged inhabiting" 2014. [Online]. Available: https://doi.org/10.13140/2.1.2561.9202

[8] "New European Bauhaus : beautiful, sustainable, together.," 2021. https://europa.eu/new-european-bauhaus/index_en (accessed Apr. 05, 2021).

[9] "EnergyPlus," *Building Technologies Office: EnergyPlus Energy Simulation Software*. http://apps1.eere.energy.gov/buildings/energyplus/ (accessed Jul. 15, 2013).

[10] "WUFI (en)." https://wufi.de/en/ (accessed Apr. 23, 2022).

[11] "Energy statistics - an overview." https://ec.europa.eu/eurostat/statistics-explained/index.php?title=Energy_statistics_-_an_overview (accessed Apr. 01, 2022).

[12] A. Bullier and C. Milin. "Alternative financing schemes for energy efficiency in buildings," in *ECEEE 2013 Summer Study proceedings*, France, 2013, pp. 795–805. Accessed: Jan. 20, 2015. [Online]. Available: http://proceedings.eceee.org/visabstrakt.php?event=3&doc=3-221-13

[13] C. F. Reinhart and C. Cerezo Davila. "Urban building energy modeling – A review of a nascent field," *Build. Environ.*, vol. 97, pp. 196–202, Feb. 2016, doi: 10.1016/j.buildenv.2015.12.001.

[14] A. Allen, G. Henze, K. Baker, and G. Pavlak. "Evaluation of low-exergy heating and cooling systems and topology optimization for deep energy savings at the urban district level," *Energy Convers. Manag.*, vol. 222, p. 113106, Oct. 2020, doi: 10.1016/j.enconman.2020.113106.

[15] "innovate - definition of innovate in English | Oxford Dictionaries," *Oxford Dictionaries | English*. https://en.oxforddictionaries.com/definition/innovate (accessed Mar. 28, 2017).

[16] CEN - Technical Bodies, "EN16883:2017 Conservation of cultural heritage - Guidelines for improving the energy performance of historic buildings," 2017.

[17] K. B. Janda. "Buildings don't use energy: people do," *Archit. Sci. Rev.*, vol. 54, no. 1, pp. 15–22, 2011, doi: 10.3763/asre.2009.0050.

[18] A. Potts. *European Cultural Heritage Green Paper*. The Hague&Brussels: Europa Nostra, 2021. [Online]. Available: https://issuu.com/europanostra/docs/20210322-european_cultural_heritage_green_paper_fu

[19] E. Paschalidou, C. Fafet, and L. Milios. "A Strong Sustainability Framework for Digital Preservation of Cultural Heritage: Introducing the Eco-Sufficiency Perspective," *Heritage*, vol. 5, no. 2, Art. no. 2, Jun. 2022, doi: 10.3390/heritage5020058.

[20] N. S. Brito. "Mismatch, exclusion and inclusion: threats/opportunities for Historic Buildings in the current Energy Efficiency paradigm," in *Energy Efficiency and Comfort of Historic Buildings - Second International Conference Proceedings*, Brussels: Belgian Building Research Institute (BBRI, Flanders Heritage Agency, 2016. [Online]. Available: https://ihbconline.co.uk/newsachive/?p=16069

# Building stock categorization for energy retrofit of historic buildings: A literature review

L. Briz

Institute for Renewable Energy, Eurac Research, 39100 Bolzano, Italy; lara.brizbejerano@eurac.edu.

**Abstract –** The analysis of a large building stock requires some generalizations, e.g., dividing it into a few statistically representative categories. Due to the inhomogeneity of historic buildings and the influence of their technical characteristics and cultural values on their energy saving capabilities, existing categorization methods for younger building stocks are not easily transferable. Yet historic buildings account for one quarter of all European buildings, offering great opportunities for energy savings. This paper analyses recent studies regarding the main features used for the categorization of a building stock, and evaluates how cultural values are considered in such processes. The results indicate that, when including historic buildings, there is a need for more comprehensive categorization methods that consider their unique characteristics and systematically balance cultural, economic and environmental factors that result from retrofit measures applied at the building stock level.

**Keywords** – building stock categorization, building typologies, historic buildings, energy retrofit, cultural value

## 1. INTRODUCTION

Buildings in the EU are responsible for 40% of energy use and 36% of greenhouse gas emissions, which mainly stem from construction, usage, renovation and demolition [1]. Nowadays, the annual growth rate of new constructions is very low; in the residential sector, it is around 0,7% [2]. Consequently, reducing existing stock's energy consumption and enhancing building operations' flexibility are essential strategies to meeting the EU's 2050 decarbonization agenda, reducing reliance on fossil fuels, and mitigating climate change risks.

A pre-assessment of the building stock is an important step in energy retrofit plans for a large number of buildings, and here the building stock categorization methodology is often applied. In practice, it is not possible to analyse all the buildings one by one, so the building stock is grouped into a limited number of categories with similar or comparable features, see Figure 1. Understanding the typical energy behaviour for each representative building will help to ensure that appropriate energy efficiency measures are selected.

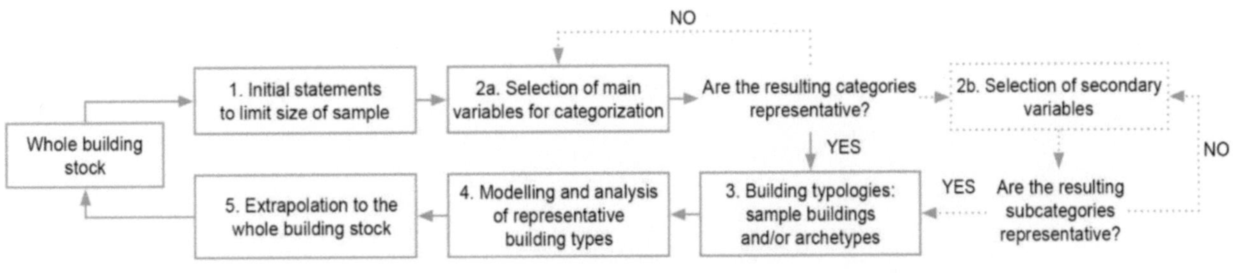


**Figure 1. Main steps involved in categorizing a building stock, including selecting variables for categorization and optionally selecting secondary variables to create subcategories.**

While there has been a surge in recent years in the use of categorizations to calculate the energy performance of a building stock, few studies have incorporated aspects regarding historic buildings. Its inherent complexity and nuanced cultural value contrast with the quantitative nature of the categorization process, and due to the nonstandard materials used prior to industrialization, providing a fair assessment of the energy performance of the resulting representative buildings and extending these results to a larger group is challenging. Consequently, historic buildings are usually left out of categorization processes or included without the appropriate caution, risking losing both their structural integrity and their cultural significance.

Yet historic buildings comprise one quarter of all EU residential buildings [3], so they represent significant energy savings potential at the building stock level. Thus, this paper examines recent literature regarding the categorization of a building stock, with a particular focus on historic building stocks. Which constructive, physical but also cultural features are usually taken into account, thus shedding light on the factors that influence the definition of building typologies when planning renovation strategies for historic buildings.

## 2. METHOD

This work is based on an extensive literature review of relevant studies published in the years 2010-2021, accessed from databases such as ScienceDirect (Elsevier), Google Scholar, or Scopus. Search terms included "building stock", "typologies", "categorization", "energy retrofit", "historic buildings", "renovation", "heritage", and "cultural value". After sorting out only the publications that considered building stock analysis for the purpose of energy efficiency, 43 articles were retrieved and classified according to the variables used for the categorization process. Although this list does not claim to be exhaustive, it provides a good overview and a basis for discussion on whether a more comprehensive approach to categorizing historic buildings for the purpose of energy retrofit is needed.

## 3. RESULTS

The development and use of categorization methodologies to calculate the energy performance of large building stocks has been a continuous focus since 2010 (Figure 2). It coincides with increasing demands for energy reductions in the built environment in Europe, which were expressed in several European directives, as well as in the launch of major research programs. Two large projects specifically targeting categorization of buildings, TABULA (2009-2012) and EFFESUS (2012-2016), seem to have contributed most to the sharp increase over these two periods.

**Figure 2. Identified articles on building stock categorization published between 2010 and 2021.**

As noted in Lidelöw et al. [4], few studies address large quantities of historic buildings, compared to those that refer to single heritage buildings or components. Though they are almost always studied as part of the building stock, their particular characteristics are rarely taken into account or analysed in more depth; much less their cultural significance. The result of the building stock analysis may exclude them indirectly precisely because of their special characteristics. Nemry et al. [5], for example, excluded 30% of the total EU25 building stock by referring to "special buildings", which include undoubtedly a large number of historic buildings.

In the process of categorizing a building stock, typologies are employed to classify objects with specific shared characteristics, thus creating unique combinations of attributes. Consequently, the importance of selecting correctly the main characteristics based on which the sample will be classified is clear. Depending on the research objectives and the data available, most articles select a few key features in the early stages of building stock analysis, which can later be supplemented by specific attributes that create additional subcategories. To avoid having too many and too diverse typologies, statements are also made at the beginning of the categorization process to narrow the scope of the sample, hence excluding or limiting a certain type of building, typically based on building uses (e.g. only residential).

In the analysed articles, two main groups of features are identified: those that relate primarily to energy use, and those that pertain more to the building's configuration. Figure 3 illustrates the results of categorizing the articles according to the proportion of features used to define the typologies of one or both groups, as well as the variables related to the cultural values. The reviewed articles most often focus on building stock energy consumption, and thus, as shown in Figure 3, the typologies are defined by energy-related characteristics; fewer studies have adopted a more comprehensive approach that considers other factors, such as local material availability or traditional uses of the space. Cultural values of buildings are included in the definitions of building typologies only in those articles that primarily focus on historic buildings. Such articles range from not mentioning it at all to identifying it as one of the main points of categorization. The two identified groups of features are presented more in-depth in the next sections, along with the various ways in which cultural values may be included in the categorization process of a building stock.

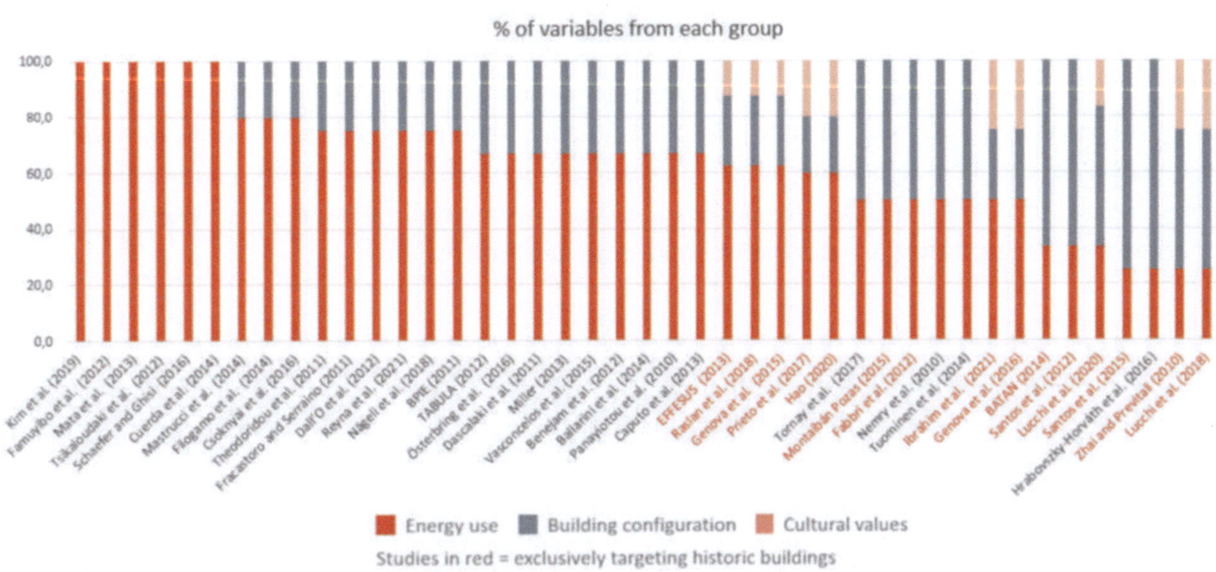


**Figure 3. On the diagram, the analysed articles are classified according to the percentage of variables from each group used for categorizing the building stock.**

## 3.1 FEATURES RELATED TO ENERGY BEHAVIOR

Studies looking at energy use characteristics link single-building energy analyses with statistical data to identify potential energy-saving measures at the building stock level. Variables most commonly used can be grouped into: building geometry, building envelope, climate zone, operation and systems, and social and behavioural aspects.

### *3.1.1 Building geometry*

Geometrical features such as volume, number of floors, or floor area are generally considered to be key factors when categorizing a building stock from an energy perspective. Beyond the obvious influence of volumetry on a building's energy behaviour, these are factors that are relatively easy to obtain today, either from building registries or from new tools such as satellite imagery or 3D city models. The major study on categorization TABULA [6] uses geometry, in addition to building age and climate, to define comparable building typologies in 20 European countries. In the years since, many studies within and outside the TABULA project have used the methodology or typologies derived from the project for further research [7]–[13].

### *3.1.2 Building envelope*

Many articles consider the building envelope in the process of classifying buildings by their energy performance, for obvious reasons. This includes both building aggregation (number of exposed facades) as well as parameters directly related to the thermal performance of the envelope, such as the U-values of the roof, floor, wall and windows, airtightness or glazing ratio. Miller [14], for example, used envelope features as the main characteristic to identify archetypes in Vancouver. Because of the nature of the sample, this parameter stood out as the most influencing one in the energy demand in an urban context, impacting such factors as glazing ratio, compactness, or solar radiation. Similarly, Cuerda et al. [15] used a façade cataloguing system to make a categorization based on envelope characteristics, though such a classification appears to be possible only because of the limited geographical and temporal scale of the sample, namely buildings built between 1950 and 1980 in a Madrid neighbourhood. In some cases, the data regarding envelope characteristics is missing or incomplete for the whole building stock. Some studies, such as Österbring et al. [16] or Fracastoro and Serraino [17], used building periods to derive typical U-values for that period, and to determine a building's envelope thermal characteristics from another building of similar age. When the scale allows it, on-site inspections can add valuable information about the building facades [18]. As evidenced by the analysis of the historic building stock of Calavino, Italy, by Lucchi et al. [8], this is of particular importance when dealing with historic buildings, as the documentation may not be complete in terms of occupancy, repair status, or subsequent undocumented renovations.

### *3.1.3 Climate zone*

According to the sample's scale and geography, climate zones will heavily influence a building's categorization. Tsikaloudaki et al. [19] suggested that climates within the EU could be the main factor when describing European building stocks based on energy performance. A typical approach looks only at heating degree days, but the authors considered both heating and cooling degree days when dividing the territory into climate zones. In accordance with other studies [20]–[23], they argued that energy used for cooling, especially in Mediterranean regions, is or could become a significant component of the energy balance. In fact, a growing number of building energy models include not only historical climate data, but also future climate scenarios. With climate change causing higher temperatures and more extreme weather events, it is becoming increasingly difficult to base energy retrofit plans on historical typical weather years. An example can be found in Hao [21], who categorizes heritage buildings of an alpine region in northern Italy based on climate change impacts. When it comes to historic buildings, the climate is even more important, due to how the local climate influenced traditional architecture. Thus, in her study, the author uses building typologies to analyse risks and adaptation plans for protected buildings.

### *3.1.4 Operation and systems*

As one of the main factors influencing a building's energy use, its operation is traditionally addressed in the early stages of the categorization process, regarding e.g. data on energy consumption or directly related features such as current building use. It seems that research is becoming increasingly interested in measuring the energy use of buildings beyond their operational use through life-cycle assessments; however, very few of the analysed studies consider it when categorizing buildings. In a study by Nemry et al. [5], the EU-25's building stock was categorized for the purpose of assessing buildings' life cycle impacts. In order to analyse representative buildings for this purpose, as well as the type of building and climatic zone, the categories were based on used materials and insulation. Another example is a study by Sartori et al. [24], which indirectly addresses this issue by utilizing construction, demolition and renovation flows to study the long term development of the Norwegian residential building stock.

When heating and cooling are considered a key variable of the categorization process, they are usually assessed in a second step of the process (See Figure 1, step 2b). As an example, studies by Dascalaki et al. [9] and Fracastoro and Serraino [17] create subcategories based on system installations, after applying the categorization method proposed by TABULA. Using the age periods, the latter calculates the efficiencies of heating systems according to the manufacturers from each period. Similar approaches can be read in [13], [16], [25], [26] or [24]. In the latter, as it occurred with the climate, the authors examine future predictions for categorizing the Norwegian building stock according to its systems and systems' efficiency, assuming that the building stock's renovation rates would evolve in accordance with European targets for $CO_2$ reduction. Since heating and cooling systems can be difficult to quantify in historic buildings, they are often excluded from categorization analysis, as was the case, for example, in Genova et al. [27] for the historic center of Palermo.

### *3.1.5 Social and behavioural aspects related to the energy use*

Studies may include variables related to social aspects if they are deemed relevant by the characteristics of the sample. For instance, Schaefer and Ghisi [31] used socioeconomic variables such as income level or family structure to obtain reference buildings in a low-income housing stock in Florianópolis, Southern Brazil. By doing so, other aspects, like occupant behaviour or heating systems were indirectly addressed. In another example, Zhai and Previtali [28] combined world climate zone boundaries with cultural heritage zones based on language families in order to identify different regions of ancient vernacular houses around the world. Though it is debatable whether language groups can be a criterion for dividing the territory into architectural typologies, the study is an important step towards incorporating energy-saving features related to vernacular architecture into current building codes.

In line with the review performed by Aguilera and Ossio [29], very few studies were found that considered the occupation patterns as well as the occupant behaviour as major factors when categorizing buildings for energy purposes. Studies like Dascalaki et al. [9] considered only the buildings that are continuously occupied (i.e. no summer vacation houses), while others like Sartori et al. [24] made a distinction for abandoned buildings since they do not contribute to the building stock's energy consumption. Some studies simply mention the existing limitations regarding occupancy of buildings and occupant behaviour since little data exists, and they are difficult to quantify. For instance, a study by Famuyibo et al. [30] revealed that 60% of the variation in energy consumption between the resulting archetypes of the Irish building stock and national databases was not explained, which the authors attributed, among other reasons, to the fact that occupant behaviour was excluded from the main variables used for categorization due to insufficient data.

Nevertheless, in the case of historic buildings, these two factors take on an even greater significance since many of these buildings are vacant or abandoned, and the behaviour of the occupant is usually an essential part of the bioclimatic design of traditional architecture. Accordingly, Lucchi et al. [31] investigated whether traditional categorization methods, in this case based on clustering analysis, would work for the historic city centre of Calavino, Italy. The results suggested that such methodologies are difficult to use for historic buildings stocks, since additional complex variables not usually considered, such as occupancy, conservation state, or heritage constraints, play a significant role.

## 3.2 FEATURES RELATED TO BUILDING CONFIGURATION

Aside from the energy assessment, studies that incorporate a greater number of variables relating to the configuration of the building will usually also attempt to improve the energy models used for simulations [32][33], or to understand the stock beyond the energy aspects, such as the structural behaviour for risk mitigation [34]. These variables are summarized into: space distribution, geographical location and materials, and construction period.

### *3.2.1 Space distribution*

Few of the identified studies considered space distribution to be a significant factor in categorization. This involves the existence of buffer spaces (areas adjacent to the building that are not heated), bioclimatic aspects in space design, seasonal heating patterns, etc., which are common in traditional architecture, but not so much in modern one. Consequently, most studies that accounted for it focused on historic building stocks. For example, Santos et al. [34] identified historic typologies in the Portuguese city of Seixal considering, among other parameters, the spatial distribution inside the houses and typical changes made to meet successive comfort and hygiene standards (e.g. additions to façade or courtyards, internal bathrooms, new coatings, larger windows...). Based on an updated building configuration, the authors could develop typologies that offered a range of levels of vulnerability and suitable interventions when it came to assessing seismic and fire risks. In a similar vein, Montalbán Pozas [33] emphasizes the importance of understanding how energy was used in traditional buildings so that improvements can be proposed accordingly. To accomplish this, additional parameters are considered in the energy simulations of the identified typologies, including the use of spaces at different times of the day and throughout the year, customs, and even traditional clothing.

### *3.2.2 Geographical location and materials*

While location is often used as an indicator of the climatic conditions that influence a building's energy behaviour, other geographical factors that, in a way, also define a building's shape and configuration are not so commonly considered. Such factors include orientation, sunlight exposure, vegetation or availability of local materials; but also cultural influences, religion, and local traditions, which as a whole help define the architecture of a place. Thus, for example, although materials are sometimes used as indicators of U-values or insulation levels, they are not generally considered as a local resource that defines construction typologies by itself. The built heritage is, however, heavily influenced by the availability of raw materials in a region. One of the few studies that consider dominant local materials as one of the main parameters to define building typologies, and without limiting the stock to historic buildings, is from Tornay et al. [7], where the whole French building stock is characterized based on urban typology, building use, construction period, and geographical location regarding building materials. The French project BATAN [32] goes one step further and defines so-called "thermal typologies" for the national historic building stock, based on three levels: aggregation at the building level, inertia of the different construction techniques at the component level and traditional materials density at the material level. Similarly, Zhai and Previtali [28] distinguish between massive and light building construction components both for walls and roofs, thus defining typologies of vernacular architecture that address different needs in terms of heat, ventilation, humidity, and extreme weather events.

### *3.2.3 Construction period*

Building age is considered by most articles as one of the most important characteristics when making building typologies. The underlying reason is the large amount of information that can be deduced from the construction period, with regard to building regulation codes, building's typology, systems and equipment used or construction practices applied, all of which are strongly connected to the building's energy performance. For example, Theodoridou et al. [35] categorized the Greek residential building stock based on the construction age only. Period shifts are normally determined by new energy regulations, if they exist. Tommerup and Svendsen [36], for example, divided the Danish residential building stock into seven construction periods; the first three periods representing a shift in architectural style and the last four representing a shift in thermal insulation requirements. In terms of heritage buildings, a quick look at the countries participating in the TABULA project reveals that there is no consensus regarding the date from which buildings are considered historic. Depending on the context, limits can be associated with historical or technological developments, or with thermal regulations. As an example, they might be referred to as buildings built before 1948, or prior to the introduction of thermal regulations in the 1970s, or making still a distinction for those built in a transitional period between 1920-1950, as the period when local materials gave way to industrial ones. As a matter of fact, Fabbri [37] suggests that "historical thresholds" do not always coincide with "historical energy thresholds", therefore categorizations where the main focus is not heritage buildings will likely ignore the former.

## 3.3 ANALYSIS OF CULTURAL VALUES

It is surprising that only studies that focused exclusively on historic buildings mentioned cultural values while categorizing a building stock, since a huge portion of the total number of studies examined samples that contained historic buildings, and the risk of destroying cultural value is always present when energy retrofits are applied at a stock level. Sometimes, this can be attributed to the purpose of the categorization. If the primary objective is not a retrofit energy assessment, discussing cultural values may not have any relevance; as in [7], [32] or [34]. Alternatively, the implicit reason may lie in the lack of information about historic buildings' cultural significance, which is available only for protected buildings, rendering categorization a difficult process. It is common for national or local directives not to apply energy standards to buildings designated as officially protected or historically valuable, so one might conclude that this omission simply results from the fact that the energy improvement measures resulting from the study will not be implemented in "valuable" historic buildings. However, the protected stock represents only a small portion of the historic stock, meaning that for the vast majority of historic buildings there is no information available in public databases about their cultural significance or structural integrity [38], making their fate difficult to determine in terms of restoration strategies, retrofit options, and energy policies.

#### *3.3.1 What is understood as cultural value*

A lack of consensus exists about what cultural value is and how it ought to be addressed in energy retrofits for historic buildings, as evidenced by both the omission of the subject and the extensive debates it induces. Lidelöw et al. [39] conducted a comprehensive analysis of how cultural values are addressed when planning energy interventions for historic buildings, dividing the projects into three groups: those omitting the topic, those that carry out implicit discussions and take into account existing charts and conventions, and those that attempt to find new ways of understanding cultural values today. Taking a closer look at the latter, a trend seems to be emerging that we no longer rely on indicators to determine cultural heritage, but rather consider it as part of social construction where "preserving by using" takes on a greater significance. Although this new, qualitative approach may be beneficial for energy renovations, it seems to conflict with the quantitative nature of the categorization process, making it difficult to strike a balance between valuing the cultural significance and uniqueness of individual historic buildings, versus the benefits of taking a stock-level approach in energy renovation strategies.

#### *3.3.2 How it is included*

Articles which consider cultural values in the categorization process are mostly confined to local contexts, where historic architecture is sufficiently homogenous and, therefore, standardized solutions that disregard the particularities and principles of conservation of individual buildings are more unlikely to be applied [40]. The additional variables that are considered usually involve protection level, traditional features, conservation states or utilization levels of the building stock. The major project EFFESUS [41] analyses the potential and consequences of energy retrofits for heritage buildings by transforming degrees of protection into actual energy retrofit measures. The typologies are mainly based on geometrical characteristics, and the main character defining elements of each typology, distinguishing between visual, material, and spatial characteristics, are identified. Using a scale of benefits and risks, each retrofit measure is then evaluated. Later studies have used the EFFESUS methodology in various contexts, such as [27], [42] and [43], or to assess a particular retrofit measure, such as Lucchi et al. [44], within the EFFESUS project, for the integration of photovoltaic systems in the historic city center of Santiago de Compostela, Spain.

## 4. CONCLUSIONS AND FUTURE DIRECTIONS

In this article the author tried to make a distinction between the variables used to categorize a building stock, distinguishing between variables focused on energy use and variables focused on building characteristics. The results indicate that, with a few exceptions, the latter are generally ignored when categorizing a building stock that is not exclusively historic. In spite of the apparent obviousness of these results, they help illuminate the characteristics that are distinctive of historic buildings and that are overlooked when these buildings are considered as part of a broader classification. It has also been observed that the features chosen for categorization are largely determined by two factors: the data available for analysis, and the classification objective itself. Regarding the latter, while energy efficiency is generally a goal in historic buildings, its level of achievement is heavily influenced by the building's cultural values and structural capabilities. Therefore, rather than trying to attain a certain energy-class or a number, for some historic buildings it might be more realistic to strive for just thermal comfort that meets current standards and keep the building in use, which is known to be the most effective preservation method, or to ensure structural integrity both from normal use as well as from new challenges, such as climate change.

It can be concluded that there is substantial uncertainty and ambiguity regarding the principles or methodologies used in identifying historic buildings typologies, and that traditional categorization methods may not always be effective due to both the nature of the sample and the objectives of the study being very different. Rather than overlooking or excluding them, a more comprehensive approach is necessary when classifying a building stock that include historic buildings, accepting the additional effort it may entail for the additional value we receive in return. While in the case of protected buildings there is a common acceptance to invest more resources or make exceptions in the name of heritage preservation, such an approach should be extended to all historic buildings that, even though lacking an exceptional individual value, when considered as a whole represent an irreplaceable heritage and symbolize the history and culture of a place.

## 5. REFERENCES


[1] European Commission. "In focus: Energy efficiency in buildings." Internet: https://ec.europa.eu/info/news/focus-energy-efficiency-buildings-2020-lut-17_en, 17 February 2020 [Apr. 02, 2022].

[2] NEED4B - New Energy Efficient Demonstration for Buildings, "Economic and Market Analysis of the EU Building Sector, Deliverable D8.1," 2015.

[3] A. Troi, "Historic buildings and city centres – the potential impact of conservation compatible energy refurbishment on climate protection and living conditions," International Conference Energy Management in Cultural Heritage, p. 10, 2011, [Online]. Available: http://www.3encult.eu/en/deliverables/Documents/2011_EM_Troi_Impact.pdf.

[4] S. Lidelöw, T. Örn, A. Luciani, and A. Rizzo, "Energy-efficiency measures for heritage buildings: A literature review," Sustain. Cities Soc., vol. 45, pp. 231–242, 2019.

[5] F. Nemry et al., "Options to reduce the environmental impacts of residential buildings in the European Union-Potential and costs," Energy Build., vol. 42, no. 7, pp. 976–984, 2010.

[6] TABULA, "Use of Building Typologies for Energy Performance Assessment of National Building Stocks. Existent Experiences in European Countries and Common Approach", 2010.

[7] N. Tornay, R. Schoetter, M. Bonhomme, S. Faraut, and V. Masson, "GENIUS: A methodology to define a detailed description of buildings for urban climate and building energy consumption simulations," Urban Clim., vol. 20, pp. 75–93, 2017.

[8] E. Lucchi, D. Exner, and V. D'Alonzo, "Building stock analysis as a method to assess the heritage value and the energy performance of an Alpine historical urban settlement," in Conference report of the 3rd International Conference on Energy Efficiency in Historic Buildings, Visby, Sweden, 2018, pp. 482–492.

[9] E. G. Dascalaki, K. G. Droutsa, C. A. Balaras, and S. Kontoyiannidis, "Building typologies as a tool for assessing the energy performance of residential buildings - A case study for the Hellenic building stock," Energy Build., vol. 43, no. 12, pp. 3400–3409, 2011.

[10] T. Csoknyai et al., "Building stock characteristics and energy performance of residential buildings in Eastern-European countries," Energy Build., vol. 132, pp. 39–52, 2016.

[11] I. Ballarini, S. P. Corgnati, V. Corrado, and N. Talà, "Improving energy modeling of large building stock through the development of archetype buildings," in Proceedings of Building Simulation 2011: 12th Conference of International Building Performance Simulation Association, Sydney, 14-16 November, pp. 2874–2881, 2011.

[12] I. Ballarini, S. P. Corgnati, and V. Corrado, "Use of reference buildings to assess the energy saving potentials of the residential building stock: The experience of TABULA project," Energy Policy, vol. 68, pp. 273–284, 2014.

[13] P. Caputo, G. Costa, and S. Ferrari, "A supporting method for defining energy strategies in the building sector at urban scale," Energy Policy, vol. 55, pp. 261–270, 2013.

[14] N. Miller, "Urban Form and Building Energy: Quantifiying Relationships Using a Multi-Scale Approach," Ph.D. dissertation, Dept. of Resour. Manag. and Envir. Studies. The University of British Columbia, 2013.

[15] E. Cuerda, M. Pérez, and J. Neila, "Facade typologies as a tool for selecting refurbishment measures for the Spanish residential building stock," Energy Build., vol. 76, no. Fig 1, pp. 119–129, 2014.

[16] M. Österbring, É. Mata, L. Thuvander, M. Mangold, F. Johnsson, and H. Wallbaum, "A differentiated description of building-stocks for a georeferenced urban bottom-up building-stock model," Energy Build., vol. 120, pp. 78–84, 2016.

[17] G. V. Fracastoro and M. Serraino, "A methodology for assessing the energy performance of large scale building stocks and possible applications," Energy Build., vol. 43, no. 4, pp. 844–852, 2011.

[18] G. Dall'O', A. Galante, and G. Pasetti, "A methodology for evaluating the potential energy savings of retrofitting residential building stocks," Sustain. Cities Soc., vol. 4, no. 1, pp. 12–21, 2012.

[19] K. Tsikaloudaki, K. Laskos, and D. Bikas, "On the establishment of climatic zones in Europe with regard to the energy performance of buildings," Energies, vol. 5, no. 1, pp. 32–44, 2012.

[20] M. Kavgic, A. Mavrogianni, D. Mumovic, A. Summerfield, Z. Stevanovic, and M. Djurovic-Petrovic, "A review of bottom-up building stock models for energy consumption in the residential sector," Build. Environ., vol. 45, no. 7, pp. 1683–1697, 2010.

[21] L. Hao, "Exploring the Performance of Historic Residential Buildings in South Tyrol : Considerations on Present and Future Climate," Ph.D. dissertation, Dept. of Archit. Built Environ. and Constr. Engin., University of Milan, 2020.

[22] U. Ali, M. H. Shamsi, C. Hoare, and J. O'Donnell, "GIS-Based Residential Building Energy Modeling at District Scale," Proceedings of BSO 2018: 4th Building Simulation and Optimization Conference, Cambridge, UK: 11–12 September 2018.

[23] A. Machard, C. Inard, J.-M. Alessandrini, C. Pelé, and J. Ribéron, "A Methodology for Assembling Future Weather Files including Heatwaves for Building Thermal Simulations from Regional Climate Models Multi-years Datasets," Energies, 13, 3424, pp. 1–34, 2020.

[24] I. Sartori, N. H. Sandberg, and H. Brattebø, "Dynamic building stock modelling: General algorithm and exemplification for Norway," Energy Build., vol. 132, pp. 13–25, 2016.

[25] É. Mata, A. Sasic Kalagasidis, and F. Johnsson, "Description of the European building stock through archetype buildings," in 8th Conference on Sustainable Development of Energy, Water and Environment Systems – SDEWES Conference, September 22-27, 2013, Dubrovnik, Croatia pp. 25–28.

[26] B. Kim, Y. Yamaguchi, S. Kimura, Y. Ko, K. Ikeda, and Y. Shimoda, "Urban building energy modeling considering the heterogeneity of HVAC system stock: A case study on Japanese office building stock," Energy Build., vol. 199, pp. 547–561, 2019.

[27] E. Genova, G. Fatta, and T. Broström, "Categorization of the Historic Architecture in Palermo for the Purpose of Energy Assessment," in CISBAT 2015 - September 9-11, 2015 - Lausanne, Switzerland, pp. 499–504.

[28] Z. (John) Zhai and J. M. Previtali, "Ancient vernacular architecture: characteristics categorization and energy performance evaluation," Energy Build., vol. 42, no. 3, pp. 357–365, 2010.

[29] F. Aguilera and F. Ossio, "Residential archetypes in urban energy simulation models in Chile: Determining factors of residential energy consumption," Rev. la Constr., vol. 16, no. 3, pp. 527–536, 2017.

[30] A. A. Famuyibo, A. Duffy, and P. Strachan, "Developing archetypes for domestic dwellings - An Irish case study," Energy Build., vol. 50, pp. 150–157, 2012.

[31] E. Lucchi, V. D'Alonzo, D. Exner, P. Zambelli, and G. Garegnani, "A Density-Based Spatial Cluster Analysis Supporting The Building Stock Analysis In Historical Towns," in Proc. Build. Simul. 2019 16th Conf. IBPSA, pp. 3831–3838.

[32] Fondation Batiment-Energie, "Garantie de performance énergétique - Typologies de bâtiments", 2014.

[33] B. Montalbán Pozas, "Rehabilitacion sostenible de la arquitectura tradicional del Valle del Jerte," Ph.D. dissertation, Dept. de arte y ciencias del territorio, University of Extremadura, 2015.

[34] T. M. Ferreira, C. Santos, R. Vicente, and R. Mendes da Silva, "Building typologies identification to support risk mitigation at the urban scale - Case study of the old city centre of Seixal, Portugal", J. Cult. Herit. 14 449–463, 2012.

[35] I. Theodoridou, A. M. Papadopoulos, and M. Hegger, "A typological classification of the Greek residential building stock," Energy Build., vol. 43, no. 10, pp. 2779–2787, 2011.
[36] H. Tommerup and S. Svendsen, "Energy savings in Danish residential building stock," Energy Build., vol. 38, no. 6, pp. 618–626, 2006.
[37] K. Fabbri, "Energy incidence of historic building: Leaving no stone unturned," J. Cult. Herit., vol. 14, no. 3 SUPPL, 2013.
[38] T. Brostrom, A. Bernardi, A. Egusquiza, J. Frick, and M. Kahn, "A method for categorization of European Historic Districts and a Multiscale Data Model for the assesment of energy interventions," in 3rd Eur. Work. Cult. Herit. Preserv. EWCHP 2013, pp. 153–158.
[39] S. Lidelöw, T. Örn, A. Luciani, and A. Rizzo, "Energy-efficiency measures for heritage buildings: A literature review," Sustain. Cities Soc., vol. 45, pp. 231–242, 2019.
[40] E. Genova, G. Fatta, and C. Vinci, "The Recurrent Characteristics of Historic Buildings as a Support to Improve their Energy Performances: The Case Study of Palermo," Energy Procedia, vol. 111, no. September 2016, pp. 452–461, 2016.
[41] B. Moshfegh, P. Rohdin, V. Milic, A. Donarelli, P. Eriksson, and T. Broström, "A method to assess the potential for and consequences of energy retrofits in Swedish historic buildings," The Historic Environment: Policy & Practice, 5 (2), 150-166, 2014
[42] H. S. S. Ibrahim, A. Z. Khan, S. Attia, and Y. Serag, "Classification of heritage residential building stock and defining sustainable retrofitting scenarios in Khedivial Cairo," Sustain., vol. 13, no. 2, pp. 1–26, 2021.
[43] E. Raslan, A. Donarelli, and E. De Angelis, "Categorization of the heritage building stock in Cairo for the energy planning purposes. A method and the typical buildings," in Energy efficiency in historic buildings 2018, 2018, pp. 503–511.
[44] E. Lucchi, G. Garegnani, L. Maturi, and D. Moser, "Architectural integration of photovoltaic systems in historic districts. The case study pf Santiago de Compostela," in Int. Conf. Energy Effic. Hist. Build. Madrid, Spain 29th-30th September 2014.
[45] A. Mastrucci, O. Baume, F. Stazi, and U. Leopold, "Estimating energy savings for the residential building stock of an entire city: A GIS-based statistical downscaling approach applied to Rotterdam," Energy Build., vol. 75, pp. 358–367, 2014.
[46] L. Filogamo, G. Peri, G. Rizzo, and A. Giaccone, "On the classification of large residential buildings stocks by sample typologies for energy planning purposes," Appl. Energy, vol. 135, pp. 825–835, 2014.
[47] J. Reyna et al., "U . S . Building Stock Characterization Study. A National Typology for Decarbonizing U.S. Buildings. Part 1 : Residential Buildings", 2021.
[48] BPIE, "Europe's buildings under the microscope. A country-by-country review of the energy performance of buildings", 2011.
[49] A. Brandão de Vasconcelos, M. D. Pinheiro, A. Manso, and A. Cabaço, "A Portuguese approach to define reference buildings for cost-optimal methodologies," Appl. Energy, vol. 140, no. January 2012, pp. 316–328, 2015.
[50] G. M. Benejam, É. Mata, A. S. Kalagasidis, and F. Johnsson, "Bottom-up characterization of the Spanish building stock for energy assessment and model validation," in Retrofit 2012 Conference, 24 - 26 Jan 2012, Manchester, UK, 2012.
[51] G. P. Panayiotou et al., "The characteristics and the energy behaviour of the residential building stock of Cyprus in view of Directive 2002/91/EC," Energy Build., vol. 42, pp. 2083–2089, 2010.
[52] I. Prieto, J. Izkara Martínez, and A. Egusquiza Ortega, "Building stock categorization for energy retrofitting of historic districts based on a 3D ctiy model," Rev. DYNA, vol. 92, no. 5, pp. 572–579, 2017.
[53] K. Fabbri, M. Zuppiroli, and K. Ambrogio, "Heritage buildings and energy performance: Mapping with GIS tools," Energy Build., vol. 48, pp. 137–145, 2012.
[54] P. Tuominen, "Assessing energy efficiency potential in the building stock. Method for estimating the potential for improvements and their economic effects", Ph.D. dissertation, VTT Technical Research Centre of Finland Ltd, 2015.
[55] T. M. Ferreira, C. Santos, R. Vicente, and R. Mendes da Silva, "Structural and architectural characterisation of old building stocks: case study of the old city centre of Seixal, Portugal, rebuilt after the Great 1755 Lisbon earthquake," Eng. Struct. Technol., 7(3): 126-139, 2015.
[56] S. Hrabovszky-Horváth, T. Pálvölgyi, T. Csoknyai, and A. Talamon, "Generalized residential building typology for urban climate change mitigation and adaptation strategies: The case of Hungary," Energy Build., vol. 62, pp. 475–485, 2013.

# Investigating stakeholders' opinion on energy retrofit of historic buildings, grasping their needs and experiences

## An overview on the use of retrofit guidelines and decisional tools

A. Buda[1]

[1] ABC Department, Politecnico di Milano, Milano, Italy, alessia.buda@polimi.it

**Abstract –** This paper aims to explore what information do refurbishment stakeholders find useful to support the decision-making and what are the key points during the retrofit design process to consider. Decisions concerning the improvement of conditions and performance of historic buildings are not only about solving technical problems: decisions revolved around balancing different objectives and value judgments. Hence, what guides different stakeholders in constructing their opinions? How are they influenced? What matters more in the process? For this purpose, in this article different stakeholders have been interviewed to understand how to activate a negotiation process that functions in real-life interactions, with the support of experts in social science. The observations of many heritage officers, architects and energy professionals were gathered to better understand how energy retrofit guidelines, the EN 16883:2017 and the Italian retrofit guidelines for historic buildings by the Cultural Heritage Ministry (MiBACT), are used in practice.

**Keywords** – historic building; energy efficiency; users' involvement; interview; guidelines

### 1. INTRODUCTION

In the last years, the energy efficiency of historic buildings has been widely discussed, being subject of EU projects [1,2] and of several research studies [3–5]. One of the goals for researchers has been the definition of a method / tool to balance the aspects which entail the retrofitting process, namely energy savings, heritage protection and indoor microclimate conditions [6–9]. In this framework, several guidance documents have been published with the intent to be used by practitioners, users, and heritage officials as a support for making decisions when selecting retrofit solutions. At European level, it is the case of the EN 16883:2017 standard „Conservation of cultural heritage - Guidelines for improving the energy performance of historic buildings" [10], developed for supporting the selection and assessment of appropriate interventions to improve the energy performance of a given historic building. At National level, it can be mentioned the „Guidelines for the improvement of energy efficiency in cultural heritage", published by the Italian Ministry of Cultural Heritage (MiBACT) in 2015 [11] for supporting retrofit stakeholders in the planning.

Despite the number of resources and efforts spent on standardization of the retrofit process in recent years, little attention has been paid to if and how these instruments are used in practice [12, 13]. Hence, through interviews and questionnaires, the observations of many heritage officers, architects and energy professionals were gathered to better understand how energy retrofit guidelines for historic buildings are used in practice. From the document pros and cons, attention is placed on how much and on what occasions energy retrofit guidelines represent a valid guiding tool but also the degree to which they can deviate from the heritage protection needs. The final aim of this survey is to understand how to modify and enrich the contents of the existing guidelines, if necessary, to support a calibrated interventions choice. Accordingly, some possible solutions will be presented as points for a future discussion.

## 2. METHODOLOGY

Interviews with stakeholders involved in the process have been conducted to have a better understanding of the methodologies for retrofitting historic buildings and to highlight the weak and the strong points of the current practice. Considerations of public and private interventions are given from different points of view, receiving feedback from people involved in the process and highlighting the main advantages and disadvantages of the various criteria and methods adopted. Thanks to the geographical variety of the origin of the interviewees, it was possible to underline differences among different Countries. At the same time, the topic on the national scale was deepened. The aim is to compare the views of experts, authorities, designers, and building end-users who interact with the process of decision-making for the retrofit of built heritage, reflecting upon the best strategies whereby a project can be developed to reach a balance between different needs.

***Table 1. List of interviewed stakeholders***

| Role | Details |
|---|---|
| ***Heritage Stakeholders*** | • 3 Northern European Heritage authorities<br>• 15 Italian heritage protection authorities |
| ***Energy stakeholders*** | • 1 Energy Manager<br>• 1 Energy Agency<br>• 1 representative of Energy Service Company (ESCo) |
| ***Designers*** | • 75 professionals (investigation through a questionnaire)<br>• 3 designers from Mediterranean Countries |

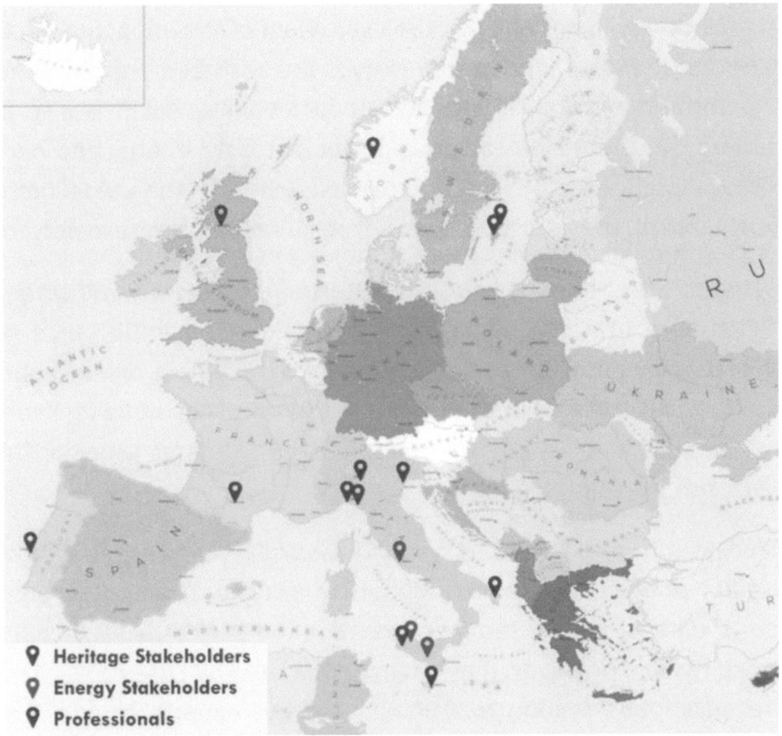


**Figure 1. European distribution of the interviewed stakeholders.**

## 2.1 HERITAGE STAKEHOLDERS

For understanding how to make a decision-making tool suitable for the real needs of heritage conservation, it was necessary to investigate the point of view of those involved in this field. It was therefore decided to compare the opinion of three Heritage authorities from EU Countries with that of 15 Italian heritage protection authorities.

### *2.1.1 Heritage Authorities in EU context*

Interviews were conducted with three different heritage managers in northern European countries (two in Sweden and one in Norway) involved in the definition of the EN 16883 guidelines. When asked what they thought of these guidelines, each interviewee pointed out that the document is too generic, with the result that it is unused. According to what was reported in the interviews, in fact, they had not received any project developed following the guidelines procedure.

Heritage Authorities' feedback is that the guidelines are not easy to be used and examples are missing. In their opinion, the guidelines are thought to be for experts, but not all applicants have expertise in the field. The discussion focuses on how to make it easier for non-expert end users, because a certain level of knowledge is required to intervene on heritage. Heritage Authorities feel the lack of guidance to make appropriate choices for the building preservation, including maintenance activities. The EN 16883 structure is considered too general; the assessment method of the measures is also criticized. The latter should be coupled with a proper understanding of the building function and of the users' needs: if everything come just from professionals, the execution will not respond to the real needs of occupants, to the behavioural use of the building. It could be hard to have an idea of risks and all the problems related to a building without an awareness of its real use, both before and after interventions. From the heritage point of view, the evaluation of the impact of measures is considered difficult: how to check that the retrofit project preserves the aesthetic historic values, the authenticity of the building? It is important to make sure to not lose values, but it is also important to make buildings that are usable. So, the main problem is to look at both energy and heritage aspects, with the right competencies. Another factor affecting the use whether the guidelines are available open access. While in some contexts they are already toll-free (Sweden), they are not, for example, in Italy.

Interviewees agrees that the guidelines are a starting point. They will simply need a review to be more detailed and more practical. Clearly it is remarked the importance of avoiding retrofit interventions which are solely energy-oriented, and ideally the whole building should be considered to define measures. When planning an intervention, the value of the building itself comes first, along with the comfort and the sustainability of the expected use. The energy aspect is therefore just part of the renovation project, not the main scope.

The interviewees were asked how they would have preferred the guideline to be structured. To facilitate the assessment of projects, the Heritage officers suggest making guidelines more practical with a sort of 'checklist' on how people should proceed in the project, a list of minimum analysis and documentation needed (which, however, may change on a case-by-case basis). The guidelines should also include a more descriptive section explaining how the experts could become involved with examples, technical papers and references to standards on other topics (diagnostics, materials analysis, accessibility, fire, safety, etc.). They further suggested to make the guidelines also adaptable to different needs, in relation to the size of the project and the economic resources for the project. In fact, public projects can afford procedures with a whole building approach, following the procedure in full, entering the merits of the project proposals with the assessment phase. In the case of smaller projects, for example those of private individuals, it can be useful to make the guidelines more like a booklet, with factsheets to make people aware of the effects of interventions on the building integrity, such as on the indoor microclimate. This would help people to understand that everything is done to any part of a building has consequences for the entire structure. Another suggestion was to integrate a repository of best practices to support the selection of solutions.

As the guidelines are thought to be adaptable to EU countries, the heritage officers suggest leaving the general structure of the documents a general instruction valuable for all European countries. It should be a guide for all EU members without specifics, to avoid misunderstandings. They are in fact sceptical about adapting the guidelines at the National level because it could result difficult to make suggestions which are valid for all buildings, so giving specific solutions could result meaningless.

#### *2.1.2 Italian Heritage Authorities*

Italian Heritage officers involved in this investigation operate in different Italian territories (Sicily, Lombardy, Liguria, Puglia and Veneto). One problem that has emerged concerns the competences in matters of protection which vary considerably at regional level. For the so-called „monumental buildings" an exception is allowed for minimum energy requirements by the Italian Energy Law 90/2013 [14]. So, the heritage officials supervise the design process to avoid that energy retrofit interventions can affect the building integrity in these monumental cases. This form of control instead fails in the case of historical centres. In fact, in minor architecture where there are no restriction decrees, the interventions follow the indications of the local planning tools; where instead specific plan forecasts (i.e. landscape plans) are in force, the Heritage Board can veto only in the case of „alteration of the state of the places and/or the external appearance of the buildings".

Furthermore, some building activities follow a simplified authorization procedure because they are considered „less impactful" according to local laws. This is translated into the impossibility for the Heritage board to exercise control over the compatibility of the interventions, especially inside the buildings. And yet, in recent years, the requests for permits have skyrocketed, thanks also to tax breaks for the rehabilitation and energy renewal of structures [15]. In general, the requests concern: the replacement of fixtures with better-performing ones, the insertion of solar and/or photovoltaic panels, the laying of external/internal insulation, the replacement and/or insertion of systems (generators, pipes and terminals) and the transformation of the horizontal sections.

As reported by the interviewees, the quality of the projects does not emerge. This is partially due to the pressure linked to the economic incentives that can only be obtained with interventions of great impact on the built environment. The Superintendents themselves complain about training gaps in the subject: their lack of knowledge in the energy field limits their ability in the request to have reliable energy diagnoses and in the evaluation of alternative solutions to those currently proposed to them by practitioners. For example, almost no one has stated that they know the difference in thermal behaviour in the case of laying insulation outside rather than inside it.

Although everyone then reports direct/indirect experiences regarding energy efficiency, not everyone knows the content (and sometimes the existence) of the MiBACT guidelines. Although the first part of the document is full of positive ideas, these do not seem to translate into the „application" section of the examples, generating more perplexity than clarification. Therefore, their major request of integrating the guidelines concerns experimental data of innovative or improvement solutions, basic knowledge of technical physics and clear and easy to use support tools.

## 2.2 ENERGY STAKEHOLDERS

### *2.2.1 Energy Planners– Assessment process and guidelines use*

Two energy planners were interviewed: one Italian energy manager and one representative of an Italian energy society (ENEA), both recently involved in retrofits of built heritage. From the point of view of those involved in energy assessments, the improvement project must optimize the building services to reduce management costs, starting from the performance data. In the general view of the interviewees, the theme of knowledge of the historical and material data of the building is not central, because it is the responsibility of the architect or of the Heritage Board officer.

If the goal is to achieve a rational cost reduction, the collection of data on the building (geometric and material survey, bills and technical data of technical installations) is functional to the determination of the energy class and to the energy simulation model (static or dynamic, the difference does not always seem to be clear), leaving out what is considered irrelevant (for the analysis, such as the state of conservation of materials or the existence of passive climate systems. This leads to a simplification of the thermo-physical behaviour of the building and to the exclusion of data relevant to the conservation project. It seems to be reduced to reading what is „valuable and what is not" in order to understand the weak points on which to act.

Among the most frequently employed solutions by the interviewed in the retrofit projects in historic buildings, we find LED lamps insertion, replacement of building services, photovoltaic insertion. Among the latter, there is also the replacement of doors and windows, considered more economically advantageous than „repairing something old" that does not ensure comfort as per standard, allowing also access to tax incentives.

The majority of those interviewed expressed their appreciation for the guidelines' procedures: contrary to what was reported by Architects and Superintendents, it is considered positive to be able to evaluate the design options according to a cost-benefit analysis based on the energy performance data. One of the major problems appears instead to be dialogue with the protection bodies, defined as „unique censors" of the design process. For this reason, technicians would consider it useful to have „ready-to-use" (even non-standard) solutions, with a declared „energy gain", „sustainability" certificate and accepted by the Heritage Board.

#### *2.2.2 Energy Service Company (ESCo)*

According to the Art. 3 of the Directive 2006/32/CE, the ESCos are companies that provide energy services for public administrations in the industrial and civil sectors with the guarantee of results in terms of energy savings [16]. The advantage of the public-private partnership with an ESCo is in the ability to duplicate the spending capacity of public administrations and therefore the resources to be used also for other projects. The construction, management and maintenance of the energy services as technical installations (mainly) or interventions on the building envelope (few cases) are under the charge of ESCo, who takes the responsibility of intervention and its maintenance for a certain number of years. Therefore, the public body has only the task of verifying the established energy savings.

The representative of the whole Sicilian ESCo, who is an Italian architect, was interviewed. He highlighted that current practice in the public sector is to opt for standard solutions that are already known and with defined energy and economic savings (relamping with LED, replacement of the boiler/heat pump, replacement of windows, insertion of BMS for managing the energetic vector, insertion of PV panels). In the case of historic buildings, non-standard interventions can be provided to the extent that it is possible to evaluate how to support them economically, even over time. However, as was remarked by the interviewed, the professionalism of those who have to deal with the diagnosis in the built heritage sector is lacking. Who should coordinate the work? An intermediary figure is needed who knows all these aspects to enhance an important heritage.

In his experience, the architect argues that the only thing that can lead to a development towards the field of historical assets is to provide an in-depth energy diagnosis. That is, the diagnosis cannot be an „industrialized" process, but must be carried out exclusively to provide information for the specific intervention. What is interesting is the degree of achievable savings, estimated by the simulation of the intervention, be it on the technical installations, or on the casing. But currently these data are not known. How convenient is it economically and energetically to replace glass rather than replacing the entire window frame?

Guidelines, in fact, are seen as generic, not related to the complex process of project development that follows a public administration and an ESCo. Technical data should therefore be enriched with examples of possible strategies. The architect also suggested to include a reward for those who adopt the guidelines, as well as the certification protocols and a section of the guidelines dedicated to economic financing strategies (like public-private partnerships).

## 2.3 DESIGNERS

### *2.3.1 A questionnaire to 75 Italian professionals about the use of decisional tools*

To analyse the criticalities and potential of the guidelines and to understand the different points of view of the actors involved, architects, engineers and superintendents were asked to complete a questionnaire during the seminar 'Energy efficiency in Cultural Heritage: guidelines, programs and experiences', held in Genoa (October 2018) [17]. The questionnaire asked for an opinion on the relationship between energy improvement and historic buildings, any experiences with the subject, and an opinion about the usefulness of the document.

A first result is that out of 75 questionnaires distributed, only 20 were returned (10 architects, 7 Heritage Board officials and 3 engineers) and only three people declared that they had used the guidelines, all for the same retrofit project (in the Palazzo Reale of Genova). When asked what the most appreciated part of the guidelines was, 45% answered theoretical contents, 35% pointed to the case studies attached and 20% said they appreciated the procedure. It should be noted that among these, the three who then used the guidelines in practice said they preferred the theoretical contents, because the procedure would need some corrections. The proposed adjustments are the addition of varied types of examples related to the world of cultural heritage (museums, libraries, etc.) with real numerical data, to make the positive impact of the interventions clear. Those who, on the other hand, have only seen the guidelines but have not used them, declared that they consider the document too generic (75%) and with a language that is not suitable for their needs (90%). Furthermore, the majority require a collection of retrofit interventions (as prepared instructions for use with pros and cons) (70%), the others an example of how to carry out the planning procedure, to be guided step by step (30%).

Hence, the questionnaires highlighted the usefulness of developing guidelines because it was considered important to find solutions compatible with the needs of protection point, it is considered essential to improve the quality of use of historic buildings through policies that are not exclusively linked to new constructions but that can encourage a more efficient use of our heritage. The poor response (as subsequently established) is linked to the lack of knowledge and dissemination of the document. It was thus decided to continue through the tool of direct interviews with three architects, experts of both energy and conservation.

### *2.3.2 Architects in the Mediterranean Climate*

Interviews were carried out with architects working in Mediterranean Countries (Italia, Malta, Portugal), to understand better how in practice the problem of balancing winter/summer conditioning is faced. Professionals primarily recognize a problem of competences: there is no full awareness of the possible effects (positive and negative) of interventions for improving energy efficiency in historic buildings during the season variability. Most of the time, other experts are involved in the decision-process, but it is not a simple dialogue if there is not proper awareness of actions.

In the EN 16883:2017, as well as in the Italian guidelines, there is an observable lack of a control system which, according to an overall logic, evaluates the consequences linked to the implementation of retrofit measures, taking into consideration the multiplicity of values involved. Hence, they admit to filling their skills gap during external trainings or courses, to be the link between the different professionals involved in the construction site. In both guidelines the designers find a certain generality in purely theoretical indications and the lack of effective examples leads almost all the interviewees not to use them. On the contrary, other guidelines (such as those for seismic vulnerability) which are also not mandatory tools, are widespread and consulted because in them the theoretical part is related to explanatory schemes and adaptable to a wide range of cases.

Among the experiences reported by the interviewees, there are many restorations works (of different importance and size) where an accurate understanding of the building, including thermal behaviour, is an indispensable condition for the choice of calibrated and compatible solutions. A good knowledge and attention to the material and constructive data, obtained by combining direct and indirect sources, makes it possible to consider limits and resources (passive ventilation systems, existing ducts, curtains or blinds, etc.) usable for the intervention. Country by country, the approach towards restoration was different: only the Italian architect talked about the need of evaluating qualitative aspects, such as compatibility, giving priority to conservation and users' needs; the others in fact talked more about 'where to intervene to reduce losses'. Accordingly, the energy improvement project should be configured as part of the restoration project to guarantee the use of the asset in the long term, contrary to a practice made by random interventions or the effect of emergency solutions such as air conditioning systems that often dot the facades. However, it often happens that the project does not affect the entire building: the funds, mostly coming from European tenders, are limited only to the intervention on the technical installations, losing as a consequence the more general dimension of the problem, neglecting other aspects related to the building preservation. It should instead push the professionals involved to consider different demands and needs (social, cultural, economic and environmental), as well as to assess the critical issues that the change to the internal microclimate can generate. Sometimes, for example, it would be enough to think of simple maintenance and interaction between building users to achieve greater results with lower costs. Therefore, an approach to restoration intervention as a „cultural act", capable of guaranteeing priority to the whole vision, should be reiterated.

***Table 2. Feedback and suggestions on the EN 16883:2017 Guidelines***

| Topic | | Stakeholders | | | Suggestions |
|---|---|---|---|---|---|
| | | Heritage Stakeholders | Energy stakeholders | Professionals | |
| ***Usability*** | Unused because too general | X | | X | |

| | | | | | |
|---|---|---|---|---|---|
| | Unused because not open access | X | X | X | |
| **Structure** | Good structure | | X | | • Add examples on the use<br>• Add a checklist to proceed in the project<br>• Make different versions for differe[illegible] countries |
| | Too much general indications | X | | X | |
| | A good starting point | X | X | X | |
| **Content** | Not connected to the planning process | X | X | X | • Add a section on pitfalls<br>• Add a repository of best practices |
| | Too much technical for not experts | X | | | |
| | Lack of examples | X | X | X | |
| | Difficulties in following the steps | | | X | |

***Table 3. Feedback and suggestions on the Italian Guidelines***

| | Topic | Stakeholders | | | Suggestions |
|---|---|---|---|---|---|
| | | Heritage Stakeholders | Energy stakeholders | Professionals | |
| **Usability** | Unused because too general | | X | X | |
| | Unused because not open access | X | X | X | |
| **Structure** | Good structure | X | X | | • Add examples on the use<br>• Add a special section dedicated to public buildings |
| | Not related to the heritage protection | X | | X | |
| | A good starting point | X | | X | |
| **Content** | Good cost-benefit analysis | | X | | • Change the Planning procedure to include Heritage evaluation<br>• Add technical papers and factsheets to express the effects of interventions on heritage with real data<br>• Add a repository of best practices on different types of building function |
| | Not connected to the planning process | X | | X | |
| | Too much theoretical | | X | X | |
| | Lack of examples | X | X | X | |
| | Difficulties in following the steps | X | | X | |

## 3. CONCLUSIONS

As result of the investigation carried out, the main points on which the research should focus to improve the existing retrofit guidelines [18] are:

*(a) Making more adherent the existing guidelines to the current practice*

Users find that the standard lacks sufficiently detailed procedures for how to complete the steps in the suggested planning process. They are lacking models to follow of what can be done, as they find difficult to understand how the standard relates to established routines and existing processes. The descriptions of each step in the suggested procedure are generic. The opinion shared by the different interviewees is that of having to integrate two existing guidelines (both EN 16883 and the Italian guidelines) to make them more adherent to the decision-making process and to the needs of current practice. Users have found it difficult to determine what is effectively required to follow the standard.

*(b) Need for a multiscale and flexible methodology*

If on the one hand we cannot think of making something detailed which by its nature is flexible as a support to the design, on the other hand it can be integrated to make it more responsive to the needs of the final beneficiaries. The structure must be designed according to the users and the design objectives. This implies a different decision complexity (residential/not residential, private/public), to be linked also to the diversity of the level of detail of the project (feasibility project, design project, final project).

*(c) Addition of best practices, factsheets and examples*

Everyone agrees on one element: it is necessary to define intervention strategies and provide material that enriches in a practical way what is already contained in the guidelines, such as explanations, examples, and factsheets. Unexperienced end-users require examples of best practice regarding both case studies and single retrofit measures. Those who protect cultural heritage are clamouring for content to be added that explains the interference between the various interventions and that shows a range of solutions with pros and cons. They also ask that the designer be assisted by a sort of building analysis checklist. For their part, the energy technicians ask that it be clarified which solutions are compatible from the point of view of conservation. The architects and the heritage boards ask for factsheets and simplified descriptions of the feasible strategies.

*(d) Adding a section on public-private partnership with best practices on heritage retrofit*

During the dialogue with energy companies and energy experts, we noted how the current practice in the public sector is to opt for standard solutions that are already known and that present defined energy and economic savings. Hence, they asked for a guidelines section where the advantages of less common retrofit measures are introduced to the public administrations from a financial point of view, explaining how to carry out the economic evaluation and showing possible examples of private-public partnerships.

*(e) Adding an example of multicriteria or a weighting system of scenarios*

How different needs coming from different disciplines should be joined remains an open question. Some suggested to compare all stakeholders' reasons and needs through a multicriteria system, weighting the different scenarios in a risk assessment chart (similar to the EN 16883 assessment method). The aim might be not to define just a classification, but to bring out the problems to be discussed in a multidisciplinary context. However, as highlighted by others, the weights can change case by case, depending on the single stakeholder's vision, as the project objectives. In fact, all the interviewees reported some problems when negotiating the right solution for a building; others confessed a lack of knowledges or skills. Some others suggested that the main part to carry an evaluation of measures is in defining the criteria to be agreed at the beginning of the project between the different actors. It is an important issue that the assessment should consider no single measures, but the whole strategy towards the adoption of a holistic approach.

***(f) Active involvement of heritage officers during the process of heritage impact assessment***

Not surprisingly, common grounds were found between heritage officers in different countries despite their different latitudes. The purpose of guaranteeing active use and conservation of cultural heritage is shared by all interviewees. However, differences do exist in the definition of heritage significance and values, appropriate to the facets of the restoration discipline – considering the Italians more 'conservative' than the others; therefore, with a different (less or more restrictive) view in considering the 'impact assessment' on heritage values of retrofit actions. Hence, it is considered that the involvement of heritage boards for evaluating the heritage impact of interventions would be very important to avoid contradictions in the evaluation.

***(g) Adding a section with common pitfalls and recommendations***

Finally, the interviewees unanimously expressed the need to highlight possible pitfalls that could be encountered throughout the procedure for defining interventions: to avoid the most frequent mistakes, some recommendations could be added in the process before proceeding with the intervention, promoting awareness of possible repercussions on the building [13].

## 4. REFERENCES

[1] IEA-SHC, „Task 59. Deep renovation of historic buildings towards lowest possible energy demand and CO2 emission (NZEB)". Internet: https://task59.iea-shc.org/ [December 19th, 2020].

[2] 3enCULT. Efficient energy for EU cultural Heritage, „D7.7 Relation historic buildings, EPBD and EPBD CEN Standards" 2011. Internet: http://www.3encult.eu/en/project/ [December 19th, 2020].

[3] Martinez-Molina, A.; Ausina, I.; Cho, S.; Vivancos, J.L. „Energy efficiency and thermal comfort in historic buildings: A review". Renew. and Sust. Ener. Rev. 2016, 61, pp. 70-85.

[4] Lidelöw, S.; Örn, T.; Luciani, A.; Rizzo, A. „Energy-efficiency measures for heritage buildings: A literature review". Sustain. Cities Soc. 2019 45, pp. 231–241.

[5] Webb, A. L. „Energy retrofits in historic and traditional buildings: A review of problems and methods". Renew. Sustain. Energ. Rev. 2017, 77, pp. 748–759.

[6] Entrada Silva H. et al. „Preventive conservation of historic buildings in temperate climates. The importance of a risk-based analysis on the decision-making process" in Energy and Buildings, 107(26-36).

[7] Gasparella A. et al. (2015, May). „Multi-objectives optimization of Energy Efficiency Measures in existing buildings" in Energy and Buildings. [On-line], 95(57-69). [September 11th, 2020].

[8] Roberti, F.; Filippi Oberegger, U.; Lucchi, E.; Troi, A. „Energy retrofit and conservation of a historic building using multi-objective optimization and an analytic hierarchy process". Energy Build. 2016, 138.

[9] Blumberga, A.; de Place Hansen, E.J. „Written guidelines for decision making concerning the possible use of internal insulation in historic buildings". Internet: https://www.ribuild.eu [October 2020].

[10] EN 16883 „Conservation of cultural heritage - Guidelines for improving the energy performance of historic buildings", June 2017.
[11] MIBACT, „Guidelines for the improvement of energy efficiency in cultural heritage". October 2015.
[12] Mazzarella L. (2014, November). „Energy retrofit of historic and existing buildings. The legislative and regulatory point of view" in Energy and Buildings. [On-line], 95(23-31). [May 2nd, 2017].
[13] Buda, A.; Pracchi, V. „Potentialities and criticalities of retrofit guidelines in their application on different case studies". Conference Proceedings: EEHB2018, Visby, Sweden, 26–27 September 2018, pp. 283-293.
[14] Legislation and Official Policy Documents of Italy, 2013. Law 90/2013, G.U. n. 181, 3rd August 2013.
[15] ENEA, Annual Report „Detrazioni fiscali", 2019. https://detrazionifiscali.enea.it/ (October 1st, 2020)
[16] Directive 2006/32/EC of the European Parliament and of the Council of 5 April 2006 on energy end-use efficiency and energy services and repealing Council Directive 93/76/EEC
[17] MiBACT and Segretariato Regionale per la Liguria, Efficienza Energetica nel Patrimonio Culturale. Linee di Indirizzo, Programmi, Esperienze, 2nd October 2018.
[18] Buda, A. Energy efficiency of built cultural heritage, guidelines and new practices of intervention in Mediterranean climate, Dissertation PhD Thesis, 23rd March 2020, Politecnico di Milano.

# Identification and thermal behaviour of traditional flat roofs in the Mediterranean using remote and in situ sensing technologies – an innovative approach.

J. Cassar[1], C. Galdies[2], E. Muscat Azzopardi[1]

[1] Department of Conservation and Built Heritage, Faculty for the Built Environment, University of Malta, Msida, MSD 2080, Malta. joann.cassar@um.edu.mt; elizabeth.muscat-azzopardi@um.edu.mt

[2] Institute of Earth Systems, University of Malta, Msida, MSD 2080, Malta. charles.galdies@um.edu.mt

**Abstract –** “Cool roofs” and “evaporative roofs” are modern construction technologies, based in part on the known behaviour of “traditional flat roofs” in the Mediterranean. This paper describes an innovative study of the behaviour of Mediterranean roofs, employing high resolution satellite data, UAV (Unmanned Aerial Vehicle) multispectral technology (visible, near- and thermal-infrared wavelengths) and direct measurements on, and underneath, the roofs of selected buildings in Malta, whilst also comparing to modified and modern flat roofs. This research is determining their real behaviour and advancing the understanding of their performance, whilst promoting their sustainable preservation and maintenance in a changing climate. An interesting outcome of this project has been that satellite (and to a certain extent) UAV technology can remotely identify traditional roofs from other roofs and understand their thermal behaviour This will be an important tool for policy makers and regulators who need to assess the presence and location of such roofs.

**Keywords –** traditional Mediterranean flat roofs; innovative methodology; satellite data; UAVs; roof identification.

## 1. INTRODUCTION

The typical landscape of the Mediterranean and Middle East consists of light-coloured or white-painted flat roofs and/or walls usually made of traditional materials which are usually also porous (brick or tile, or limestone and lime-based overlying layers) and thus “breathable”, being capable of absorbing and releasing moisture resulting in a cooling evaporative effect on drying. These materials have been commonly used materials in hot climates for hundreds of years but this passive and sustainable technique has lost its popularity.

This paper is the result of an innovative research project which is a first attempt to quantify the behaviour of such roofs in one particular Mediterranean country (Malta), gathering baseline data which will help promote the maintenance and even reinstatement of such traditional roofs on historic and traditional buildings.

If a country is to promote widely the use of such traditional “breathable” roofs, not only does their behaviour need to be understood and quantified, but public bodies also need to be able to estimate the number (and hence area) and location of these roofs. This project also evaluates the use of UAV and satellite technology for locating and quantifying these roofs.

## 2. THE PROJECT

The research which is the focus of this paper centres on the innovative use of satellite data, combined with visible and thermal wavelength data obtained from UAVs, and *in-situ* (external and internal) monitoring of rooftops in buildings with four different roof types. The main aim of the project was to identify existing temperature and moisture gradients for specific traditional and/or historic buildings in Malta, as representative of a Mediterranean typology, and comparing them to traditional roofs which have been modified[1] as well as modern roofs[2]. Hybrid roofs[3] have also been targeted in this study. This methodology has been described in detail in [2].

A second aim which came into focus as the rich data on such roofs started being collected, was the potential identification of different types remotely (by UAV and/or satellite), to quantify this resource and thus enable policy on these roof types to be formulated.

## 3. THE PROJECT'S DATASTREAM

Data was collected by remote sensing (via satellite and UAV) and *in-situ* data (inside and outside the building). This was considered essential as remote sensing gives accurate information on large areas with the use of both satellite images and images obtained by sensors mounted on a UAV complementing each other. These data are however indirect. The indirect data has thus been supplemented with direct *in-situ* measurements consisting of air temperature and relative humidity (RH) and also immediately above and below the studied roofs, by appropriate dataloggers. Direct surface and sub-surface measurements have also been recorded on each roof type.

### 3.1. MULTISENSOR DATA FUSION

The actual data which have been collected are the following:

- Satellite data – very high-resolution satellite data derived from KOMPsat 3A[4] (discussed below).
- UAV – A combination of two sensors have been used: a multi-spectral sensor (Blue (446nm), Green (548nm), Red (650nm), Red Edge (720nm), NIR (840nm) and a thermal camera system.
- The *in-situ* data are explained above.

### 3.2. REMOTE SENSING

### 3.3. VERY HIGH-RESOLUTION SATELLITE DATA FOR THE PROJECT

KOMPsat 3A is a sun synchronous, low Earth orbit with an altitude of 528 km, an inclination of 97.5o, an orbital period of 98.5 minutes and repeat cycle of 28 days. It is South Korea's first EO with two imaging systems onboard that have the aim of obtaining infrared and very high-resolution images for geographical information systems applications in many fields, including natural disasters. The payload of this satellite includes (1) a high-resolution electronic optical camera AEISS-A giving a 55cm class optical photography, and (2) an IR sensor capable of detecting heat on the ground. The

---

[1] In our case by the placement of an overlying impervious membrane

[2] In our case, roofs which have been totally replaced by modern concrete planks with overlying membrane

[3] Composed of a limestone structure (including roof) covered with cement

[4] This is an Earth Observation satellite that was launched in January 2019 by the Korean Aerospace Research Institute to provide high resolution imagery.

spectral bands of the AEISS-A used for the present study are the 450-520 nm MS1 (MultiSpectral), blue; 450-520nm MS2, green; 630-690nm MS3, red and the 760-900 nm MS4, NIR (Near-infrared). The ground sample distance is 2.2m for MS at nadir (downscaled to 50cm pansharpened imagery) and 5.5m for IR data.

### 3.4. UAVs

The technological evolution of UAVs (drones) is nowadays much used in the building industry for roof insulation inspections, investigation of thermal anomalies in the building fabric, technical condition of flat roofs and documentation of building performance.

This technology has here provided a closer look at the thermal and moisture-related properties of the roofs, utilising also satellite data over a wider area. The use of aerial surveillance of selected rooftops using a UAV equipped with cameras scanning the RGB-NIR-TIR range of wavelengths has further provided this study with the highest spectral and spatial resolution information needed. The drone was flown co temporally with the satellite coverage. This was in addition to the localised *in-situ* data collected, as well as specific information on the real composition of the said roofs.

### 3.5. DATA PROCESSING

Some studies have presented interesting results on data integration and fusion by employing quantitative means for multi-sensor remote sensing data sources. The synthesis and interpretation of the overall results from the different remote sensing sensors and platforms used have led this study to identify the behaviour of traditional roofs under different environmental conditions, and how these compare to modified and modern roofs, as well as "hybrid" roofs. This has been done following precise photogrammetric registration of UAV multispectral images (acquired in the same multispectral range as those acquired by the KOMPsat 3A satellite) taking note of radiometric and atmospheric correction of satellite and the co-location and correlation with seasonal *in-situ* data at pixel level. The *in-situ* data consists of both air and surface temperature, as well as RH data.

This three-tiered data acquisition approach has provided this study with opportunities to conduct multi-source pixel data fusion, analysis, correlation and seasonal modelling using standard analysis at pixel level. Specifically, this has allowed (1) statistical correlation between very high remotely-sensed data and *in-situ* measurements, (2) the spectral characterisation of the different target roofs, including traditional, historical ones, (3) detection and quantification of flux anomalies (i.e. reflectances (NGBIR), emissivities (TIR) and their calibration with *in-situ* data) into quantifiable material properties of traditional historical, and other roofs, and (4) derivation of seasonal spectral models of KOMPsat data that is able to separately distinguish traditional, modified, modern and hybrid roofs in a cost-effective, and rapid manner.

## 4. RESULTS & DISCUSSION

### 4.1. SATELLITE AND DRONE DATA ANALYSIS FOR ROOF TYPE IDENTIFICATION

Following atmospheric and radiometric correction of the satellite data, statistical analysis of each of the four bands derived from KOMPsat 3A was conducted so as to test whether specific bands can be more suitable than others to characterise roof types. Figure 1 shows representative radiometric data derived from KOMPsat 3A and UAV carrying similar sensor payload for the target roofs taken during the entire surveying period.

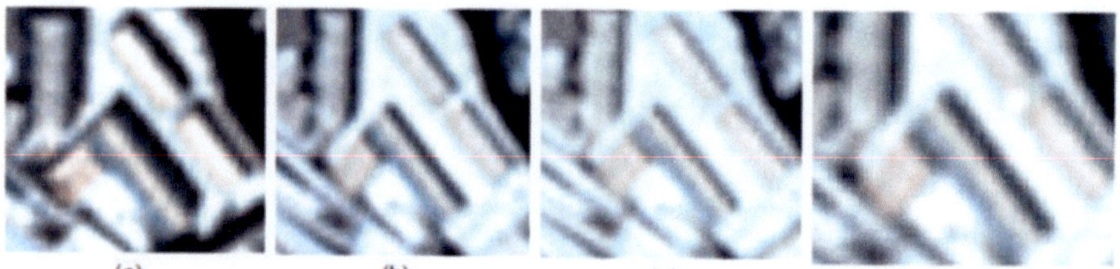

(a) (b) (c) (d)

RGB-NIR Satellite images over Fort St.Angelo: (a) February 16 2021; (b) June 10, 2021; (c) July 6, 2021; (d) August 2, 2021

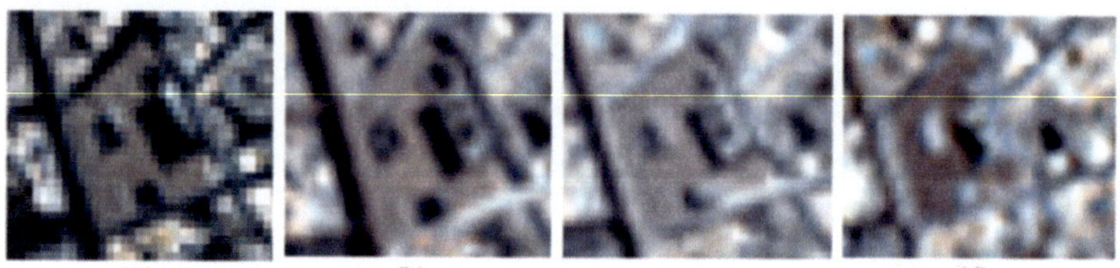

(a) (b) (c) (d)

RGB-NIR Satellite images over Inquisitor Palace: (a) February 24, 2021; (b) June 19, 2021; (c) July 6, 2021; (d) August 19, 2021

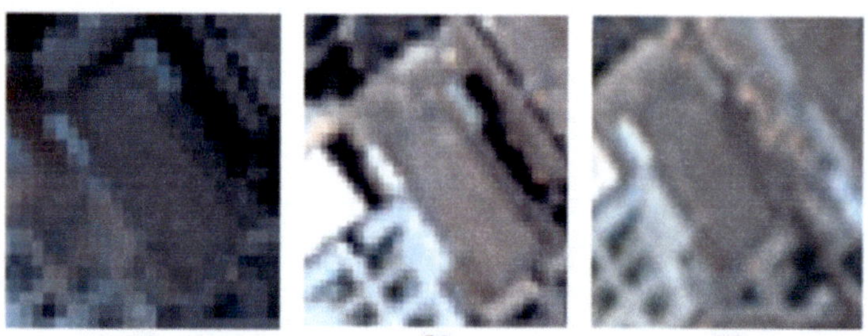

(a) (b) (c)

RGB-NIR Satellite images over Bighi: (a) March 14, 2021; (b) June 19, 2021; (c) August 19, 2021

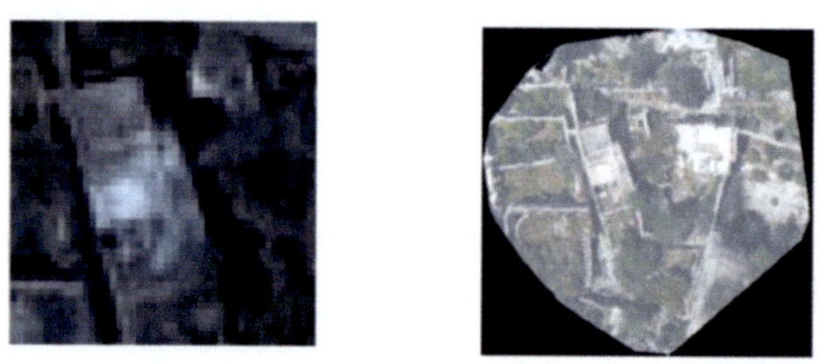

(a) (b)

RGB-NIR Satellite image (a) and drone (b) images over St. Mark's Convent, May 12, 2022

(a) (b)

RGB-NIR Satellite (a) and drone (b) images over Villa Frere. June 8, 2022

***Figure 1. Selected RGB-NIR satellite and drone images of the target roofs observed during the surveying period.***

A statistical cluster analysis of all the roof types based on their satellite radiometric median and standard deviation values was conducted. All the predicted roof clusters are labelled from 0 to 5, while the true labels are represented with symbols (Fig. 2). Results shows that K-means clustering analysis of the satellite radiometric data was able to clearly identify traditional from hybrid roof types. Traditional roof types were identified primarily when using the reflectance in the blue waveband, while hybrid roofs were most identifiable in the red waveband. On the basis of the K-means clustering a Silhouette clustering score of 0.46 was derived, indicating sufficient cluster separation on the basis of multispectral reflectance from these roof types. Particularly prominent is the cluster separation of the traditional (Villa Frere and St. Mark's Convent) roof types from the other clusters. It is interesting to note that statistical analysis clustered together modified traditional and modern roofs, whereas hybrid roofs were individually grouped into two clusters. This result applies for the entire time period used in this study.

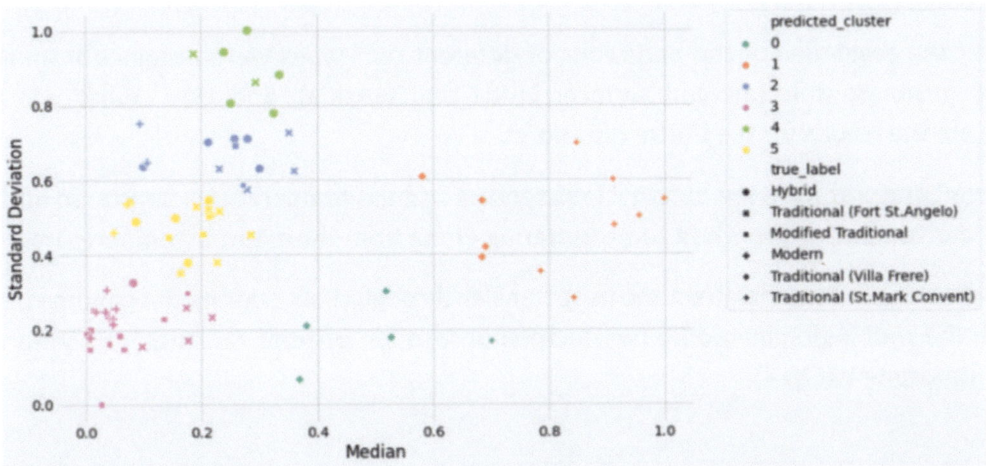


***Figure 2. KOMPsat-3A derived spectral reflectance clustering using K-Means clustering for Grids.***

Using a similar procedure, the K-Means clustering results for the multi spectral reflectances derived from the UAV sensors showed that this approach was also sensitive enough to be able to clearly distinguish between traditional, hybrid and modern roofs (Fig. 3). The derived Silhouette clustering score was equivalent to 0.49, which gave a clearly separated clustering for the traditional roofs, especially noticeable for Villa Frere and St Mark's Convent, but with a bit less separation between hybrid and modified traditional roofs clusters. This result applies for the entire time period used in this study.

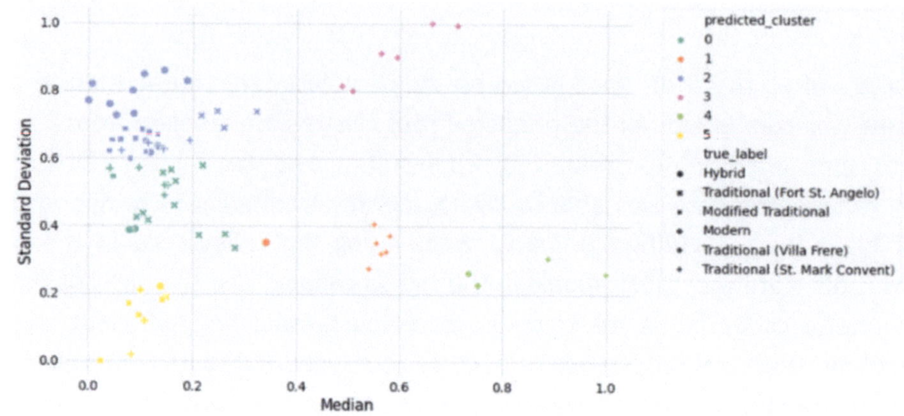


***Figure 3. UAV-derived spectral reflectance clustering using K-Means clustering for Grids.***

## THERMAL ANALYSIS FOR ROOF TYPE BEHAVIOUR – INITIAL RESULTS

The data used for the evaluation of the behaviour of different roof types were obtained from the weather station, *in-situ* sensors (air and surface, inside and outside), and UAV. What will be considered here are the results for the Cottonera roofs[5].

In all cases the temperatures from the external (air) sensors and the temperatures registered at the weather station showed that all roofs had some buffering effect from the external temperatures.

It is known that different factors can affect the heat transfer through roofs - during the daytime heat transfer through the roof is dominated by two factors: absorption of solar radiation and infrared emission to the atmosphere (Fig. 4).

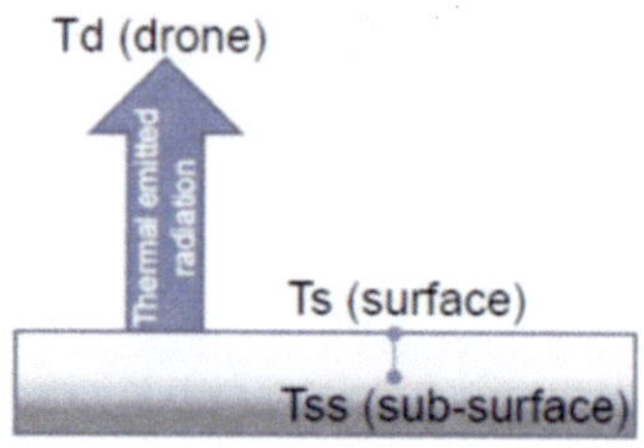


***Figure 4. Factors affecting thermal insulation in a roof.***

This study found that assessing the surface insulation directly, using the difference between the surface (Ts) and the sub-surface temperature (Tss): $|Ts-Tss|$, depends on the deployment configuration of the sensors[6] and is almost completely uncorrelated with seasonal conditions and therefore not useful to characterize the roof top characteristics. What is useful to use is the surface radiative assessment, namely the difference between the drone-acquired temperature Td and the surface temperature Ts: $|Td-Ts|$, characterising the roof top material based on its emissivity capabilities. This conclusion is presented in Table 1; the roof with high insulation is the traditional

---

[5] For the additional two roofs (St Mark's Convent and Villa Fere) UAV data only were used, due to time limitations; this study is still under way and results will not be reported here.

[6] It must be remembered that traditional roofs tend to be applied manually, and therefore the laying and compaction of the material is usually very inhomogeneous

roof part of the Fort St. Angelo, followed by the modified traditional roof type. Only entries with Ts > 30°C were considered to take into account summertime conditions. More results being processed for the other traditional roofs studies will enrich these conclusions.

**Table 1. Summary table of the roof temperatures, their differences (in °C) thermal insulation results for all roof types.**

| Roof type | Ts | Tss | Td | \| Ts – Tss\| | \|Td-Ts\| | Thermal Insulation |
|---|---|---|---|---|---|---|
| C (hybrid) | 15,25 | 13,40 | 13,04 | 1,85 | 2,21 | 0,837 |
| C (hybrid) | 30,70 | 29,47 | 38,99 | 1,23 | 8,29 | 0,148 |
| C (hybrid) | 33,91 | 32,02 | 42,61 | 1,89 | 8,70 | 0,217 |
| C (hybrid) | 36,31 | 34,04 | 41,20 | 2,27 | 4,89 | 0,464 |
| D (traditional) | 15,21 | 13,45 | 15,78 | 1,76 | 0,57 | 3,088 |
| D (traditional) | 32,22 | 30,46 | 43,54 | 1,76 | 11,32 | 0,155 |
| D (traditional) | 34,47 | 32,99 | 45,98 | 1,48 | 11,51 | 0,129 |
| D (traditional) | 35,26 | 33,34 | 42,42 | 1,92 | 7,16 | 0,268 |
| E (traditional border) | 31,45 | 32,62 | 42,63 | 1,17 | 11,18 | 0,105 |
| E (traditional border) | 35,11 | 35,20 | 46,23 | 0,09 | 11,12 | 0,008 |
| E (traditional border) | 36,57 | 36,49 | 42,81 | 0,08 | 6,24 | 0,013 |
| G (Modern) | 19,24 | 18,66 | 21,87 | 0,58 | 2,63 | 0,221 |
| G (Modern) | 34,34 | 33,85 | 42,10 | 0,49 | 7,76 | 0,063 |
| G (Modern) | 31,70 | 32,30 | 41,88 | 0,60 | 10,18 | 0,059 |
| F (Modern) | 20,67 | 18,30 | 22,12 | 2,37 | 1,45 | 1,634 |
| F (Modern) | 37,13 | 31,62 | 41,50 | 5,51 | 4,37 | 1,261 |
| F (Modern) | 33,42 | 37,56 | 43,36 | 4,14 | 9,94 | 0,416 |
| H (Modern) | 18,23 | 18,15 | 20,74 | 0,08 | 2,51 | 0,032 |
| H (Modern) | 32,80 | 33,55 | 42,87 | 0,75 | 10,07 | 0,074 |
| H (Modern) | 33,29 | 32,05 | 42,15 | 1,24 | 8,86 | 0,140 |
| J (Modified Traditional) | 35,05 | 24,26 | 27,34 | 10,79 | 7,71 | 1,399 |
| J (Modified Traditional) | 41,96 | 33,72 | 29,96 | 8,24 | 12,00 | 0,687 |
| J (Modified Traditional) | 50,64 | 39,53 | 31,28 | 11,11 | 19,36 | 0,574 |
| J (Modified Traditional) | 47,33 | 38,91 | 42,26 | 8,42 | 5,07 | 1,661 |

## 5. CONCLUSION

This study has demonstrated a dual approach towards roof top material characterisation by means of satellite and collocated and co temporal UAV remote sensing. Radiometric analysis showed that very high-resolution satellite data derived from KOMPSAT 3A allows the identification of traditional and hybrid roofs using the blue and red wavelengths. This technologic synergy augurs well on the basis of a future, wider application of this technique towards a comprehensive detection of traditional roofs at a national scale. This will be a tool of great use to heritage professionals, as well as urban planners and regulators.

As regards to roof behaviour, thermal insulation is one of the most important characteristics of a roof which can help reduce heating and cooling energy costs. During the daytime heat transfer through the roof is dominated by two factors: absorption of solar radiation and infrared emission to the atmosphere. Differences occur because solar radiation hits roofs at a much higher angle during the hours of greatest intensity. Roofs are also greatly exposed to the sun, and therefore they will lose a much greater amount of heat to the atmosphere through infrared radiation.

These results have shown that the traditional roof type gives the best compromise between potential dispersion and insulation assessment in thermal performance characterization; more research and data analysis is however needed to better understand this process.

Another primary output of the project has been the development of a methodology to consolidate the use of multispectral information coming from diverse space platforms (including commercial data), aerial technology (sensors operated from UAVs) and *in-situ* sensors to analyse roof properties (thermal and moisture-based). In the long term this information will be used to develop recommendations and guidelines to safeguard buildings with these traditional roofs and to develop a way forward to apply this methodology in the wider Mediterranean. All of this is closely linked to potentially regaining the sustainable use and reuse of historic buildings, including reducing energy consumption and ensuing $CO_2$ emissions, and possibly also affecting in a positive way the Urban Heat Island effect [3].

## 6. ACKNOWLEDGEMENTS

Project EO4HBCS has been financed by the Malta Council for Science & Technology through the Space Research Fund 2019, CONTRACT NUMBER: SRF-2019-2S1. Project data merging and analyses were carried out by SISTEMA GmbH – the great input of Dr Stefano Natali and Alexandra Bojor is gratefully acknowledged. The authors also wish to thank Heritage Malta, the Restoration Directorate, Architect Edward Said and Fabio Scicluna (Research Support Officer II), and Pilots Jason Gauci and Kevin Theuma (Institute of Aerospace Technologies, University of Malta) and Mario Demanuele (Aerial Malta) for their tangible support extended.

REFERENCES

[1] L. Foroudi for Financial Times: Architecture. "The architecture of heat: how we built before air-con." Internet: https://www.ft.com/content/839d4ccf-269f-44fe-914b-544644a4c819, Aug. 28, 2020 [Dec. 20, 2020]

[2] Cassar, J.; Galdies, C.; Muscat Azzopardi, E. A New Approach to Studying Traditional Roof Behaviour in a Changing Climate—A Case Study from the Mediterranean Island of Malta. *Heritage* 2021, *4*, 3543-3571. https://doi.org/10.3390/heritage4040196

[3] C. Galdies and S.L. Hok. "Urban heat island effect, extreme temperatures and climate change: a case study of Hong Kong SAR" in Climate Change, Hazards and Adaptation Options: handling the impacts of a changing climate. 1st ed., vol. 1, W. Filho, G. Nagy, M. Borga, D. Chávez Muñoz and A. Magnuszewsk, Eds. Switzerland AG: Springer Nature, 2020, pp. 369-388.

# Internal insulation – a preliminary assessment tool based on probabilistic simulations

E. J. de Place Hansen[1], E. B. Møller[2], T. K. Hansen[1]

[1] Department of the Built Environment, Aalborg University, Copenhagen, Denmark. deplace@build.aau.dk; tessa@build.aau.dk

[2] Department of Civil Engineering, Technical University of Denmark, Kgs. Lyngby, Denmark. ebmo@byg.dtu.dk

**Keywords –** Internal insulation; Preliminary assessment tool; Hygrothermal simulations; Probabilistic approach; Risk

## 1. ABSTRACT

Historic buildings contribute heavily to the energy consumption of the European building stock, and internal insulation offers a possibility to improve energy performance and indoor thermal comfort, without compromising the buildings' architectural appearance. However, due to the complexity and many potential risks and uncertainties less experienced planners and building owners need a preliminary assessment tool and simple, practical guidance in a specific situation. This paper presents and discusses such a tool developed as part of RIBuild (www.ribuild.eu); the content, preconditions and limitations.

User input encompasses specific location and orientation of a building, and wall thickness and type, by using checkmarks and sliders. In each case, the tool delivers a list of solutions selected from a database of pre-calculated simulations made in the hygrothermal simulation tool DELPHIN with Quasi-Monte Carlo based repetitions of simulations; a probabilistic assessment. The user can prioritize solutions based on e.g. risk of mould growth or maximum thickness of insulation. The output describes the heat loss through 1 m$^2$ wall with and without internal insulation, as well as risk of mould growth behind the insulation and algae growth at the façade. An improvement of the reliability of the tool requires more simulations and the inclusion of other failure mechanisms than mould and algae growth.

## 2. INTRODUCTION

Internal insulation offers a possibility to improve energy performance and indoor thermal comfort of historic buildings, reducing their contribution to the energy consumption in buildings in Europe, without compromising their architectural appearance. However, internal insulation entails several risks as the original wall becomes colder and more humid. The RIBuild project (www.ribuild.eu) identified the need for a preliminary assessment tool and simple, practical guidance targeted at less experienced planners and building owners to overcome their restraint and to avoid mistakes when internal insulation of historic buildings is considered. The RIBuild website guides on how to determine if the building is suitable for internal insulation and it offers the Insulation Calculation Tool (ICT).

ICT is a web tool developed to be used by building owners or building professionals with limited knowledge on building physics, when they want to predict the implications of adding internal insulation to a solid wall made of brick or natural stone, by entering a few data on the building. Although energy savings are often the initiator for renovating a building, the development also focused on moisture related issues, e.g. mould and algae growth. The needed input data should be easy to find and not require laboratory tests. Before using the ICT, the user should have checked the building e.g. according to the guidelines at www.ribuild.eu and found it suitable for internal insulation. The ICT is a preliminary assessment tool; it cannot replace professional assessment by experts, but gives the owner an idea of whether it might be a good idea to proceed with plans of applying internal insulation.

Section 3 describes the main features of the ICT. The result of a test of the ICT is presented in Section 4. The ICT has some limitations, as at present combinations of relevant orientations, locations and insulation systems are few, and more failure modes would be relevant to include. This is discussed in Section 5. Finally, conclusions and suggestions for improving the tool are given in Section 6.

## 3. MATERIAL AND METHODS IN THE PRELIMINARY ASSESSMENT TOOL

The core of the ICT is a database consisting of 275,000 hygrothermal 1D simulation results based on DELPHIN [1] that provide a probabilistic assessment of the hygrothermal conditions at specific points in an internally insulated solid wall. Although important development has been achieved in RIBuild concerning the numerical efficiency of a probabilistic assessment approach [2], [3], real-time simulations that consider uncertainties, performed by a non-specialist, are still not realistic. Therefore, simulations have been performed beforehand, as a fast response has been a high priority. The ICT finds the simulations that come closest to a specific case described by an average U-value of the insulated wall, heat loss through 1 $m^2$ wall with and without a specific insulation system, and the internal surface temperature. Further, it assesses the risk of mould growth at the interface of insulation and existing wall, and the risk of algae growth at the external surface.

### 3.1 SIMULATIONS FOR A PROBABILISTIC ASSESSMENT

To reduce the number of simulations for a probabilistic approach, a Sobol sampling was chosen as the most effective Quasi-Monte Carlo method [2]. The probabilistic approach resulted in 16 uncertainty layers defining the wall configuration and the parameters that were to be varied in the simulations [4]. The RIBuild project showed that solid walls made of brick or natural stone by far are the most common external wall types in historic buildings [5]; in most cases they have an internal rendering and, in some cases, also an external rendering. As the internal rendering might be removed before applying internal insulation, four configurations of the existing wall were relevant to include: 1) Without any rendering, 2) Rendering on both sides, 3) Only external rendering, 4) Only internal rendering.

The selected insulation systems were also based on investigations made in RIBuild [6]. Each of them is a complete system, i.e. with an intersection layer (glue mortar), an insulation material and a finishing layer. As an exception mineral wool was included as well, as this is common without being a complete system. For each system, simulations were based on standard thickness of the insulation material.

To avoid having unrealistic combinations of material parameters in each of the historic or insulation materials, the materials were treated as discrete parameters, i.e. for the probabilistic assessment different bricks were used, each with fixed material parameters. The different materials in each category (e.g. bricks) were chosen from the DELPHIN database included in [1], supplemented by a few historic materials, tested as a part of RIBuild and added to the DELPHIN database.

To cover Europe from North to South, weather data files were obtained from 152 weather stations. The data was future data developed in the project Climate for Culture (www.climateforculture.eu). Simulations were run for five years with a discrete starting year between 2020 and 2045. As indoor climate the two climate classes A and B were chosen based on EN 15026 [7]. The sources of the basic input data and the number of materials are listed in Table 1 .

**Table 1: Basic input used for simulations and sources for these**

| Input | Variations | Source |
|---|---|---|
| Wall configuration:<br>Solid wall (brick or natural stone)<br>Plaster on either side is optional | 52 brick types<br>33 natural stone types<br>11 plaster types | RIBuild investigations, www.ribuild.eu |
| Insulation systems:<br>11 different systems | Standard thickness given by the manufacturers | RIBuild investigations |
| Weather data | 158 European weather stations | Climate for Culture |
| Indoor climate | Climate class A or B | EN 15026 |

Ten of the uncertainty layers were uniform parameters, i.e. could be any value within a range, many of these were coefficients describing the boundary conditions, e.g. heat or moisture transfer coefficients. Also, the orientation was a uniform parameter, between 0° and 360°, and the thickness of rendering and masonry were uniform parameters in the range of 1-2 cm and 10-90 cm, respectively.

## 3.2 USER INPUT

To filter out the relevant simulations, user inputs as simple as possible are needed. Only parameters that would be fairly easy to assess were chosen, described in Table 2. Based on this, the ICT delivers a list of possible solutions concerning internal insulation systems.

**Table 2: User input to filter out the relevant simulations**

| Parameter | Type of input | Range |
|---|---|---|
| *Location* | Coordinates, address or click on map | - |
| *Distance to weather station* | Slider to maximize distance | 10 km – 500 km |
| *External plaster* | Check box | Yes or No |
| *Wall material* | Check box | Brick or Natural stone |
| *Internal plaster* | Check box | Yes or No |
| *Thickness of wall* | Slider from both sides | 100 mm – 900 mm |
| *Orientation* | Slider from both sides | 0° – 360° |

ICT includes two features to ensure that the most relevant solutions among those selected, based on the mandatory user inputs, will be on top of the list. Firstly, sliders can be used to prioritize the five parameters: U-value, heat loss, internal surface temperature, risk of mould growth, risk of algae growth. Secondly, the user can choose to restrict the thickness or type of internal insulation, within a range of 10 – 150 mm and between 11 insulation systems, respectively.

### 3.3 OUTPUT

The result of each of the performed simulations is a 5-year series of temperature, moisture content and relative humidity (hourly based) at five different points, illustrated in Figure 1. Points 1-4 were chosen as the most relevant places to investigate different failure modes. Point 5 was included as the internal surface temperature is important for the comfort close to the wall. The user does not see the 5-year series, these are used for the post processing, where failure modes are assessed together with the heat loss (per year), U-value, and minimum internal surface temperature.

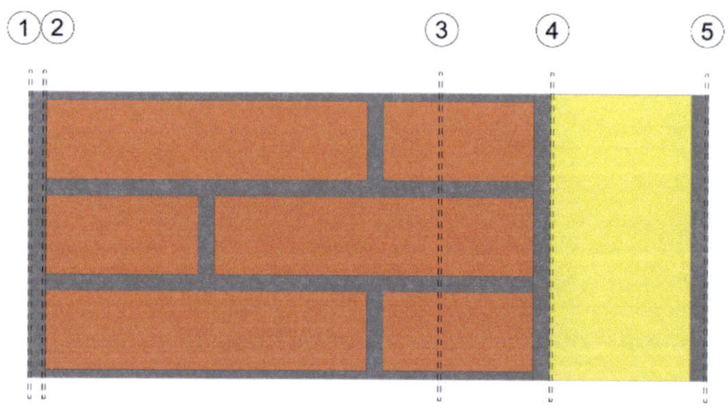


**Figure 1. Cross-section of an internally insulated solid masonry wall indicating five locations relevant to four different failure modes and the internal surface temperature: 1) External surface; algae growth 2) 5 mm from the external surface; frost damage 3) 50 mm from the original internal surface; wood rot in wall plate 4) 0.5 mm from the interface between insulation and existing wall; mould growth 5) 0.5 mm from the internal surface; surface temperature**

Originally it was planned to test for four different failure modes: Algae growth, frost damage, wood rot and mould growth. However, as the existing models for frost damage and wood rot were not fully developed to give realistic results, the ICT only tests for algae growth at the external surface (point 1) and for mould growth at the intersection between the existing wall and the insulation system (point 4). Minimum internal surface temperature and heat loss are shown as direct output in the ICT.

The risk of algae growth was determined using the method described in [8], giving an algae index between 0 and 1. For mould growth the VTT model [9] was used, depending on the material the scale goes from 0 to 6. For mould growth, the sensitivity class of the materials in the intersection has been set to 'Resistant', due to the initially high pH at the intersection caused by alkaline glue mortar, and the materials being inorganic. There are no general accepted threshold values for algae growth or mould growth; they may vary on a national scale or depend on the risk the specific user is prepared to take. Consequently, the ICT includes no threshold values, however, the results are coloured, indicating a high (red), medium (yellow) or low (green) number. The heat loss is given as a number (kWh/m$^2$/year) and as a reduction compared to the uninsulated case. Furthermore, the calculated U-value and the minimum internal surface temperature is shown.

## 4. TEST OF THE PRELIMINARY ASSESSMENT TOOL

To compare the outcome of the ICT with the potential achievement of putting more effort into simulating a specific case, a test was made involving six case studies from RIBuild, containing in-situ measurements of temperature and relative humidity in the insulated walls. The cases represent brick and stone wall with and without external rendering, different systems for internal insulation, and different locations and orientations [10]. U-value, annual heat loss through 1 $m^2$ of external wall, Mould Index and minimum internal surface temperature were used as output parameters. Further, deterministic hygrothermal simulations in both WUFI and DELPHIN have been carried out using data on wall thickness etc. from the case studies. Post-processing tools were used, including WUFI Mould Index VTT [11] and the VTT Mould Model included in DELPHIN [1]. Figure 2 illustrates the procedure.

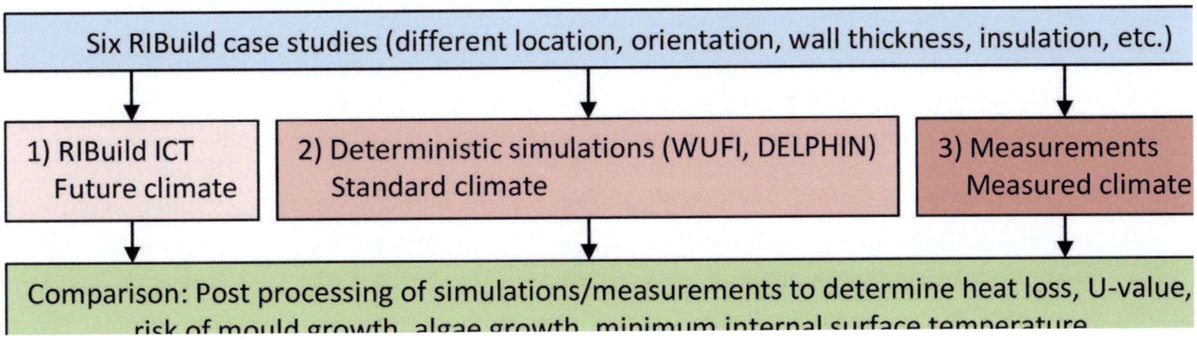


**Figure 2. Procedure for testing RIBuild ICT by comparing outcome using 1) ICT, 2) deterministic simulations, and 3) measurements. Climates were chosen as close to the location as possible.**

As the RIBuild ICT is based on previously performed simulations involving a probabilistic approach, ICT models do not necessarily match the case studies directly. Deviations are present both in relation to orientation, wall thickness, insulation thickness and thermal conductivity of the insulation material. Further, the ICT and the simulation tools used to support measurements in the case studies make use of different sets of climate data. All the deterministic simulations (WUFI, DELPHIN) were carried out for five years with local climate data from a location close to the case building in question.

Comparison of internal surface temperature simulated in ICT, WUFI and DELPHIN, shows in general good coherence. In most cases, the ICT appears to underestimate the heat loss when compared to simulation models and reported values from the case studies, which is surprising as is tends to overestimate the U-value.

The Mould Index is generated for the interface between internal insulation and the existing wall, as this is an area of risk, due to possible condensation. This parameter showed large differences between ICT, DELPHIN and WUFI. In some cases, WUFI and DELPHIN agreed, in other cases ICT and WUFI agreed. In general, the ICT generates no mould growth (Index of 0), except for one case 0.1, while DELPHIN in general gave a Mould Index around 2.

## 5. DISCUSSION

### 5.1 SIMPLIFICATIONS AND LIMITATIONS IN THE NUMBER OF SIMULATIONS

In regard to historic building materials, the user was given two options only; brick or natural stone. The different kind of brick (or natural stone) were handled as a parameter with a relatively high uncertainty, representing the many different kinds of bricks or stones in the pre-calculations. This choice was based on analyses made within RIBuild, showing that other factors such as location or orientation of the building were more important than the properties of the core material [12].

Aspects such as solar gains, ventilation losses, etc. are not included in the ICT, focusing on the external wall only. Irregularities such as thermal bridges and building details (e.g. around windows and beam ends), are also not accounted for as the simulations were made in 1D. However, such irregularities influence both heat loss and potential risk of moisture related damage and must therefore be assessed by professionals. 2D or 3D simulations would have prolonged the computing time enormously.

The database holds 275,000 simulations, however, in many cases the user will only find a few simulations to fit the specific case, as the wall thickness of the actual case might not be covered, and/or the orientation of the wall in the presented solutions, and the weather data that they are based on, might not be relevant for the case. Roughly 10 million simulations are needed to ensure that the user in most cases could find several simulations based on a relevant wall thickness and orientation, and weather data from a location close to the building. This would take years with the current rate that the simulations have been made with, despite queueing systems and powerful computers.

Using a meta-model with a machine learning algorithm, such as a neural network, would therefore be an obvious choice. Unfortunately, hygrothermal simulations sometimes fail, often rainfalls can have a fatal influence on the simulations, making them crash. This makes it difficult to introduce machine learning especially if the outcome is supposed to be time series of several parameters. Simpler things like heat loss might be possible. However, it is expected that the development of more advanced machine learning algorithms in time might solve the problem of time consuming calculations.

## 5.2 LIMITATIONS CONCERNING FAILURE MODES

Originally it was planned to include frost and wood rot as failure mechanisms. However, frost in masonry is a complex mechanism and no reliable model was found. Therefore, frost was omitted. Wood rot models exist, although, when applied to e.g. wall plates in old masonry, these models seem to overestimate the risk of wood rot [13]. According to the models most beam ends in historic buildings would have suffered from a fatal mass loss after a much shorter time (e.g. 10 years), than they already have existed (> 100 years). Consequently, wood rot was also omitted. Besides, the moisture and temperature threshold for mould growth based on the VTT model [9], is lower than what is expected for wood rot, and although wood rot is expected to occur at a slightly different place (point 3 in Figure 1) than mould growth (point 4) and there might be cases where mould may be acceptable but wood rot not, the mould growth threshold is most likely to be the decisive threshold.

The algae model was included although there is limited experience with it outside the lab. It might overestimate the risk of algae growth as the model will continue algae growth whenever the conditions are favourable, i.e. there is no decline. Consequently, if there are shorter periods favourable for algae growth at specific times of the year, the algae index will reach 1 given enough time.

The mould growth model does not have this problem, as it can decline. However, the VTT model was developed for surfaces, not interfaces where the pH value is high, and without direct contact to interior air. As the high pH value may inhibit mould growth [14], the sensitivity class was changed. Simply changing the sensitivity class from 'Medium resistant' to 'Resistant' may be too simple a solution. The consequence has been that the mould index is very low in most cases. Further investigation on how pH can be integrated into e.g. the VTT model is needed.

When better failure models are available, an update of the tool is quite simple, as the failure models are post processors. The heavy work load are the simulations, the time series from these can be used for any failure mode, and by having outputs at point 2 and 3, frost and wood rot can also be included.

## 5.3 FEASIBILITY

Overall, RIBuild ICT appears to be a helpful tool, for a fast analysis of a given construction. It is based on a probabilistic approach, and the validity is therefore dependent on the number of simulations that have been run previously. The ICT generates probabilistic results within the given ranges of input, and can therefore include uncertainties on e.g. material parameters and climate. WUFI and DELPHIN operate deterministically, and do not address the uncertainties to the same extent. Further, parameters such as orientation, wall thickness and in some cases insulation thickness vary significantly in the ICT compared to the actual cases, which influences the results. With more simulations, the ICT will only improve. Furthermore, the ICT makes use of forecasted weather, while simulations in WUFI and DELPHIN are in generally based on historic climate data integrated in these tools. This is also expected to have an impact on results achieved.

RIBuild ICT does seem to overestimate the U-value in most cases, which may be explained by the use of a probabilistic approach, involving different types of brick and wall thicknesses, opposed to the case studies and the created WUFI and DELPHIN models. With an overestimation of the U-value, one would expect the ICT to overestimate the heat loss as well; surprisingly, in most cases the opposite can be seen. This may be due to different locations used for the Heating Degree Hours (HDH), used to generate the heat loss; for the manual processing, HDH has been chosen for the physically nearest location, while the ICT may interpolate between the locations included for climate determination. Furthermore, the ICT uses future climate, which might mean lower HDH.

The Mould Index yielded significant discrepancy between simulations, largely due to the variety in input parameters in the different programs, including the climate files. It is a general problem, not only in the ICT, that mould prediction models may not correspond to what is seen in reality [4]. Apparently, the models do not take into account that the pH behind insulation applied with a cement containing glue mortar inhibits mould growth [14]. That is why the sensitivity class in the intersection between wall and insulation was chosen to be 'resistant' in the ICT rather than 'medium resistant', although this may not be the right choice in all cases, depending on the pH level in the actual case. Correspondingly, a lower sensitivity class (higher resistance) than the default values were chosen for the post-processing tools when simulating in WUFI and DELPHIN (deterministic simulations). Mould index values based on the case studies show the effect of using a higher sensitivity ('medium resistant'), as they in two cases are much higher (higher than 2) than the output of ICT (zero).

# 6. CONCLUSION AND PERSPECTIVES

The Insulation Calculation Tool (ICT) is based on a large number of pre-calculated simulations, using a probabilistic approach, and a few user input, using checkmarks and sliders. The user input filters the simulations and the results are shown as different solutions describing risk of mould and algae growth, minimum internal surface temperature, U-value and heat loss. This makes it accessible for people not necessarily having any pre-existing knowledge about internal insulation, but who are interested and able to investigate the building renovation options if the information available is not too technical.

Based on a comparison with deterministic simulations performed with WUFI and DELPHIN and with measurements from case studies, the overall impression is that the ICT, in its current state, gives valuable information to the user. Discrepancies are largely accounted for by missing simulations in RIBuild ICT, some of the results differ considerably and the algae growth may be overestimated. The user must be aware of the shortcomings of the tool at its current state, included in a disclaimer at the website. The ICT is meant as a help in the planning phase of a renovation possibly involving internal insulation and it should only be used as a preliminary assessment tool, giving an overview of possible solutions, of which analyses that are more detailed should be made by building professionals.

To improve the feasibility of the ICT, further development should be based on applying a meta-model with a machine-learning algorithm to the already performed simulations and through machine-learning fill in the gaps in the simulations. When reliable models for frost damage or wood rot are available it is possible to extend the ICT further, as this would only require simple post processing.

## 7. ACKNOWLEGDGEMENTS

RIBuild has received funding from the European Union's Horizon 2020 research and innovation programme under grant agreement No 637268.

## 8. REFERENCES

[1] Bauklimatik-Dresden. DELPHIN. Simulation program for the calculation of coupled heat, moisture, air pollutant, and salt transport [online]. Dresden: TU Dresden. Available: http://bauklimatik-dresden.de/delphin/index.php?aLa=en
[2] H. Janssen, P. Freudenberg, A. Tijskens and T. Hou. Basic probabilistic analysis of hygrothermal performance of interior insulation (RIBuild Deliverable D4.1, www.ribuild.eu). 2018.
[3] H. Janssen, A. Tijskens and T. Hou. Metamodelling approaches for hygrothermal performance of interior insulation (RIBuild Deliverable D4.2, www.ribuild.eu). 2019.
[4] C. Kongsgaard, S.P. Bjarløv, T. Perkov and E. Møller. Description of the simulations (RIBuild Deliverable D6.5, www.ribuild.eu). 2020.
[5] A. Blumberga, D. Blumberga, E. Kamendere, A. Kamenders, K. Kass, R. Purvins, G. Zogla, Report on historical building types and combinations of structural solutions (RIBuild Deliverable D1.1, www.ribuild.eu). 2015.
[6] A. Blumberga (ed.). State of the art on historic building insulation materials and retrofit strategies. (RIBuild Deliverable D1.2, www.ribuild.eu). 2015.
[7] EN 15026. Hygrothermal performance of building components and building elements. Assessment of moisture transfer by numerical simulation. 2007.
[8] L. Graziani and E. Quagliarini. "On the Modelling of Algal Biofouling Growth on Nano-TiO2 Coated and Uncoated Limestones and Sandstones". Coatings, vol. 8, no. 2, 2018. https://doi.org/10.3390/coatings8020054.
[9] T. Ojanen, T. Viitanen, R.H. Peuhkuri, K. Lähdesmäki, J. Vinha and K. Salminen. "Mold growth modeling of building structures using sensitivity classes of materials," in Proc. Thermal Performance of the Exterior Envelopes of the Whole Buildings XI Int. Conf., 2010.
[10] E.B. Møller (ed.). Web tool including feasibility study of possible input and output data (RIBuild Deliverable D6.1, www.ribuild.eu). 2020.
[11] D. Zirkelbach, T. Schmidt, M. Kehrer and H.M. Künzel. WUFI Pro -Manual. 2007.
[12] E.J. de Place Hansen and E.B. Møller. "How to estimate material properties for external walls in historic buildings before applying internal insulation," in Proc. EEHB 2018, Visby, Sweden, http://eehb2018.com/conference-report/. 2018.
[13] T.K. Hansen, N.F. Jensen, E. Møller, E.J. de Place Hansen and R.H. Peuhkuri. "Monitored conditions in wooden wall plates in relation to mold and wood decaying fungi," in E3S Web of Conferences, vol. 172, 2020. https://doi.org/10.1051/e3sconf/202017220004
[14] N.F. Jensen, C. Rode, B. Andersen, S.P. Bjarløv and E.B. Møller. "Internal insulation of solid masonry walls – field experiment with Phenolic foam and lime-cork based insulating plaster," in E3SWeb of Conferences, vol. 172, 2020. http://doi.org/10.1051/e3sconf/202017201003

# Does age matter? How building age influences energy use in the Swedish residential building stock

A. Donarelli, T. Broström

Department of Art History, Conservation, Uppsala University, Visby, Sweden
(anna.donarelli@konstvet.uu.se, tor.brostrom@konstvet.uu.se)

**Abstract –** The energy use in existing buildings must be reduced and historic buildings need special attention in terms of practice and policies. Data on energy use in the building stock, and which factors influence it, can be used to develop national policies for energy saving. This study uses data from Swedish energy performance certificates (EPCs), representing almost 90 % of apartment buildings and 15 % of single family houses. EPCs are based on measured energy use.

The objective of this study is to describe how energy use relates to building age in residential buildings. The energy use in buildings built before 1945 stand for a significant part of the energy use, whereas buildings built before 1845 account for a very small part of the energy use for buildings. Most energy is used in buildings built 1965–1974. This is a segment where, generally, the impact on heritage significance could be limited.

**Keywords** – Energy performance; EPC; Historic buildings; Built cultural heritage; Building stock

## 1. INTRODUCTION

With an increasing global focus on reducing energy use in the existing building stock, it has become clear that historic buildings need special attention, both in terms of practice and policies [1] [2] [3] [4]. The recent European standard on improving energy performance in historic buildings points at the main challenge: "to reduce energy demand and greenhouse gas emissions without unacceptable effects on the heritage significance of the existing built environment". The same guidelines define historic buildings as buildings "of heritage significance due to their historical, architectural or cultural values". This definition is not limited to buildings with a formal heritage protection ("listed buildings") but applies to a much wider range of buildings [5]. In the general discourse on energy performance in historic buildings there are two parallel and contradictory anecdotes; on the one hand it is often argued that their energy performance is worse than modern buildings, on the other hand, there is a conception that older buildings perform better than one would expect [6].

The data used in this study concern the existing building stock. Parts of the building stock would be considered historic buildings according to the definition in the European guidelines. Heritage significance is not strictly related to age, modern buildings may be defined as historic buildings and older buildings may have little or none heritage significance. However, within the older building stock we will find most of the historic buildings with heritage significance. Furthermore the Swedish Planning and Building Code (PBL) stipulates that all changes to all buildings should be performed cautiously, with regard to the character of the building, maintaining its technical, historical, environmental and artistic values [7].

In the present study we have analysed the building stock based on different limits for building age. One limit is 1920 and based on the Swedish building legislation, which states as a general advice that "Buildings from the pre-1920s building expansion, which have preserved their main character, constitute such a limited part of the building stock today that most of them can be expected to meet the criteria for a heritage building" [8]. Another age limit for our study is 1945. Materials and construction techniques used in buildings built before 1945 are different from more modern buildings. During the late 1940's building construction became more industrialised and materials such as aerated concretes became more commonly used. The first national building standards in Sweden are from 1946 where standards for heat transmission through the building envelope were set [9].

### 1.1 RESEARCH OBJECTIVE

The general objective of this study is to provide an improved evidence base for policy development regarding energy targets and renovation practices in the historic building stock. In this study we use the data available in the Swedish EPC data base to answer the following questions:

- How does energy use and energy performance in the building stock relate to building age?
- How can this information be used for planning and polices on a national level?

### 1.2 ENERGY CERTIFICATION OF BUILDINGS IN SWEDEN

The Swedish law on energy certification of buildings (SFS 2006:985) is based on the European Parliament's directive on the energy performance of buildings (EPBD) [10]. The Swedish system for energy performance certificates is managed by the Swedish National Board of Housing, Building and Planning (Boverket). The energy performance certificate (EPC) is meant to provide the owner, buyer or tenant with correct information on how to improve the building's energy performance. By law, all apartment buildings should have EPCs. Single family house are required to get an EPC only when they are sold. Two factors set aside the Swedish EPCs from the practice in many other European countries. Firstly the EPCs are made on whole buildings, not apartments. Secondly the Swedish EPCs are based on measured, not calculated values of annual energy use [11].

### 1.3 PREVIOUS RESEARCH

Aksoezen et al. (2015) studied the relation between building age and energy use in the City of Basel, Switzerland, where they had data on building type, year of construction, footprint area, number of floors, number of apartments, number of residents and type of heating system [12]. Volume and gross floor area turn out to have the highest correlation with energy consumption among the studied variables. The authors went on to analyse the shape factor (the ratio of outer surface to volume), energy use and building age. Results showed that shape factor of buildings influenced the energy performance of the studied buildings, more compact buildings consumed less energy. Shape factor in turn, correlated to building age; buildings built before 1921 in the study were more compact than modern buildings and thus consumed less energy for heating.

Webb made a critical assessment of the general conception that older buildings perform worse when it comes to energy use, but also the idea that older buildings perform better than we expect due to what is referred to as inherent energy efficient features [6]. In her dissertation she used data from two surveys of the U.S building stock; Commercial Buildings Energy Consumption Survey (CBECS) and Residential Energy Consumption Survey (RECS). The data from the surveys was analysed through different statistical methods, variables and their correlations were compared. The results showed that other variables than age were more important in influencing the buildings' energy performance.

Michelsen and Müller-Michelsen made an analysis of the relation between building age and energy use in multi-family buildings in Germany [13] [14]. The data contained records of whether or not the buildings had been refurbished during the last 15 years. Their sample contains around 157 000 buildings, where 19 500 had been fully refurbished within the last 15 year and 26 000 buildings were not refurbished. The results show that the effect of refurbishment for buildings before 1918 in this study is limited. The authors argue that one reason could be the complex task of refurbishing Wilhelminian style buildings. They conclude the greatest potential for energy savings is for buildings built between 1918 and 1945.

A statistical survey of the Swedish building stock, called BETSI, was carried out in 2009 [15]. Data on U-values and areas of building envelopes, building services systems, indoor temperature and air exchange rate was collected through building surveys and questionnaires on around 1800 buildings. One aim with the study was to investigate what measures would be needed and how much it would cost to fulfil the targets set for building energy use by the Swedish government. Buildings were selected statistically to represent the whole building stock, both commercial and residential buildings. Even though the data give an indication on how energy performance varies with the age, the sample is too small to draw any statistically sound conclusions.

Previous research has contributed to an improved general understanding of how energy performance varies with age, but it does not fully answer the research questions at hand. This study is based on a large amount of building data. The focus of the study is on buildings built before 1945. Data on buildings built between 1845 and 1945 are presented with quite high detail and the parameters that generally are expected to have an impact on the energy performance of the buildings are analysed. The study compares the policies for energy targets with the real energy performance of the older buildings in the building stock.

## 2. DATA

### 2.1 BUILDINGS AND ENERGY USE IN THE EPC DATA BASE

The data used in this study comes from the national data base of Energy Performance Certificates (EPCs) [16]. The Swedish EPC data base contains about 650 000 EPCs (January 2017), out of which around 530 000 refer to residential buildings. 407 000 of those are single family houses and 125 000 are apartment buildings (table 1). In some cases, more than one building is covered by a single EPC. This is often the case if they have the same technical characteristics and are connected to the same centralised heating system. Thus the 530 000 individual EPCs in the data base cover more than 560 000 residential buildings. In the present study the data selection includes all individual EPCs, but not all individual buildings.

The whole Swedish residential building stock consists of almost 3 000 000 residential buildings divided between the following categories [17]. There are 2 750 000 single family houses, 160 000 apartment buildings, and 80 000 unspecified residential buildings. Around 89% of all apartment buildings in the Swedish residential building stock are represented in the EPC data base, but only about 15% of the single family houses.

The building area reported in the EPCs is called $A_{temp}$, which is the total heated floor area. It includes the whole area within the outer walls, including staircases, shafts and the footprint area of inner walls. The EPCs report the annual measured final energy consumption used to heat the building and the domestic hot water, including electricity for ventilation and heat distribution systems. The measured energy consumption for heating is adjusted in reference to a climatically normal year (1981–2010). The energy performance (kWh/m$^2$ year) in the EPC is calculated by dividing energy use for heating, domestic hot water, and the property electricity with the $A_{temp}$.

**Table 1: Number of EPCs, $A_{temp}$ and energy use for residential buildings in the EPC data base.**

| | Number of EPCs | $A_{temp}$ ($10^6$ m$^2$) | Energy use (TWh) |
|---|---|---|---|
| *Apartment buildings* | 124 659 | 213 | 30 |
| *Apartment buildings built before 1945* | 27 852 | 37 | 5.3 |
| *Apartment buildings built before 1920* | 6 362 | 9 | 1.3 |
| *Single family houses* | 406 802 | 69 | 7.5 |
| *Single family houses built before 1945* | 94 602 | 17 | 2 |
| *Single family houses built before 1920* | 40 070 | 7 | 0.9 |

### 2.2 DATA UNCERTAINTY

The calculations needed to register the energy use in the buildings are not standardised, but based on assessments made by the energy auditor [11]. The energy data regarding single family houses is more uncertain than those concerning apartment buildings. The single family houses are often heated with different fuels and heat pumps, which requires the use of conversion factors, calorific values and division between electricity use for heating and household electricity, where the latter is subsequently removed from the final calculation on energy performance. The data on energy use in apartment buildings is more reliable as most of them are heated with district heating.

The heated floor area in the building is often recalculated from the residential floor area (BOA) and the non-residential floor area (LOA) by using conversion factors recommended by Boverket. A study shows that these factors generally result in an underestimation of the heated area [18]. The discrepancies in $A_{temp}$ impact the calculation of energy performance of the buildings. Hence, if $A_{temp}$ is generally underestimated the energy performance will in reality be better than reported the EPC data base.

The year of construction is a key parameter in the present study. In the data there are obvious peaks every first year of a decade (i.e. 1840, 1850 etc.) probably indicating that the building was constructed during that decade rather than an exact year. There are peaks of single family houses with the year of construction 1909 and among both building types with the year of construction 1929, this is due to registration in the national cadastre, where the year of the first taxation has been registered as year of construction. In reality many of these buildings are older.

## 3. RESULTS

As a first step, the residential buildings were divided in two groups; apartment buildings and single-family houses. Looking at the energy use (fig.1) in all the residential buildings represented in the

EPC data base we find that almost 18 % of the energy use is in apartment buildings built before 1945 and around 4 % in buildings built before 1920. In single family houses around 27 % of the total energy used is in buildings built before 1945 and around 12 % in buildings built before 1920. The energy use in buildings built before 1845 is 0,9 % in single family houses and 0,2 % in apartment buildings.

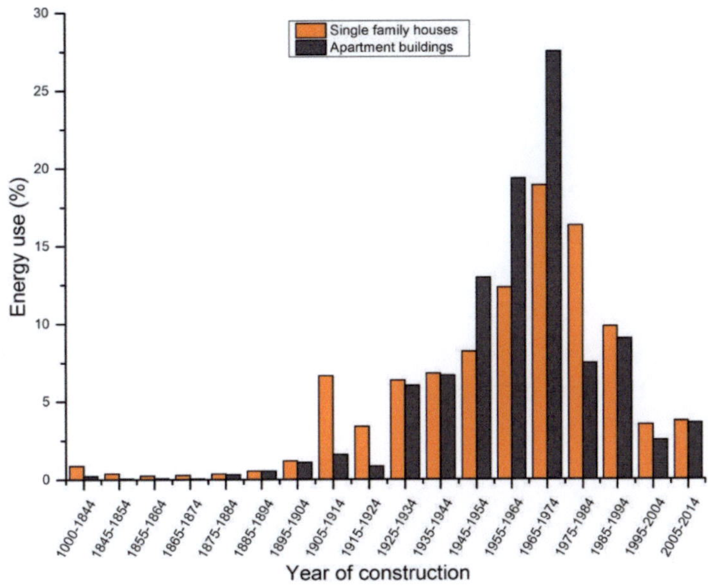


**Fig. 1: Energy use in the residential building stock, related to year of construction.**

The energy performance (kWh/m$^2$ year) in all residential buildings in the EPC data base was also analysed (fig. 2). The average energy performance is lower in the single family houses. The energy performance is more or less independent of building age until around 1965–1975, after that there is a distinct nick-point from where energy performance improves steadily. This is based on buildings in all Swedish climate zones and with all types of heating. In the next step, data was broken down in further categories.

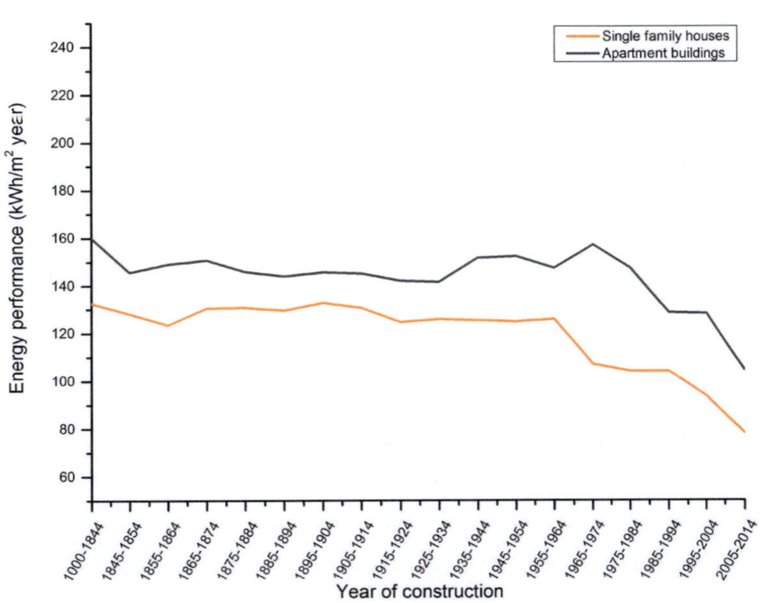


**Figure 2: Average energy performance for the residential buildings in the EPC data base.**

The effect of climate zone, adjacent walls and heat source on the buildings' energy performance was analysed next in order to create a sample of comparable buildings to better determine the effect of building age on energy performance.

1) **Climate zone:** Sweden is divided in four different climate zones, where 1 is the coldest and 4 is the warmest [8]. Climate zone 1 stood out with higher energy use, whereas buildings in climate zones 2, 3 and 4 were on the same level. Climate zone 1 was removed for the following analysis. Climate zones 2, 3, and 4 account for 93% of the buildings.
2) **Adjacent walls:** Whether a building was detached, semi-detached or terraced turned out to have little or no influence on energy performance and all buildings regardless of this factor were kept for further analysis.
3) **Heat source:** In the EPC delivered energy is registered. We selected two heat sources to make a comparison; district heating and heat pump (fig. 3). The analysis showed that the buildings heated with district heating use around twice as much as the buildings heated with heat pumps, which is expected since the Coefficient of Performance (COP) of the heat pumps typically is around 2.

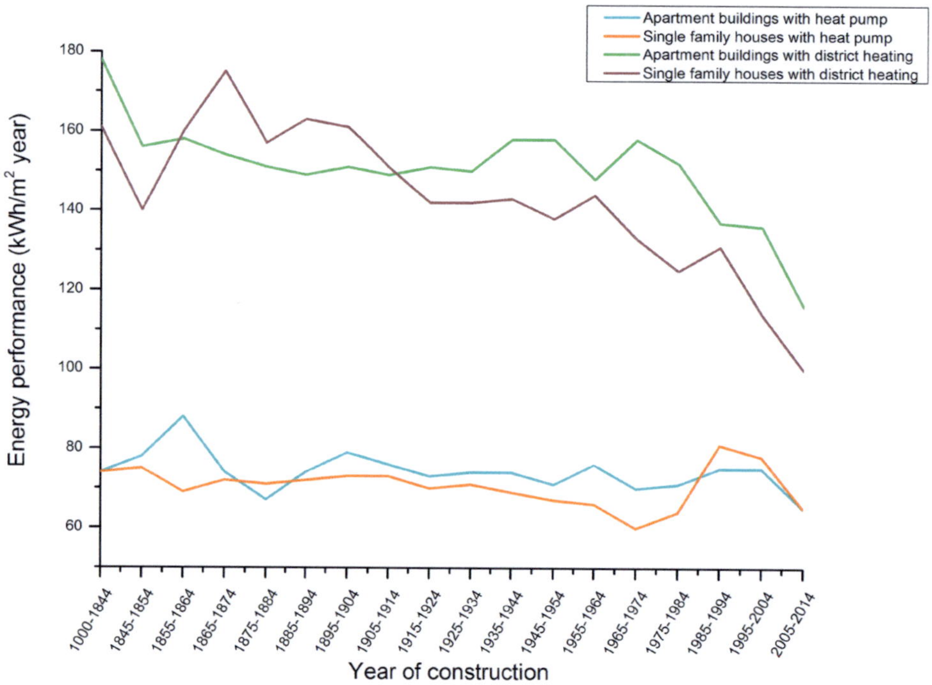


**Fig. 3. Average energy performance of residential buildings in climate zones 2, 3 and 4 related to heat source.**

## 4. DISCUSSION

The energy use in buildings built before 1945 stand for a significant part of the energy use in the Swedish building stock. It is around 27% of all energy used in single family houses and 18% of energy used in apartment buildings. This is an important part of the national energy use and the buildings built before 1945 should not be automatically exempted from societal demands towards lower energy use. The energy use in buildings built before 1920 is 4% in apartment buildings and 12% in single family houses. This is still a significant part of the energy use, especially for single family houses, but it is reasonable to be more restrictive in terms of targets and measures. On a national level, if we look at buildings built before 1845, we can see that they account for a very small part of the energy use for buildings, 0,9 % in single family houses and 0,2 % in apartment buildings. These buildings would generally have higher heritage values and thus be more vulnerable to renovation measures than the rest of the building stock. Thus, from a national point of view this should not be a prioritised group of buildings when it comes to achieving national targets for energy saving. This does not mean that these buildings do not have a potential for energy savings, rather that targets and measures should be based on the character and context of each building. Policy makers should be aware that buildings built 1965–1974 is by far the segment with the highest energy use. This is also a segment where, generally, the impact on heritage significance could be limited.

It should be noted that year of construction for buildings built before 1929 to some extent is unreliable, given the uncertainties in the registration of year of construction. For the purposes of this explorative study, data uncertainty is not a limiting factor. However for future investigations with higher resolution, this must be addressed.

The steady improvement of energy performance for modern buildings is most likely an effect of national building regulations. The fact that the building stock from the second half of the 19th century has the same energy performance as buildings from the first half of the 20th century is unexpected. It is beyond the scope of this paper to offer an explanation, but this should be an area of future research. The buildings we selected are comparable in terms of age, use, climate zone and heat source, but other factors, not available to us in this data base, influence energy performance as well. For example, we do not have data on type of construction, technical status and previous renovations.

This paper shows a method and a national case study of how an improved knowledge base of a building stock can facilitate more precise and realistic targets for policy planning and renovation aiming to improve energy performance in the Swedish building stock. Broad and blunt political targets for energy saving are not well suited to historic buildings as they can result in serious negative effects on the heritage values of the buildings. By grouping comparable buildings, in terms of type of use, climate zone and heat source, realistic and differentiated targets can be set and a potential for energy savings can be assessed. The magnitude of energy use for different age segments of the Swedish building can be estimated showing policy makers where there is little to gain in terms of national energy savings – buildings built before 1845 – and where the biggest potential for energy savings is – buildings built between 1945 and 1974.

Recently the energy requirements for new buildings changed and new EPCs show primary energy use, calculated from measured delivered energy. The energy use for the production of the energy, from the extraction to the delivery is now included. All new buildings in Sweden will also be required to have a climate declaration from 2022, which will show the climate impact of the building during its life cycle. There are also initiatives to make calculations on climate impact for the existing building stock, to show how renovation and reuse can reduce the production and distribution of new materials, and thereby carbon emissions. If and how this more holistic approach to energy use and climate impact in the life span of the building will affect the targets for existing buildings remains to be seen.

## 5. ACKNOWLEDGEMENT

This work was supported by the Swedish Energy Agency as part of the national research programme on energy efficiency in historic buildings – Spara och Bevara.

## 6. REFERENCES

[1] A. Troi, Energy efficiency solutions for historic buildings. A handbook., Birkhäuser, 2014.
[2] A. L. Webb, "Energy retrofits in historic and traditional buildings: A review of problems and methods," Renewable and Sustainable Energy Reviews, vol. 77, pp. 748-759, 2017.
[3] L. Mazzarella, "Energy retrofit of historic and existing buildings. The legislative and regulatory point of view," Energy and Buildings, vol. 95, pp. 23-31, 2015.
[4] A. Martínez-Molina, I. Tort-Ausina, S. Cho and J.-L. Vivancos, "Energy efficiency and thermal comfort in historic buildings: A review," Renewable and Sustainable Energy Reviews, vol. 61, pp. 70-85, 2016.
[5] European Committee for Standardization, Conservation of cultural heritage – Guidelines for improving the energy performance of historic buildings, EN 16883:2017, 2017.
[6] A. L. Webb, Evaluating the energy performance of historic and traditionally constructed buildings, 2017.
[7] Plan- och bygglag (2010:900) c. 8 § 17.
[8] Boverket, Boverkets Byggregler (Föreskrifter och allmänna råd), BFS 2011:6.
[9] Kungliga Byggnadsstyrelsen, Byggnadsstadgan, BABS 46, 1946.
[10] European Parliament, Council of the European Union, "Directive 2010/31/EU of the European Parliament and of the Council of 19 May 2010 on the Energy Performance of Buildings," Official Journal of the European Union, no. 153, p. 13, 2010.
[11] Boverket, Boverkets föreskrifter och allmänna råd om energideklaration för byggnader, BFS 2007:4.
[12] M. Aksoezen, M. Daniel, U. Hassler and N. Kohler, "Building age as an indicator for energy consumption," Energy and Buildings, vol. 87, pp. 74-86, 2015.
[13] C. Michelsen, "C. Michelsen, Im Fokus: Der lange Weg zur Energieeffizienz von Immobilien – Ergebnisse des ista-IWH-Energieeffizienzindex," 2010.
[14] C. Michelsen and S. Müller-Michelsen, "Energieeffizienz im Altbau: Werden die Sanierungspotenziale überschätzt? Ergebnisse auf Grundlage des ista-IWH-Energieeffizienzindex," 2010.
[15] Boverket, Energi i bebyggelsen – tekniska egenskaper och beräkningar, resultat från projektet BETSI, 2010.
[16] Boverket, "Energideklarationsregistret, Gripen," Boverket, 2017-01-04.
[17] Lantmäteriet, "Fastighetsregistret, helårsstatistik 2016," 2016.
[18] M. Mangold, M. Österbring and H. Wallbaum, "Handling data uncertainties when using Swedish energy performance certificate data to describe energy usage in the building stock," Energy and Buildings, vol. 102, pp. 328-336, 2015.

# "Mechanization Takes Command". Indoor Environmental Control in 20th-Century Historic Buildings

Anna Bonora, Kristian Fabbri, Giulia Favaretto, Marco Pretelli

Department of Architecture, University of Bologna, Italy (anna.bonora3@unibo.it, kristian.fabbri@unibo.it, giulia.favaretto2@unibo.it, marco.pretelli@unibo.it)

**Abstract** – Before industrialization, architecture itself was designed for ensuring specific microclimatic indoor conditions. When "mechanization takes command" – citing the title of a volume by Sigfried Giedion –, this task was transferred to the Heating, Ventilation and Air Conditioning (HVAC) systems. Starting from a reflection on this pivotal passage, the present paper focuses on the indoor environmental control in 20th-century historic buildings. In so doing, it outlines an investigation concerning the building-plant system of some significant case studies: Villa Tugendhat in Brno, the Salk Institute for Biological Studies in California and the French Cité de Refuge. Attention is focused on both the plant solutions implemented during the realisation of these iconic buildings and the restoration interventions that have been carried out for their preservation. The aim is to underline the relevance of restoration interventions capable of not removing the historic plants and, at the same time, not implementing an uncritical musealisation.

**Keywords** – Indoor Environmental Control; Plants; Microclimate; Preservation; 20th-Century Historic Buildings.

## 1. BUILDINGS' DESIGN AND MICROCLIMATE BEFORE AND AFTER THE INDUSTRIAL REVOLUTION

For thousands of years men have built public and private buildings mainly considering their purpose and the environmental boundary conditions, such as the local climate, the available materials, and others.

As we read in early modern treatises (e.g. [1-5]), before the 19th century one of the priorities for architects was to obtain a specific microclimate inside a building. Indeed, particular attention was paid to the building envelope, the internal layout of the rooms, and the favourable and adverse environmental characteristics, for example.

According to the above-mentioned treatises, the relationship between form, function, and the characteristics of the place was fundamental: material and formal features determine how the outdoor conditions will influence the indoor microclimate of a building. The pre-modern architects suggested how to proceed to design a building, with the aim to exploit the benefits of climate and location, obtaining optimal natural lighting, warmth in the winter, breeze during the summer, and other.

The introduction of the so-called "systems" has progressively changed this approach. Between the 19th and 20th century, it managed to determine an imbalance between architectural and systemic solutions. The functions of the two are different and not always coordinated.

Le Corbusier's five points of architecture [6], which are the outcome of the application of the industrialization principles to the field of buildings' construction, are the example of the new logic of the 20$^{th}$ century. Le Corbusier believed that the introduction of concrete drew a line between the pre-modern and the modern architecture, because it allows the architect to be free from many constructive constraints, which are no longer dependent on the outdoor climate or the geographical location. The use of concrete, indeed, followed by the use of lighter and lighter materials, the reduction in thickness of the external walls and the widespread and massive use of glass surfaces, has introduced a new necessity: the use of the Heating, Ventilation and Air Conditioning (HVAC) systems.

After "mechanization takes command" [7], the presence of those systems has made it possible to build architectures with the same characteristics (shape and materials) in completely different locations: nowadays, thanks to the HVAC systems, buildings can guarantee the same temperature at the equator and at the pole. This approach has in turn generated new needs: one related to a huge change of the perception of thermal comfort; the other related to the strong increment in energy demand.

In this paper we will present three case studies of 20$^{th}$-century historic buildings: Villa Tugendhat in Brno, the Salk Institute for Biological Studies in California and the French Cité de Refuge. The aim is to illustrate and compare their HVAC systems, as well as the approaches of both the architects who fitted those systems and those who decided if and how to preserve them.

## 2. VILLA TUGENDHAT, BRNO

Villa Tugendhat in Brno is a 20$^{th}$-century iconic architecture designed by Ludwig Mies van der Rohe since 1928. Realised between 1929 and 1930, this building is an emblematic work also from a plant-engineering point of view. It constitutes an interesting example of dialogue between its architect, Mies van der Rohe, and the plant engineer, J.L. Bacon, for the functioning study of the building [8].

### 2.1 THE INDOOR ENVIRONMENTAL CONTROL

Focusing attention on the air-conditioning system, Villa Tugendhat is an architecture that presents both a hydronic and aeraulic distribution system.

The heat generator consisted of two coal-fired cast iron Strebel boilers and a smaller boiler, the latter probably used only for sanitary water. The hydronic system started from the two Strebel boilers and arrived at the plant terminals that are of two typologies. Six-column cast iron Hahn radiators are situated in rooms with small cubic capacity and window frames, while chromed or painted tubular profiles are located in spacious rooms at the base of extensive glass walls. Such devices allowed not only the heating of the related rooms, but also – according to the thought of the period – the generation of rising flows of warm air capable of counteracting the phenomenon of condensation.

The above-mentioned hydronic system, presented in manuals of the early 20$^{th}$ century [9], constitutes a standard plant solution for that time in Europe, at least for valuable buildings. The novelty of Villa Tugendhat lies in the presence of an aeraulic system with a masonry Air Handling Unit (AHU). Here, air treatment was guaranteed by a precise pathway: an air intake was connected to a chamber with nozzles to humidify the incoming air; a vertically sliding steel panel regulated the air exchange with a mixing chamber, where there was the encounter with the recycled air; an adjacent filter chamber was equipped with a double filter, the first one with oil and the second one with wood shavings. The masonry AHU presented heat exchangers connected to the boilers and a low-pressure radial ventilator. This is where the canalisation network started. In the living area, canalisations arrive to metal grilles that supply air, contributing to heating in winter and allowing ventilation in summer. Quality and quantity of the injected air were manually managed by a control panel situated in the technical floor of the building, while in the living area a wall metal grille allowed air to be drawn in and reintroduced it into the masonry AHU. A vertically sliding steel panel thus regulated the transition from the extracted air chamber to the mixing chamber, where the encounter with the incoming air took place and restarted the pathway.

In Villa Tugendhat, the indoor environmental control is also ensured by grilles and vertical ducts that allow the natural ventilation of the service rooms. Furthermore, if extensive curtains screen the building glass walls, the transparent surfaces of the winter garden produce – perhaps not by design will – an inner tube useful in the winter season.

### 2.2 THE PRESERVATION APPROACH

The latest restoration of Villa Tugendhat, carried out between 2010 and 2012, chose to reuse the existing ducts and terminals. Other components have instead become an integral part of the building museum route. Case in point is the boiler room that has become a "technical monument" [10, p. 219] in which the musealisation of the plant system is combined with new elements for the heat generation. The innovation of the plant conservation project therefore lies in placing at the heart of the process what is often simplistically removed (Figg. 1-2).

**Figure 1. Villa Tugendhat, Brno: outdoor of the building [11, p. 150].**

**Figure 2. Villa Tugendhat, Brno: solutions for the indoor environmental control of the building [photo G. Favaretto, 2017].**

## 3. THE SALK INSTITUTE FOR BIOLOGICAL STUDIES, CALIFORNIA

The Salk Institute for Biological Studies was built between 1960 and 1965 for Jonas Salk, by Louis Kahn: one of the most influential architects of the 20th century. This building, designated as National Historic Landmark in 1991, is located in the La Jolla community of San Diego, California. For its construction, Kahn used traditional and industrially produced materials, such as teak, stone, concrete, glass and steel. His aim was to design an architecture, which could last for many generations, as maintenance-free as possible. Moreover, the laboratories' configuration had to be easily adaptable to any change of science practised there. To accomplish this, the labs' design is free from columns, open and unobstructed: all the systems, such as electricity, piping systems and ventilation ducts, are set inside the Vierendeel trusses, in the interstitial floors between the laboratories.

### 3.1 THE INDOOR ENVIRONMENTAL CONTROL

The virtuous case of the Salk Institute for Biological Studies shows the noteworthy approach that the architect had towards systems, by designing the building. Here, indeed, the design of plants results as important as the architecture itself [12]. Its construction has been developed hand in hand with all the traditional buildings' components.

This architecture comprises two symmetrical buildings, six stories tall and a travertine courtyard between them (Fig. 2). Kahn decided to confine the building systems in horizontal pipe spaces between each laboratory floor: August Komendant, the structural engineer, used 13 concrete Vierendeel trusses to separate each floor above each laboratory. The trusses are 9 feet deep and 65 feet wide. Between them there are *"columns arranged at 20-foot centers along the side walls"* [13, p. 38]. Electrical lines, ventilation ducts and piping systems are hosted inside the trusses. Through that system, the air is supplied to the laboratories and offices at the desired temperature, whereas the return air duct transfers it to the central chamber. Chillers, fan exhausts, etc., are hosted in the mechanical room (Fig. 3).

The idea to create spaces between each laboratory floor begun from Kahn with the Richards Medical Research Laboratories, in Philadelphia [14]. Jonas Salk, the client, contributed to the decision to separate served and servant spaces in the Salk Institute, suggesting to Kahn: *"give the pipes a floor their own"* [15, p. 125].

The "pipe spaces" allow keeping the areas of laboratories unhindered and flexible. Moreover, the dual function of the trusses (structural and of utilities containers) allows the access for the systems' maintenance without interfering with the use of laboratories.

**Figure 3. Salk Institute for Biological Studies, California: outdoor and indoor of the building [12, pp. 183, 249].**

### 3.2 THE PRESERVATION APPROACH

Today, after more than five decades, the Salk Institute for Biological Studies strongly "looks like" in the 1960s. In 2013, it was mainly the need to preserve the teak window wall assemblies that called for a partnership between the Salk Institute and the Getty Conservation Institute (GCI) [16-18]. If the 70% of the original material has been preserved, the systems of the building haven't been replaced, in conformity with the Conservation Management Plan [12] which presents the conservation policy for the Salk Institute, in order to guide the future conservation, helping to "*identify areas where change is appropriate, and where it is not*" [13, p. 83]. One of the policies mentioned in the Plan invites to "*respect the integration of the structural and building services systems in the building and their coordination by the architect*" [13, p. 249].

Here below, two standards [19] which promote practices that can show the kind of preservation approach adopted for this building:

"*2. The historic character of a property will be retained and preserved. The removal of distinctive materials or alteration of features, spaces, and spatial relationships that characterize a property will be avoided*" [13, p. 85];

"*6. Deteriorated historic features will be repaired rather than replaced. Where the severity of deterioration requires replacement of a distinctive feature, the new feature will match the old in design, color, texture, and, where possible, materials. Replacement of missing features will be substantiated by documentary and physical evidence*" [13, p. 86].

The major intervention that concerned the systems dates back to 2012, when solar panels were placed atop of the laboratory buildings.

## 4. CITÉ DE REFUGE, PARIS

The Cité de Refuge is an architecture by Le Corbusier and Pierre Jeanneret realised in Paris in 1933. Differently from the previous case studies, this building highlights the refusal of a close relationship between designers and plant engineers [8].

### 4.1 THE INDOOR ENVIRONMENTAL CONTROL

The Cité de Refuge was designed by applying the two concepts that Le Corbusier considered essential for solving the issue of the indoor environmental control: *la respiration exacte* and *le mur neutralisant*. The first concept corresponds to a forced ventilation responding to the Lecorbusierian dogma of the eighteen degrees everywhere. The second concept, instead, consists in the realisation of a double glazing with hot or cold filtered air circulating between the two glazing slabs to prevent the air at 18 °C from been affected by external influences.

The above-mentioned solutions have been praised by the Saint-Gobain company, producer of glass material, for the use of a double quantity of glass. Nevertheless, the ideas proposed by Le Corbusier and Pierre Jeanneret have been set aside for economic reasons. This led to the use of simple glass walls with the result of comfortably warm rooms in winter that were similar to an unbearable greenhouse in summer.

Discussion with plant experts would probably have allowed the application of the principles formulated by Willis Haviland Carrier, the inventor of "air conditioning" that was born in the United States of America in 1902. As a matter of fact, Carrier had already arrived at more practical and economic solutions which would have made it possible to undertake a less unsuccessful course in terms of indoor comfort [20-21].

### 4.2 THE PRESERVATION APPROACH

Object of major transformations over time, the Cité de Refuge has witnessed the rearrangement not only of the HVAC systems, given the unsuccessful choices from this point of view, but also of the building curtain wall.

Restored between 1948 and 1952 by the authors of the architectural work themselves, with the involvement of Iannis Xenakis and Vladimir Bodiansky, the façade with glass walls was modified with the insertion of glass slabs sustained by wooden frames on panels acting as parapets. Only in 1961, the polychromy of the primary colours blue, yellow and red – originally circumscribed to the entrance coating – was introduced. Between 1973 and 1977, Philippe Verrey directed works that not only led to the adoption of new tones, but extended the polychromy on the originally monochrome back façade. In the 1980s, instead, wooden elements have been replaced by aluminium ones. Finally, the intervention carried out between 2014 and 2015 has maintained the changed image of the building, even if adopting colours considered similar to those of the 1950s [22] (Fig. 4).

The components documenting the original functioning of the Cité de Refuge are no longer present. Nevertheless, this building constitutes an emblematic case study useful to highlight how most of the time books and articles focus on the description of iconic 20th-century architectures neglecting the aspects relating their plant-engineering component. The Cité de Refuge, indeed, is widely known for its architectural value, but much less well known for the failed solutions in terms of plant engineering and microclimate control studied by some of the best-known architects of the Modern Movement.

**Figure 4. Cité de Refuge, Paris: outdoor of the building over time [22-23].**

## 5. CONCLUSIONS

Comparing what emerged from each case study, some conclusions can be drawn: on the one hand, regarding the choices which characterised the construction of these buildings; on the other hand, about the approach adopted at the time of the following interventions.

With reference to the first point, the case studies of Villa Tugendhat and the Salk Institute for Biological Studies strongly represent how important is to integrate fully the systems project in the building project: both architects – Ludwig Mies van der Rohe and Louis Kahn – gave equal importance to the design project and to the plant project for the construction of these 20$^{th}$-century iconic buildings. It proved to be a successful choice. On the contrary, for the Cité de Refuge, Le Corbusier and Pierre Jeanneret refused a consultation with the plant engineers since the beginning, and this decision proved to be a failure.

With respect to the second point, the plants preservation approach embraced for both Villa Tugendhat and the Salk Institute for Biological Studies has been characterised by the intent to document the original functioning of the plant systems. Rather, the plant solutions adopted for the Cité de Refuge have not been preserved due to their failure. Nevertheless, this latter case study results emblematic because it clearly shows how 20$^{th}$-century historic buildings are often well known for their architectonical value, but not for the choices, even unsuccessful, adopted for their indoor environmental control. Whereas, these buildings should be considered as «resources of knowledge» [24, p. 56] also in terms of plant-engineering solutions, hopefully to be preserved, and indoor microclimatic conditions, eventually to be improved.

## 6. REFERENCES


[1] M. Vitruvio Pollione. I dieci libri dell'architettura. Translated and commented by Daniele Barbaro (1567). Milano: Il Polifilo, 1987.

[2] L.B. Alberti. De Re Aedificatoria, voll. 1-2. Milano: Il Polifilo, 1966.

[3] A. Palladio. I Quattro libri dell'Architettura. Electronic edition, 15 September 2009, anastatic copy. Milano: Hoepli, 1980.

[4] P. Pellegrini. L'Architettura. Milano: Il Polifilo, 1990.

[5] G. Valadier. L'Architettura Pratica, voll. I-II-III-IV-V. Roma: Sapere 2000, 1992.

[6] Le Corbusier. Verso una nuova architettura. Milano: Longaresi, 1923.

[7] S. Giedion. Mechanization Takes Command: a contribution to anonymous history. New York: Oxford University Press, 1948.

[8] A. Bonora, K. Fabbri, G. Favaretto, M. Pretelli. "Le diverse consistenze dell'acqua. Dotazioni impiantistiche e controllo del microclima a Villa Tugendhat a Brno", in Gli effetti dell'acqua sui beni culturali. Valutazioni, critiche e modalità di verifica. G. Biscontin, G. Driussi, Eds. Marghera-Venezia: Arcadia Ricerche, 2020, pp. 513–524.

[9] C. Rumor, H. Strohmenger. Riscaldamento – Ventilazione e impianti sanitari. Milano: Hoepli, 1911.

[10] D. Hammer-Tugendhat, I. Hammer, W. Tegethoff. Tugendhat House. Ludwig Mies van der Rohe. Basel: Birkhäuser, 2015.

[11] P. Pelčak, I. Wahla. Mies van der Rohe. Villa Tugendhat in Brno. Brno: Brno City Museum, 2016.

[12] M. Pretelli, K. Fabbri, L. Signorelli. "Interventi sugli impianti storici. Un patrimonio dal significato in evoluzione". Materiali e Strutture. Problemi di conservazione, vol. 11, pp. 9–24, 2017.

[13] D. Slaton, T. Penich, T. Olson, Eds. Conservation Management Plan. Salk Institute for Biological Studies. La Jolla, California. Los Angeles, California: The Getty Foundation, 2016. [Online]. Available: https://www.getty.edu/foundation/pdfs/kim/salk_cmp.pdf [Dec. 16, 2020].

[14] C.M. Shih, F.J. Liou. (2010, Nov.). "Louis Kahn's Tectonic Poetics: The University of Pennsylvania Medical Research Laboratories and the Salk Institute for Biological Studies". Journal of Asian Architecture and Building Engineering. [Online]. Vol. 9:2. Available: https://www.tandfonline.com/doi/pdf/10.3130/jaabe.9.283?needAccess=true [Dec. 16, 2020].

[15] T. Leslie. Louis I. Kahn: Building Art, Building Science. New York: G. Braziller, 2005.

[16] S. Lardinois. Salk Institute for Biological Studies Conservation Project: Teak Window Wall Assemblies. Phase 1: Research and Investigation Results and Preliminary Conservation Proposals. Project Report. Los Angeles, California: The Getty Conservation Institute, 2017. [Online]. Available: https://www.getty.edu/conservation/publications_resources/pdf_publications/pdf/salk_report.pdf [Dec. 16, 2020].

[17] S. Lardinois, K. Normandin. "Conserving the Teak Window Wall Assemblies at the Salk Institute for Biological Studies". Louis I. Kahn. The Permanence. Docomomo Journal, vol. 58, pp. 30–39, 2018/1.
[18] K. Normandin, S. Lardinois. "Salk Institute for Biological Studies: Teak Window Wall Conservation Project", in Metamorphosis. The Continuity of Change. Conference Proceedings of the 15th International Docomomo Conference (Ljubljana, 28-31 August 2018). A. Tostões, N. Koselj, Eds. Lisboa: Docomomo International. Ljubljana: Docomomo Slovenia, 2018, pp. 410–417.
[19] K.D. Weeks, A.E. Grimmer. The Secretary of the Interior's Standards for the Treatment of Historic Properties with Guidelines for Preserving, Rehabilitating, Restoring & Reconstructing Historic Buildings. Washington, D.C.: U.S. Department of the Interior, National Park Service, Cultural Resource Stewardship and Partnerships, Heritage Preservation Services, 1995. [Online]. Available: https://www.thempc.org/docs/lit/hist/interiorstandards.pdf [Dec. 16, 2020].
[20] R. Banham. The Architecture of the Well-tempered Environment. London: Architectural Press, 1969.
[21] M. Ingles. Willis Haviland Carrier – Father of Air Conditioning. Garden City: Country Life Press, 1952.
[22] F. Rosa. (2017, Oct.). "Cité de Refuge, restauro esemplare per un'opera «viva»". Il Giornale dell'Architettura. [Online]. Available: https://ilgiornaledellarchitettura.com/web/2017/10/03/cite-de-refuge-restauro-esemplare-per-unopera-viva [Dec. 16, 2020].
[23] Fondation Le Corbusier. Armée du Salut, Cité de Refuge, Paris, France, 1929. [Online]. Available: http://www.fondationlecorbusier.fr/corbuweb/morpheus.aspx?sysId=13&IrisObjectId=4593&sysLanguage=fr-fr&itemPos=4&itemCount=78&sysParentId=64&sysParentName= [Dec. 16, 2020].
[24] I. Hammer. "The Tugendhat House: Between Craftmanship and Technological Innovation. Preservation as Sustainable Building Policy". Modern and Sustainable. Docomomo Journal, vol. 44, pp. 48–57, 2011/1.

# Hygrothermal building simulation from a 3D point cloud - the Edo-Wiemken-Monument in Jever

L. Feldmann[1], S. Bichlmair[2], R. Kilian[2], M. Winkler[2]

[1] Ingenieurbüro INDICAMUS PartG, Cologne, Germany. Email: l.feldmann@indicamus.de

[2] Institute for Building Physics (IBP), Holzkirchen, Germany

**Abstract** –For hygrothermal building simulations the correct building geometry is required, among other things. This is necessary due to the moisture buffer processes, heat storage effects and heat fluxes for building components as well as for the indoor air volume. For the as-built survey, a procedure such as 3D laser scanning provides a modern method for the fast and accurate recording of complex building geometries. The result of 3D laser scanning is a 3D point cloud that contains the geometric properties of the building. The current state of the art is to extract conventional 2D drawings out of the laser scans, i.e. 3D point cloud, and to redraw the 3D geometry in a CAD software which is able to generate a suitable export format for the hygrothermal building simulation software. A direct export out of 3D point cloud in a suitable export format would simplify the work flow. The aim of this paper is to explore the possibilities and restraints of direct modelling in 3D point clouds and requirements of an export/import interface. The exemplary workflow to extract a suitable building model for hygrothermal simulation is developed on the example of the historical building of Edo-Wiemken monument in Jever (Germany).

**Keywords** – historic buildings, hygrothermal simulation, laserscanning, 3D geometries, preventive conservation

## 1. INTRODUCTION

### 1.1 BACKGROUND OF THE INVESTIGATIONS

The indoor climate of the case study Edo-Wiemken monument, which was found to be unfavourable by conservators, has to be improved by a technical measure. The monument was built around 1560 as a tomb of the East Frisian chieftain Edo Wiemken in Jever (northern Germany) [1]. In preparation of restoration and technical measures, laser scans were compiled and 3D models and 2D plans created. To assess the effect of technical measures on indoor climate and on energy consumption it was decided to perform a whole building hygrothermal simulation with WUFI® Plus [2, 3]. The results serve as basis for evaluation and decision making.

One principal problem of modelling for building simulation is the reduction of the complexity of real building geometry in a sufficient simplified 3D model. Especially building components of historic buildings are mostly uneven or have slight changes in direction. For example walls may have varying thicknesses, stuccos and other irregularities like small heels, wall cornices, protrusions, inward and outward indentations and so on. For historic buildings typically deformation correct plans are available which reflects the complex structure. The 2D drawings already simplify the real structure even if those plans are based on deformation-compliant allowance. Even more accurate are laser scans and descriptions of a building structure in 3D point clouds which are nowadays state of the art for historic buildings. The typical workflow for building simulation is still to extract 2D drawings out of the 3D point cloud and to redraw simplified those extracted plans for import into the simulation tool. The simplification is necessary to reduce the effort for modelling, reduction of possible errors by mistake and to reduce computational processing time. This means reduction of modelling time from weeks to days and reduction of computational processing time from days to hours (depending of the size of the building). But how to simplify? This work has to be done by a user who is able to decide the simplification work and judge the possible effect on the simulation result. In doubt of eligibility of simplification, simulation variants have to be performed. To do so it is necessary that the simulation user performs the 3D modelling based on 2D drawings which are redrawn in a CAD-software in order to get a sufficient simplified 3D model. The largely digital approach offers significant advantages in terms of efficiency compared to other methods of taking inventory [4]. The focus here is on a new approach to transfer the building geometry from 3D laser scans (i.e. 3D point cloud) via modelling software and description of interface characteristics necessary for export and import of data into the simulation software. For an evaluation of the results the simulation results are compared with the results of a hygrothermal simulation independently modelled in a conventional way by another user.

**Figure 1. Exterior view (left), interior view (middle) and the point cloud based 3D-model (right) of the Edo-Wiemken-Monument in Jever (source 3D-model: DhochN-Nord Digital Engineering GmbH).**

## 2. 3D MODELLING OF HYGROTHERMAL SIMULATION IN WUFI® PLUS

### 2.1 EXISTING TOOLS

#### 2.1.1 3D-EDITOR

There are several options to transfer geometric data into the simulation software WUFI® Plus. The simplest and quickest variant is the acquisition with the help of the "Building Wizard". Basic 3D geometries can be generated from a set of typical building-geometries via this wizard. The building created in this way is then visualized in the software's 3D editor where further adjustments can be made. The 3D editor also serves as an independent option for modelling geometries in WUFI® Plus. With the help of the simple functions offered by the 3D Editor, more complex geometries can be created compared to modelling with the Building Wizard. Entering the individual corner points via the 3D Editor is relatively time-consuming and only recommended for simply structured or small buildings.

#### 2.1.2 GBXML-IMPORT OF 3D GEOMETRIES

Since the 3D model provided by the engineering firm was created with the CAD/BIM software Autodesk Revit [5], the implementation of the geometric information in WUFI® Plus using the gbXML import function is attempted first. This software supports the gbXML format (Green Building XML) [6], which has been developed for the transfer of various building data. A gbXML file can store and transfer geometric and semantic information similar to an IFC file (Industry Fondation Classes) [7]. In the context of Revit, the gbXML format is one of the main export options for energy analysis software. The provided 3D model in Revit is a model with a high level of detail (figure 2, image left and in the middle). This LOD (level of detail) was chosen by the engineering firm because, in addition to the derivation of as-built plans (floor plans, sections and elevations), it should be possible to process them in virtual reality with the model. Especially for the latter, a high level of detail is absolutely necessary.

The gbXML file derived from the existing 3D model is prepared for an import with the WUFI® Plus XML project file converter. In the converter, the parameters for the import can be defined. Furthermore, basic information relevant for processing the data is listed there. These include, for example, the number of polygons or zones. In this case, the exported model contains 542 points, 162 polygons and one zone that was previously defined in Autodesk Revit as a room with the name 1 Interior. Information such as the number and type of component types is also listed. The file generated from the converter has the abbreviation *.xml.WUFIproject* and can be opened with the function gbXML-Import in WUFI® Plus. Here, only the components adjacent to the room are imported. By the representation of the 3D model in figure 2 (image right) it can be seen that a large part of the building is not imported. In addition, the model apparently has several deficiencies that lead to a representation of the geometry that is not true to reality. In addition to the transfer of areas that do not correspond to the actual sizes in the model, the geometry has gaps that make simulation difficult or even impossible without post-modelling. Furthermore, some components are not transferred. The absence of these components can lead to a falsification of the simulation results. Finally, the exemplary import carried out with the model provided on the basis of the gbXML import function shows that the implementation of a BIM-based model in WUFI® Plus is difficult in view of a large number of default settings. Although the geometric information is basically available in the model, it cannot be exported due to the settings made during modelling.

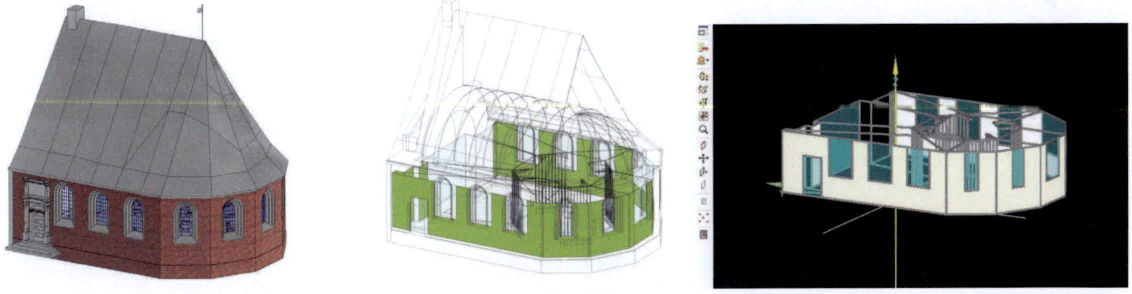

**Figure 2. Exemplary view of the model in *Autodesk Revit* (left), gbXML-export in *Autodesk Revit* (middle) and visualization box of the 3D Editor of WUFI® Plus (right) after import.**

### 2.1.3 SKETCHUP-PLUGIN FOR IMPORT OF 3D GEOMETRIES

The import of 3D geometries from SketchUp [8] to WUFI® Plus is enabled via plugin. A plugin is an optional software extension, also called an add-on module. The CAD software offers a user-friendly interface with elementary drawing and modelling tools. With the help of these tools, basic geometric shapes can be created in 2D and then drawn up into solids using the Extrude function. Depending on their position and orientation in the coordinate system, the solids represent, for example, vertical components such as walls or horizontal components such as ceilings or floors. In this way, geometries of buildings can be quickly represented as 3D models in the software. The additional module (plug-in) for SketchUp makes it possible to assign properties to modelled components or the component surfaces, respectively, which are necessary for an import into the simulation software. The information is saved in WPS format (WUFI® Plus - SketchUp geometry file). This format can be opened by WUFI® Plus and contains the previously defined geometric and semantic information. The structure of a WPS file is explained in Figure 3 using the geometry shown in the same figure. A WPS file can be opened and edited with an ordinary text editor.

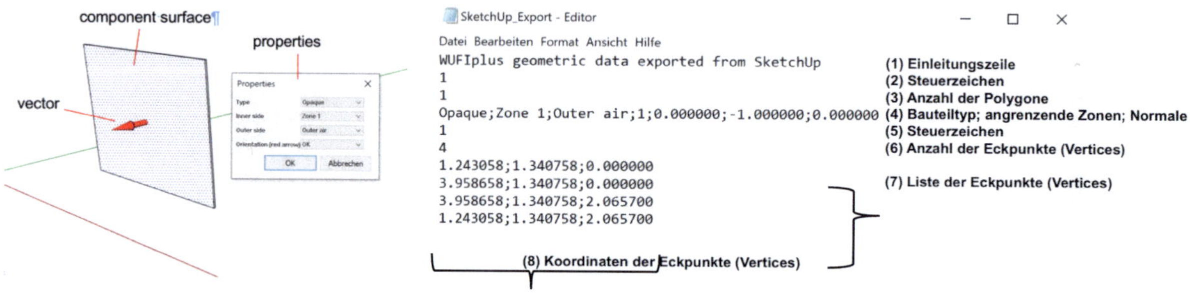


**Figure 3. *SketchUp* user interface with the WUFI® Plus Plugin (left) and the structure of a WPS file (right).**

When importing the existing 3D model to SketchUp, a grouping of the individual components is performed automatically. To edit the components, the grouping must be removed and the model must be divided into individual parts. In this case, editing the components means assigning the properties required for the import into WUFI® Plus. Due to individual properties of the components, it must be possible to select them individually. Using the example of the south-facing exterior wall, the grouping was removed and the component was divided into individual elements (Figure 4).

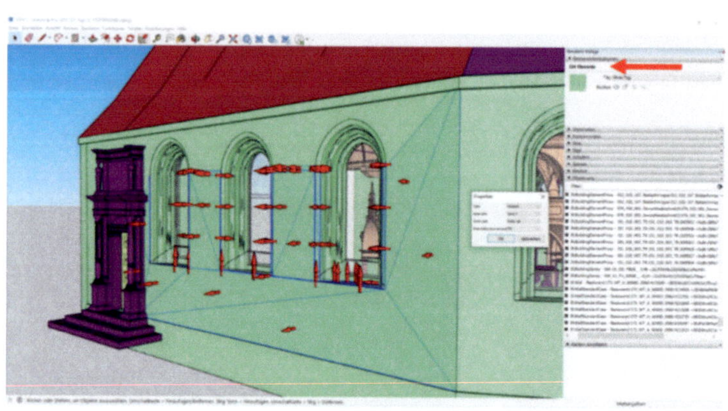

**Figure 4. Exemplary view of the model in *Autodesk Revit* (left), gbXML-export in *Autodesk Revit* (middle) and visualization box of the 3D Editor of WUFI® Plus (column right in the figure).**

When looking at the red vectors, which indicate the orientation of the component, it becomes clear how many partial surfaces (triangles) are involved. With the help of the triangles a triangulated surface description takes place. This effect is called meshing and is a common way of representing 3D geometry. These surfaces are also oriented differently. The orientation of the vectors determines the direction of heat and moisture transport in relation to the adjacent zones. Even if it were possible to assign properties to all elements of the model, a large amount of effort would have to be spent to post-process the model in order to regroup the components and provide them with further information.

In summary, it is determined in the scope of the described import attempts that an import of the existing 3D model is possible with the help of the currently available functions (gbXML import and SketchUp plugin), but the results are only conditionally usable with WUFI® Plus. While the import of a gbXML file failed due to the default settings made in the 3D-model, the representation of the surfaces as triangles (meshing) proved to be problematic because of generation of to many building components by the meshing process.

## 2.2 DEVELOPMENT OF A NEW WORKFLOW

Because of the problem of meshing and unfavourable pre-sets, a survey is conducted with the aim of finding a file format that works both without pre-sets from BIM processes and without meshing of surfaces. In the course of the research, the point cloud processing software PinPoint from the developer Scasa [9] showed advantageous properties, as it generates polygon models - similar to a WPS format - by defining surfaces using multiple vertices.

### 2.2.1 POLYGON MODEL WITH PINPOINT

Before the modelling of the laser scan data (point clouds) takes place, the requirements for the model are basically defined. The choice of boundary conditions in the form of the necessary accuracy and the level of detail depends on the requirements for the model. Since the scope of this case study is initially a general investigation of the workflow, a low level of detail is defined as a requirement for the model. The selected level of detail is described with an LOD 100 - 200. The level of detail is similar to that of a preliminary design or draft model. As can be seen in Figure 5, the software PinPoint (version 2.2.0) offers different representation options of the point cloud. Through pre-calculations, the software recognizes individual planes and colours these surfaces accordingly. Due to the more complex surface geometry in connection with the unevenness typical of the building age, a large number of planes are created, which the program colours differently (figure 5 images in the middle).

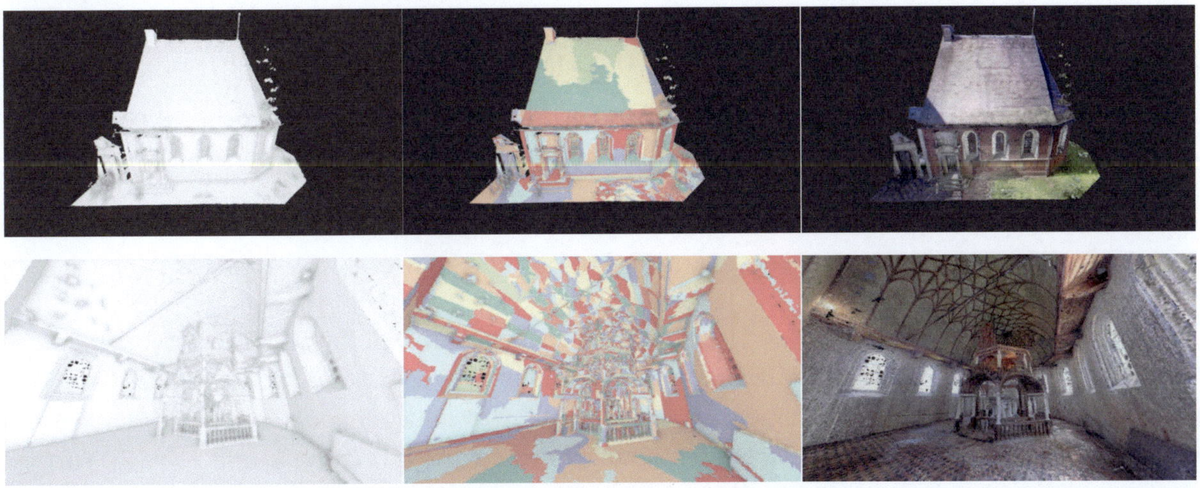

**Figure 5. Different display options of the point cloud in *Scasa PinPoint*.**

The point cloud is pre-processed in the scope of the registration and loaded into the software (Scasa PinPoint). The most important tool for modelling geometries is the "add surface" function. With the help of the surfaces recognized by pre-calculations and marked in colour, vertices (corner points) are captured and defined with the mouse pointer. With this and other functions of the software, a simplified polygon model is created based on the geometric data of the point cloud. Although it is possible to create files that describe polygons by multiple vertices using existing storage options in Scasa PinPoint, these file formats cannot be directly imported into WUFI® Plus; processing must first take place using the SketchUp plugin. Here, as previously described, problems such as the triangulated surface description (meshing) occur, which either requires extensive post-processing or the use of the data is not even given. For this reason, the developers of Scasa PinPoint, in cooperation with the developers of the WUFI® Plus Software, implemented a function in Scasa PinPoint, which enables the saving of the modeled polygons in the WPS format.

### 2.2.2 POSTPROCESSING IN WUFI® PLUS

The so created WPS-file can directly be imported into WUFI® Plus. During geometry-processing in WUFI® Plus, semantic properties such as the component type (opaque or transparent) or the adjacent zones are assigned to the individual polygons. As soon as the geometry-processing in WUFI® Plus is finished, the actual simulation model can be defined. This requires the input of further parameters such as component structures, materials and boundary conditions [10].

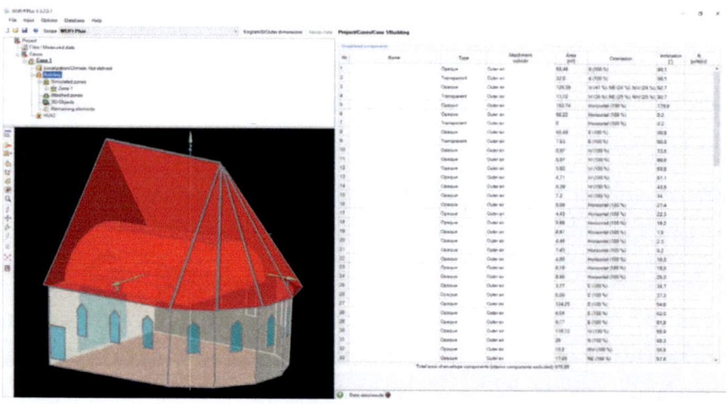

**Figure 6. View of the imported model (before the postprocessing) and the generated parts list in WUFI® Plus.**

## 3. HYGROTHERMAL SIMULATION AND EVALUATION OF RESULTS

After working out the methodology, it is now applied to the case study Edo-Wiemken monument. The results of the hygrothermal simulation with the new work flow are compared to measured indoor climate data and to a calibrated hygrothermal simulation performed by another user.

### 3.1 HYGROTHERMAL SIMULATION RESULTS

The climate of the interior (Zone 1) and the roof space (Zone 2) of the Edo-Wiemken monument are simulated with WUFI® Plus. The following diagrams show the calculated indoor climate (Sim-F) in comparison with the indoor climate measurements (MW) over the course of a full year from January 24, 2017 to January 24, 2018, see Figure 7 two diagrams on the left (zone 1 and zone 2). Further comparison of the results is made with the already calibrated simulation model (Sim-B), which serves as a benchmark, see Figure 7 two diagrams on the right (zone 1 and zone 2).

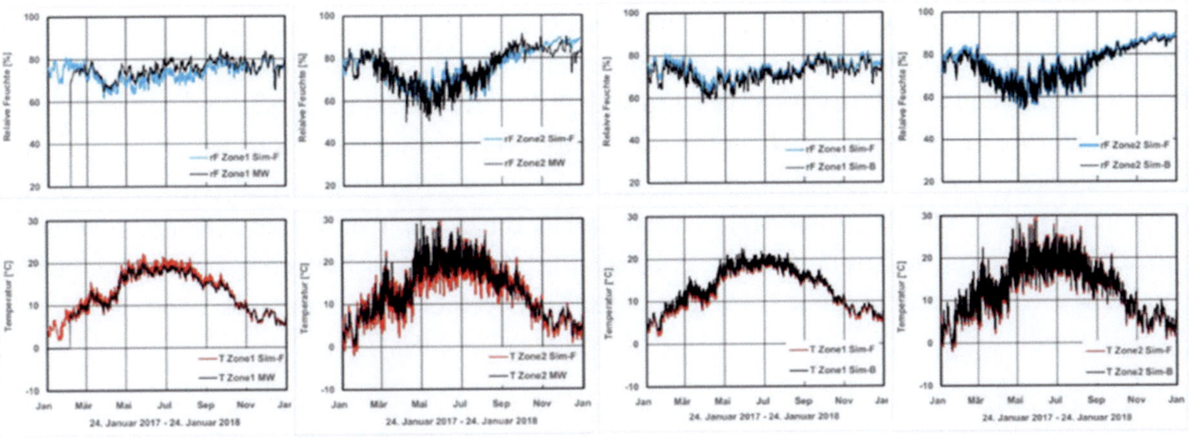


**Figure 7. Comparison of the calculated indoor climate of the case study (Sim-F) with the measured values (left two columns) and the calibrated model (Sim-B) in zone 1 (indoor) and zone 2 (attic) (right two columns). Course of relative humidity is shown in the upper diagrams, temperature in the diagrams below.**

### 3.2 EVALUATION SIMULATION RESULTS

The quality of the results of the simulation model generated in the case study is determined on the basis of statistical characteristic values and the evaluation scale according to Kilian [11]. This is an abbreviated evaluation, since a representative comparison of the results is not possible due to the lack of measured values. As the comparison of the simulation results in the above diagrams shows, the results of the newly created model for both zone 1 and zone 2 are very close to the results of the indoor climate measurements and those of the calibrated model (benchmark). In accordance with the evaluation standard according to Kilian [11], almost all of the characteristic values listed there are classified as excellent. Only the range of the calculated relative humidity for zone 2 can be rated as acceptable in comparison with the measured values. Evaluating the benchmark model due to [11] gives the same rating compared to the new model.

### 3.3 EVALUATION OF NEW DEVELOPED WORKFLOW

In contrast to the existing import functions, the newly developed workflow is based on the use of a polygon model. This describes surfaces on the basis of several vertices. In addition, a polygon model has analogies to the representation form and the storage format (WPS) of a WUFI® Plus model. For the modelling in the polygon model, however, a software suitable for this is necessary. In the case study, the modelling of the polygon model was performed using the point cloud processing software Scasa PinPoint. Compared to other software solutions, the modelling in Scasa PinPoint could be performed in a short time without prior knowledge of the software. For the transfer of the geometric data, a file format readable by WUFI® Plus is required. For this purpose, the software manufacturer developed a new interface that allows the model to be saved in WPS format. Since only a few semantic properties can be assigned to the polygons during modelling in this first version of the newly developed interface, post-processing is required in WUFI® Plus. Compared to the developed workflow, however, the existing import functions in WUFI® Plus are not judged to be insufficient. The fundamentally different requirements assumed for modelling by architectural and engineering firms ensure that the models can only be used for hygrothermal building simulation in the context of extensive post-processing.

Essentially, the models differ in terms of the method used to create the building geometry. The benchmark model was created using 2D plans (floor plans, sections and elevations) provided in the CAD program SketchUp and then imported into WUFI® Plus via the SketchUp plugin or edited directly in the 3D editor. The model described in the case study, on the other hand, is based on a point cloud of the building. Modelling polygons within a point cloud offers various advantages over the conventional creation of 3D geometries, as illustrated by the example of the Scasa PinPoint software. In addition to the intuitive operation of the program, the advantages include time savings and the high accuracy of the modelled polygons. The newly created model therefore requires less editing.

## 4. CONCLUSION

Using the example of an already performed laser scan and 3D model of the Edo-Wiemken monument in Jever (Germany), the workflow for implementing the data in WUFI® Plus to perform a hygrothermal simulation is examined. The focus of the work is on the geometric registration of the surfaces based on the existing point cloud. The correct representation of the surfaces and volume of the building or rooms is of great importance for the hygrothermal simulation due to moisture buffer processes and heat storage effects as well as heat fluxes of the different building materials. While an import attempt using a gbXML file derived from the 3D model failed due to default settings in the model and a high level of detail, the approach using SketchUp plugins also did not result in a successful import into WUFI® Plus. The latter was complicated by the triangulated surface description (meshing) used by many common file formats to describe three-dimensional geometries.

Due to this problem a new simplified modelling is carried out with the point cloud processing software Scasa PinPoint in order to generate a polygon model that shows analogies to the 3D editor of WUFI® Plus with regard to the structure as well as the display form. In cooperation with the developers of the PinPoint software a function was implemented in the program that allows 3D geometries to be saved in polygon description format which correspond to WPS file format of WUFI® Plus. After post-processing with respect to the semantic information on the model, a hygrothermal simulation is carried out. The results of the new simulation model is then compared with the results of room climate measurements and an already calibrated simulation model, which was modelled in a conventional way using 2D plans. The evaluation of the results is positive due to the achieved quality of results in combination with a lower processing effort and comparable to the benchmark model.

The advantages of an as-built survey using a laser scanner can be made available for hygrothermal simulation software. For this a 3D Point cloud processing software is necessary which is able to simplify the complex structure as-built and to export the derived building components in a polygonal model description format instead of triangular meshing.

## ACKNOWLEDGMENT

The project was supported by Scasa GmbH and ingenieurbüro heimsch GmbH. We also thank the staatliches Baumangement Ems-Weser and the Schlossmuseum Jever for their great support.

## 5. REFERENCES

[1] https://www.schlossmuseum.de/forschung/das-edo-wiemken-grabmal
[2] www.wufi.de
[3] Künzel, H. M.: *„Verfahren zur ein- und zweidimensionalen Berechnung des gekoppelten Wärme- und Feuchtetransports in Bauteilen mit einfachen Kennwerten."* Dissertation, Universität Stuttgart, 1994.
[4] F. Kern.: „Terrestrisches Laserscanning kurz & bündig." i 3 mainz Fachhochschule Mainz. 2007 [Online]. Available: https://docplayer.org/docview/27/10393891/#file=/storage/27/10393891/10393891.pdf
[5] www.autodesk.com/products/revit
[6] www.gbxml.org
[7] https://technical.buildingsmart.org/standards/ifc
[8] www.sketchup.com
[9] www.pinpoint-suite.com
[10] Antretter, F.; Pazold, M.; Künzel, H.M.; Sedlbauer, K.: *„Anwendung hygrothermischer Gebäudesimulation."* In: Bauphysik Kalender 2015. Simulations- und Berechnungsverfahren. pp 189-225. WILEY-VCH Verlag.
[11] R. Kilian. "*Klimastabilität historischer Gebäude*." Dissertation, Univ. Stuttgart. Fraunhofer Verlag, 2013.

# Preserving and Improving Energy Efficiency in Modern Architecture. Efficacy of Digital Tools.

G. Franco[1], S. Bergero[1], D. Desideri[2], B. Filograna[3]

[1] Architecture and Design Department, University of Genoa, Italy, giovanna.franco@unige.it; stefano.bergero@unige.it.

[2] Architect, expert in BIM, davide.desideri@gmail.com.

[3] Architect, expert in energy diagnosis, beafilo@yahoo.it.

**Abstract –** Modern Movement architectures, built in the first half of 20th Century, constitute a heritage where possible conflicts among conservation principles and energy improvement aims are most evident. Built in a lighter way and with poorly energy performing materials, this heritage is more fragile to environmental actions and presents greater problems of material degradation and functional inefficiency. The contribution concerns a research conducted on a private villa, Villa Domus in Sestri Levante, designed by a nationally renowned rationalist architect (Luigi Carlo Daneri), located in a highly prestigious environmental position in eastern Ligurian coast. The research, in accordance with the owners, had two main objectives: a) testing digital 3D parametric modeling tools (BIM "Autodesk Revit") to verify its interoperability with an energy performance calculation software; b) once the energy behaviour has been understood, identify and evaluate the "admissible" interventions, in accordance with the criteria adopted by the protection bodies (Italian Ministry of Culture).

**Keywords –** Energy Efficiency; Cultural Heritage; Digitalization; Interoperability; Interdisciplinarity.

## 1. INTRODUCTION ON MODERN ARCHITECTURE AND ITS DESTINY

The architectures of the Modern Movement, built in the first half of the twentieth century, are a heritage where one can very clearly see how conflicts can arise between principles of material conservation and goals of regeneration, renewal and improvement of their performance and duration in time. Many modern architectures were born from a planned quest for infinite adaptability to increasingly fast changes in the way of living, typical of recent times. Architecture in those days was imagined as a dress: always adaptable and/or replaceable, when it no longer suited the needs or the fashion of time. At other times, modern architecture arose convinced that it was precisely new materials and new building techniques which would ensure indefinite duration and resistance through time, eliminating the bothersome need for constant maintenance which, for centuries, had characterised the lives of ancient buildings. This seems to have been one of the most significant consequences, for example, of the generalised and widespread use of reinforced concrete. However, many modern architectures are actually affected by a precarious state of conservation, also because they are the outcome of a new way of building, heedless of the trial of time or of the thousand repairs, of the coordinated or random macro and micro-adaptations which buildings have to undergo after construction, at times at a very short interval after they were made. Due to the perennial pursuit and idealisation of experimentation, the avant-garde spirit and also contingent political difficulties (one need only think of the autarchic period of the Fascist regime in Italy), many buildings put up between the end of the nineteenth century and the 1950s are almost unknown to us, as much or even more so than the ancient medieval or baroque factories.

Also for these reasons, the scientific-architectural community, focussed on issues of historiography and composition-design, is now dealing with the fate of the vast architectural production resulting from the urbanisation, modernisation and industrialisation of the so-called "short age" [1]. The expansion of the range of interest and of the meaning of the term "heritage" [2] has led the experts to pose questions about the values (in terms of testimony, history, economy, society...) purveyed by recent and very recent architectural production, about its destiny and hence about how to protect, preserve, promote, enhance, regenerate it, or, on the contrary, forget or finally destroy it.

## 2. IMPROVING PERFORMANCE OF MODERN HERITAGE. GOALS AND METHODOLOGY ADOPTED

In this spirit - a curious and "reverent" attitude towards the masters of the Modern Movement and their works but which is also aware of the needs of contemporary society (saving resources, improved energy behaviour, the ecological footprint of constructions, new ways of dwelling) and of the inevitable conflicts which they may lead to - a research project was launched, in the context of a master's degree thesis in Architecture at the University of Genoa.

The study set itself several goals. First of all, to undertake an in-depth investigation of a Modern architecture, not only in its spatial and volumetric arrangement but also in its essential and material nature, in order to understand its liveability and layout flexibility, its relationship to its surroundings, the behaviour of the exposed parts also in terms of indoor comfort, its durability. In second place, to understand whether a clearly rationalist work, essential in its construction features, affords any margin of flexibility for suggesting technical solutions aimed at improvements also in the field of energy use, while not betraying, at the same time, the protection and preservation of the work in line with the protection guidelines of the Ministry for Cultural Assets and Activities [3]. Finally, being aware of the inevitable process of digitalisation which, though slowly, is also affecting the building industry, the study sought to test the validity and efficacy of the use of parametric Building Information Modelling software (BIM "Autodesk Revit") for operations involving built assets, and interoperability with building energy performance calculation software in compliance with the Italian legislative framework on energy saving in buildings (checking the efficacy in moving IFC files among different kinds of software). All of this was tried out studying a single-family residential building, Villa Domus in Sestri Levante, located in a place of great prestige on the eastern coast of Liguria. The working group featured multidisciplinary skills, in the fields of architecture, regeneration and construction technology, energy systems for buildings and of facilities.

## 3. VILLA DOMUS ON THE PROMONTORY OF SESTRI LEVANTE (GENOA)

The villa, designed in 1938 by Luigi Carlo Daneri, a rationalist architect from Rome who had moved to Genoa, presents features of orientation and position which are unique in their kind and it is still used as a dwelling today, though not continuously [4]. The building is protected by the Ministry for Cultural Assets and Activities. Villa Mantelli, also known as Villa Domus, stands atop the cliff of a promontory rising steeply from the sea (surrounded by the sea on 270°), in one of the most beautiful stretches of the Liguria coastline, but at the same time quite fragile in environmental terms (Fig. 1). The irregular escarpment, the presence of tall trees, the need for exposure to the south, have determined the layout of the house, built on different floors: two on the north, four on the east, one each on south and west. The Villa has a C-shaped plan, open towards the sea and surrounded by volumes with single-pitched roofs covered with slate slabs, which converge towards the centre in a kind of *impluvium*. From the clean and unitary horizontal line of the eaves, which confers order on the design, the building descends until it settles down on the ground, adapting to different elevations, and centring on two patios, one on the sea front, on the ground level, and another on the opposite front, on the first floor. Indoor life takes place on the ground floor for the halls, living rooms (Fig. 2) and hearth room, the dining room, the kitchen and on the first floor, with the bedrooms and sanitary facilities. The large glass windows of the ground floor disappear into the masonry by means of an electrically operated device; with them, there vanishes the already subtle diaphragm separating indoors and outdoors, between the exposed areas and the patio overlooking the sea, which suitably filters sun and wind.

In the 1970s, after ownership changed hands, important indoors and outdoors works were carried out which changed the original appearance. Glass brick walls exposed to the north were eliminated and replaced by windows and by an opaque masonry wall. The interior appearance of the house, featuring clean lines and well-defined spaces, was modified by rounding off the partition walls of the stairwell and of the hearth room. A few years later, and with new owners, the villa underwent new intervention to restore the original parts which had been rashly removed (with the recovery of some elements or the introduction of other ones). The villa is in good state of repair; however, it has never undergone energy efficiency improvement. For this reason, and for the interest of the owners, the following study has been undertaken.

**Figure 1. View of the Villa Domus, Sestri Levante, eastern Liguria, Italy. Figure 2. View of the living room.**

## 4. ENERGY ANALYSIS

European Standard UNI CEI EN 16247-1 [5] defines the energetic audit procedure as a "systematic inspection and analysis of energy use and energy consumption of a system or organization with the objective of identifying energy flows and the potential for energy efficiency improvements". In the present paper a limited energy audit is presented, based on the procedure provided by UNI CEI EN 16247-2 [6]. Both thermal envelope and plants for heating and domestic hot water services are considered in the actual state to validate the energetic model, carried out by the commercial software EC700 of EDILCLIMA® software-house [7], according to technical UNI/TS 11300 Italian standards [8]. Therefore, only a few improvements have been studied regarding the building envelope compatible with the historical-artistic character of the building. The parameters used to compare the performance of the before and after scenarios are described below.

The building energy need for heating $Q_{H,nd}$ is calculated on the basis of ventilation $Q_{H,ve}$ and transmission $Q_{H,tr}$ heat exchanges, taking into account internal $Q_{int}$ and solar $Q_{sol,w}$ heat gains weighed through the utilization factor $\eta_{H,gn}$:

$$Q_{H,nd} = (Q_{H,ve} + Q_{H,tr}) - \eta_{H,gn}(Q_{int} + Q_{sol,w})$$

The $EP_{H,nd}$ index provides the heating energy need per building useful area $A_u$:

$$EP_{H,nd} = \frac{Q_{H,nd}}{A_u}$$

The summer performance of the building was expressed in terms of the mean periodic thermal transmittance $Y_{ie}$ for opaque components, provided by the UNI EN ISO 13786 [9], and the equivalent solar area referred to useful area $A_{sol,est}/A_u$ for glazed components, provided by DM 26.06.2015 [10] as follows:

$$A_{sol,est} = \sum_k F_{sh,ob,k} g_{gl+sh,k} (1 - F_{F,k}) A_{w,k} F_{sol,est,k}$$

- $F_{sh,ob,k}$ = reduction factor for shading related to fixed external obstacles for the k-th glazed component, referring to the month of July;
- $g_{gl+sh,k}$ = total solar energy transmittance of the k-th glazing component calculated in July, when mobile solar shading is used;
- $F_{F,k}$ = frame factor of the k-th glazed component;
- $A_{w,k}$ = area of the k-th window component;
- $F_{sol,est,k}$ = correction factor for the incident irradiation of the k-th glazed component, obtained as the ratio between the average irradiance in July, in the location and for the considered exposure, and the annual average irradiance of Rome, on the horizontal plane.

Fig. 3 shows the plans of the different floors, highlighting the heated and unheated areas. Table 1 contains the main characteristics of the site and the main geometrical and thermal features of the building envelope.

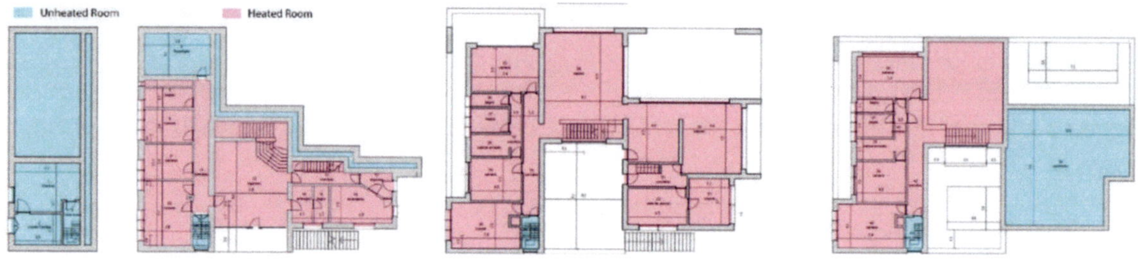


**Figure 3. View of thermal distribution.**

The main features of the building envelope are listed below.

- The external load-bearing walls are in stone on the ground floor, in solid bricks on the upper floors; at the first floor a solid brick wall separates the heated area from a cavity in contact with the ground.

**Table 1. Site and building envelope properties.**

| | | | |
|---|---|---|---|
| Sestri Levante, IT (Elevation 4 m, Climatic zone D) | | Latitude 44° 16′ Longitude 9° 23′ | |
| Legal heating season November 1 to April 15 | | S/V ratio = 0.69 | |
| Useful area $A_u$ = 480 m$^2$ | | Leaking gross area S = 1396 m$^2$ | |
| Net volume $V_u$ = 1459 m$^3$ | | Gross volume V = 2022 m$^3$ | |
| **Vertical and horizontal components** | *Thickness [m]* | *Thermal transmittance [W/(m$^2$/K)]* | *Surface mass [kg/m$^2$]* |
| Load-bearing external wall | 0.40 ÷ 0.50 | 1.49 ÷ 2.51 | 700 ÷ 1200 |
| Cavity external wall | 0.40 | 2.55 | 352 |
| Sub-window wall | 0.16 | 2.67 | 284 |
| Partition wall towards unconditioned spaces | 0.12 ÷ 0.40 | 1.30 ÷ 1.57 | 104 ÷700 |
| Inter-floor slab | 0.29 | 2.51 | 642 |
| Pitched roof | 0.30 | 1.65 | 420 |
| Doors | 0.05 | 2.55 | 43 |

- The external cavity walls, where the large windows are present, favour the disappearance of windows and shutters; the cavity was assumed to be strongly ventilated.
- Internal partition walls towards unconditioned spaces are in solid bricks.
- The inter-floor slabs are in reinforced concrete.
- The pitched roof is in reinforced concrete structure and bricks, covered with slate slabs.
- Windows are made of wood frame and single glazing ($U_w$ = 4.6 W/m$^2$K); the large windows of the living area on the seaside are made of iron frame without thermal break and single glazing ($U_w$ = 5.9 W/m$^2$K); roller shutters with non-insulated boxes are present (U = 6.0 W/m$^2$K).
- All buildings elements both horizontal and vertical, which separate the heated zone from the outside environment or from unconditioned spaces are not insulated.

The details of the building elements and thermal bridges have been obtained from visual inspections and analysis of design documents. Thermal conductivities of construction materials are provided by the standards UNI 10351 [11], UNI EN ISO 10456 [12] and UNI/TR 11552 [13]. Thermal properties of doors and windows are derived from UNI EN ISO 10077-1 [14]. Thermal bridge analysis is carried out according to UNI EN ISO 14683 [15], using the abacus provided by the software.

## 5. ENERGY EFFICIENCY IMPROVEMENT AND ASSESSMENT

The improvement project focused exclusively on interventions of a passive kind on the construction system, due to a greater interest by the owners, merely proposing, for technical equipment, the placement of thermostat valves for the radiators. Given the architectural values of this building, not all the energy improvement interventions on the building envelope have been considered in an undifferentiated way. The experiences accumulated over the years by the authors have allowed to select the operations considered compatible with the cultural and heritage significance and therefore more easily admissible by the Ministry of Culture. When working on architectural heritage, in fact, the main objective cannot be the maximization of energy saving, but rather a balance between different aims, among which of primary importance is the formal and material preservation of the work.

The first principle in choosing possible energy efficiency improvement solutions was, in fact, the maximum material preservation of the villa, not just of its "outside appearance" (or the so called "aesthetic value"). This is why applying layers of insulating material on opaque walls was ruled out, either indoors or outdoors. A second principle was to work on "balancing" systems, seeking to optimise and prevent one system from "maximising" over another. This meant identifying the parts of the construction which have the greatest heat dissipation (roof and glazing) and acting in a way compatible with their preservation, without betraying the architectural ideas of the designer. A third principle was to make proposals which considered the permissibility criteria laid down by the protecting agencies and, at the same time, easy to be undertaken even without face a huge refurbishment operation. The improvement work analysed in the study therefore focused on:

- insulating the roof (which represents a large surface of energy loss and closes an unliveable space);
- insulating the basement-underground spaces used as cellars and storage rooms, both in the vertical walls, against the ground and on the horizontal floors. Both floors towards the attic and the cellars were insulated with 14 cm thick polyurethane panels, leading their transmittance to U = 0.17 W/m$^2$K, below the limit value ($U_{limit}$ = 0.26 and 0.32 W/m$^2$K respectively). The solid brick wall towards the cavity has been insulated with 14 cm thick polyurethane panels, leading their transmittance to U = 0.16 W/m$^2$K, below the limit value ($U_{limit}$ = 0.36 W/m$^2$K);
- replacing the glazing with better performing materials, low-emissivity vacuum glasses ($U_g$ = 1.2 W/m$^2$K, $g_{gl,n}$ = 0.65; in this way the transmittance of the windows reaches values of $U_w$ = 1.3 W/m$^2$K and $U_w$ = 1.7 W/m$^2$K for wood and metal frames respectively, in compliance with the limit value established by current legislation ($U_{limit}$ = 2.1 W/m$^2$K) [10]. This is a particular intervention that is encouraged by the Superintendence; from the technical point of view, it is affordable due to the high structural resistance of the iron frame of the large windows but less sustainable form the economic point of view (even too high costs);
- preserving and restoring the existing doors and window frames (iron and wood);
- insulating the roller shutter boxes up to U = 0.9 W/m$^2$K.

Introducing the new transmittance values into the calculation software and updating the thermal bridges made it possible to check how far the new solutions approach compliance with the parameters set by law (compliance with the values in Ministerial Decree 26.06.2015). The gross exchanging surface affected by the improvement interventions is about the 35% of the entire exchanging surface of the building; the intervention is therefore configured, based on the Decree 26.06.2015, as an important second level renovation. In this condition, the main requirement to be respected is that of the mean thermal transmittance of the components subject to intervention; this requirement is therefore satisfied. The compliance with the average heat exchange coefficient for transmission $H'_T$ [16] is difficult to achieve due to the choice not to act on the vertical opaque components.

Table 2 shows the parameters relating to the whole building before and after the improvements and the relative percentage variation. Both the floors insulation and glazing replacement contribute to the improvement of winter performance based on the $EP_{H,nd}$ index. The small reduction in periodic transmittance $Y_{ie}$ is due to the impossibility of adopting external insulation on vertical walls, while the glazing replacement involves a significant reduction in equivalent solar area due to the lower value of solar transmittance $g_{gl,n}$.

**Table 2. Parameters concerning the entire building.**

| Parameter | Before | After | Variation [%] |
|---|---|---|---|
| $EP_{H,nd}$ [kWh/m²] | 123.14 | 95.63 | -22.3 |
| $Y_{ie}$ [W/(m²K)] | 0.62 | 0.56 | -9.7 |
| $A_{sol,est}/A_u$ | 0.081 | 0.061 | -24.7 |

## 6. INTEROPERABILITY AMONG BIM AND ENERGY SOFTWARE AND CONCLUSIONS

The interest, in the world of Cultural Heritage, for ways of understanding heritage and its values (historic, urban, environmental, functional/spatial, morphological) must pay greater attention to digitalisation and to the various ways to file and use information data. This also, and especially, in order to set up and carry out correct interventions for managing an asset (preservation, maintenance, regeneration, improvement), from technical/economic planning to implementation and later management of the asset, with an integrated approach. The team involved in planning and execution must in fact be able to share not only general goals of preservation and improvement, but also tools which can guarantee efficacy in managing information throughout the whole process. Reference here is also made to the so-called "scientific review (*consuntivo scientifico)*" phase of the intervention on the protected asset and on implementation of the plans of maintenance/conservation to be carried out on it during its useful life cycle. The description of an integrated process, that is characterised by a great facility of dialogue and sharing of information among actors, has the practical aim of identifying the needs of information exchange during the various phases of a virtuous process, as the premise for designing interoperable platforms. Many IT products and services currently used in the field, generated with a mental attitude focusing on self-referential sub-processes (according to limited and specific goals) are not able to import or export information from or to other sub-processes, generating a punishing diseconomy, due to the useless multiplication of data survey campaigns, of formats, of ways of filing.

Besides the identification of the most suitable interventions to improve energy efficiency of the building envelope characterised by architectural principles of the Modern Movements (but still built recurring to load masonry walls on the rocks), another result of this work was, in fact, the verification of the efficacy of parametric software (BIM "Autodesk Revit") for modelling an existing, non-standardised building and its interoperability with energy calculation software (Edilclima) working on IFC standard. The IFC (Industry Foundation Classes) standard, which is the foundation of interoperability of BIM, was born for industrial production. BIM itself was born for planning new products, on the basis of standardising the construction elements of industrial production. Over time, however, they were progressively transferred to the whole building process, and hence to managing existing buildings and to working on historic buildings. Application of BIM to intervention on existing buildings, especially those for which it is difficult to track down construction drawings, are too often forced to descend to oversimplifications which damage their real efficacy (LOD). The construction of a BIM model for a historic building, in the absence of a precise description of its constituent nature, was rather difficult; moreover, it required the adaptation and construction of new, non-pre-constituted families.

A third issue we met concerns the fact that Revit does not provide an internal energy model useful for drawing up energy models aligned with Italian regulations. This requires integration with internal plug-ins or other items created by software houses independent from Autodesk, for example Edilclima, which also affords the opportunity of using dedicated internal plug-ins for Revit. As far as we have been able to see in our work, interoperability between the two kinds of software too can be greatly improved, as can their use in existing buildings. In Edilclima, exporting IFC files calls for a long work of adaptation and correction so as to use the model for energy calculations. In fact, the workflow proved to be slow and difficult, often requiring some operations to be repeated more than once in the two software. In addition, it was necessary to create a simplified model in Revit, intended only for energy calculation, for better compatibility between the two software. In any case, this work has been useful to add a tile to the great mosaic of building a framework of information needs which can perform an environmental quality assessment (temperature and visual comfort, air quality etc.) compatible with the requirements for material preservation of the asset, as well as a complete diagnosis and energy assessment of the building-facilities system.

Finally, something which was especially interesting and related to the specific case, was the energy analysis associated with the architectural design and its environmental context (choice of volumes and shielding systems, organisation of interior spaces, environmental location and orientation, organisation of closures partly opaque and partly glazed, materials) which revealed the environmental awareness of the decisions made in the design, on the façade exposed to wind, solar radiation and sea aerosol. This also helps dispel the myth of rationalist design as something far from, independent of and relatively uninterested in its environmental context.

## 7. REFERENCES


[1] F. Choay. L'allegorie du patrimoine. Paris: Éditions du Seuil, 1996.

[2] E. Hobsbawm. The Age of Extremes: The Short Twentieth Century, 1914–1991, London: Michael Joseph, 1994.

[3] MIBACT. Linee di indirizzo per il miglioramento dell'efficienza energetica nel patrimonio culturale, Roma, 2015.

[4] L.C. Daneri. Una piccola casa sulla scogliera di Sestri Levante. Genova: Bozzo, 1958.

[5] UNI CEI EN 16247-1, "Energy audits. Part 1: General requirements", 2012.

[6] UNI CEI EN 16247-2, "Energy audits. Part 2: Buildings", 2014.

[7] EDILCLIMA® Engineering & Software, EC 700 (version 9.19), https://www.edilclima.it8.

[8] UNI/TS 11300-1, "Energy Performance of Buildings, Part 1: Evaluation of Energy Need for Space Heating and Cooling", 2014.

[9] UNI EN ISO 13786, "Thermal performance of building components. Dynamic thermal characteristics. Calculation methods", 2018.

[10] Decree 26.06.2015 of the Ministry of Economic Development, "Applying the methods of calculating energy performance and definition of requirements and minimum standards of buildings", 2015.

[11] UNI 10351, "Building materials and products. Hygrothermal properties. Procedure for determining the design values", 2015.

[12] UNI EN ISO 10456, "Building materials and products. Hygrotherrnal properties. Tabulated design values and procedures for determining declared and design thermal values", 2008].

[13] UNI/TR 11552, "Opaque envelope components of buildings. Thermo-physical parameters", 2014

[14] UNI EN ISO 10077-1, "Thermal Performance of Windows, Doors and Shutters. Calculation of Thermal Transmittance. Part 2: General", 2018.

[15] UNI EN ISO 14683, "Thermal bridges in building construction. Linear thermal transmittance. Simplified methods and default values", 2018.

[16] P. Cavalletti, S. Bergero. "The mean heat transmission coefficient as a new parameter to control heat transmission and mould growth in building refurbishing", in Journal of Civil Engineering and Architecture Research, vol. 3, no. 6, 2016, pp. 1495-1502.

# Future energy performance and overheating risks in retrofitted historic buildings of South Tyrol

L. Hao[1], D. Herrera-Avellanosa[2], Alexandra Troi[2]

[1] Department of Architecture and Art design, Hebei University of Technology, Tianjin, China; 2020107@hebut.edu.cn

[2] Institute for Renewable Energy, Eurac Research, 39100 Bolzano, Italy; Daniel.Herrera@eurac.edu (D.H.-A.); Alexandra.Troi@eurac.edu (A.T.)

**Abstract** – The Alpine region of South Tyrol in the north of Italy is characterized by its mountainous topography and diverse climatic conditions. Around 22% of the residential buildings in South Tyrol was built before 1945, and the energy retrofit of the building stock could achieve great energy saving. To investigate future energy performance and overheating risks of the retrofitted historic buildings, this paper proposes a methodology based on the analysis of local weather conditions and identification of homogenous climatic zones. For each climatic zone, the typical buildings are defined and tailored climate projections are created. Finally, the future energy performance and thermal comfort state of the reference buildings are simulated. The results demonstrate that the retrofit interventions could significantly improve energy efficiency of historic buildings in both present and future scenarios. A change in climate together with retrofit interventions will, however, result in higher risk of indoor overheating in the warmer climate zone of South Tyrol.

**Keywords** – historic buildings; energy retrofit; climate change; internal climate; overheating

## 1. INTRODUCTION

Studies carried out in the Alpine context have confirmed the serious challenges that climate change will impose in the region. Compared with other regions in Europe, South Tyrol already suffers a more severe temperature increase. The 2018 South Tyrolean Climate Report [1] indicates that the temperature in summer have risen 2.2°C from the 1960s and the temperature increase will be up to 5°C under the most pessimistic scenario by 2100. Along with the temperature rise, days with extreme temperature would be more frequent. For instance, the number of summer days (Days during which the maximum temperature is above 20°C) in the capital of South Tyrol has grown from around 100 in the 1960s to around 115 in 2018, and it will reach to 175 by 2100. Similarly, the number of days when the minimum temperature remains above 20°C rises significantly. The previous highest record of South Tyrol was 24 days in 2015, while there would be an average of more than 60 days a year by 2100.

The changes in climate may influence the pattern of energy use and the thermal comfort inside historic buildings [2]. Studies in different countries observed a decreasing trend in heating load in winter, and the dilemma of an increasing cooling load or uncomfortable conditions in summer. Since most historic buildings are not equipped with modern cooling systems and working on "free-running" mode in summer, the "cooling system" combining thermal mass and natural ventilation is commonly used as a passive cooling strategy. The effectiveness of this passive cooling is dependent on buildings' thermal mass, user behaviour, and climate factors such as outdoor temperature daily swing, solar radiation, etc [3, 4]. With outside temperature changing in future, the passive cooling system may fail to ensure a comfortable thermal condition any longer [5]. Energy retrofit may further exaggerate the overheating problem. To keep the original outlook of historic buildings, internal insulation is broadly used in energy retrofit. However, the application of internal insulation is observed to increase the indoor temperature in different climate conditions [6-8], therefore, aggravate the risk of overheating.

Considering the possible challenges in future, this study aims to assess the combined impacts of climate change and energy retrofit on the energy use and thermal comfort of the historic buildings in South Tyrol.

## 2. METHODS

### 2.1 REFERENCE BUILDINGS

To select robust and reliable building references, the correlations between local climate and building inventories are analyzed [9]. The results highlighted the necessity of using different reference buildings to represent the typical buildings in different climate zones. The whole climate in South Tyrol is divided into three homogenous sub-climate zones and building categories are defined for each climate zone. In this study, two reference buildings that represent the "Portici house" in Climate zone I and II are used (Figure 1, Figure 2).

**Figure 1 Reference building of Climate zone I (Portici house-I), Piazza Erbe 11, Bolzano**

**Figure 2 Reference building of Climate zone II (Portici house-II), Schallerhaus, Glorenza**

"Portici house" is a typical trading-residential building model, with shops occupying the ground floor and apartments located on the upper floors. In the reference building of Climate zone I (Portici-I), the shop and apartments extend toward the back, with an inner courtyard. In the reference building of Climate zone II (Portici-II), on the other hand, a small yard is located behind the shop, originally leading to stables for livestock, with access from the back for staff and animals. These two reference buildings are constructed with masonry walls (0.58m thick), and plastered with lime plaster both internally and externally. Table 1 presents a summary of the envelope U-values before and after retrofit.

**Table 1 Construction characteristics of the reference buildings**

| | Type | Retrofit solutions | U-value before retrofit | U-value after retrofit |
|---|---|---|---|---|
| External wall | Masonry wall | Wood fiberboard,12cm | 2.59 W/m$^2$K | 0.28 W/m$^2$K |
| Roof | Timber rafters with wooden casing and roof tiles on top | Wood fibreboard, 20cm | 5.6 W/m$^2$K | 0.17 W/m$^2$K |
| Foundation | Tamped earth and concrete pavement | Polystyrene, 10cm | 4.19 W/m$^2$K | 0.25 W/m$^2$K |
| Window | Single glazed window | Double glazed window | 3 W/m$^2$K | 1.1 W/m$^2$K |

### 2.2 FUTURE CLIMATE PROJECTION

To predict the future climate data, climate projections are taken from four combination of general circulation models (GCMs) and regional climate models (RCMs) of EURO-CORDEX initiative [10]. These four climate model combinations (M1,2,3,4) simulate the most likely change in future climate (Table 2). Representative Concentration Pathway (RCP) 8.5 is adopted in this study which is a business-as-usual scenario with GHG emissions continuing to rise in the 21st century. The future climatic data from the four future projections is bias-corrected and downscaled for the climate zones, and each projection offers multi-year climate data of near future scenario (F1: 2041-2050) and far future scenario (F2: 2091-2100) to reflect the long-term climate conditions in simulations.

**Table 2 GCM and RCM combination, RCP of the selected climate models**

| Acronym | GCMs | RCMs | RCP |
|---|---|---|---|
| M1 | ICHEC-EC-EARTH | DMI-HIRHAM 5 | 8.5 |
| M2 | ICHEC-EC-EARTH | SMHI-RCA 4 | 8.5 |
| M3 | IPSL-IPSL-CM5A-MR | SMHI-RCA 4 | 8.5 |
| M4 | MPI-M-MPI-ESM-LR | CLMcom-CCLM 4 | 8.5 |

### 2.3 NUMERICAL SIMULATION

Energy demand and indoor climate are calculated using EnergyPlus 8.7.0 [11]. The heating energy use is calculated on the basis of the temperature set point during heating period, while indoor temperature and relative humidity (RH) in summer are calculated in free floating conditions, without any mechanical cooling system. The heating period of Climate zone I and II is defined according to the Italian requirement on Heating Degree Days (HDD). The occupancy, lighting, and electric appliances profiles are based on ISO 17772-1 [159] and the 2014 Building America House Simulation Protocols [160]. Airtightness of the building envelope before retrofit is defined according to literature review: 10 ac/h, at 50 Pa [12]. The value after retrofit is defined according to CasaClima standard A [13]: 1.5 ac/h, at 50 Pa. Natural ventilation happens only when the room is occupied, the indoor temperature is higher than 24°C, and the difference between indoor and outdoor temperature is higher than 3°C. It is modelled by simplified ventilation calculations in EnergyPlus' Wind and Stack Open Area model.

### 2.4 INDOOR COMFORT ASSESSMENT

The adaptive thermal comfort model proposed in EN 15251 [14] is used in the evaluation of indoor overheating levels in the present study. It suggests that occupants can adapt the indoor thermal conditions through window operation or clothing arrangement. It was developed from extensive field studies and defined the comfort temperature range in free-running buildings as a function of the outdoor running mean temperatures. Its upper and lower limits used in this study are:

$$\boldsymbol{\theta}_{\max} = \mathbf{0.33}\boldsymbol{\theta}_{\mathrm{rm}} + \mathbf{18.8} + \mathbf{3} \qquad \textbf{Equation 1}$$

$$\boldsymbol{\theta}_{\min} = \mathbf{0.33}\boldsymbol{\theta}_{\mathrm{rm}} + \mathbf{18.8} - \mathbf{3} \qquad \textbf{Equation 2}$$

where $\theta_{rm}$ is the running mean outdoor temperature.

The approach is applicable when $10°C < \theta_{rm} < 30°C$ for the upper limit and $15°C < \theta_{rm} < 30°C$ for lower limit. However, the outdoor running mean temperature of South Tyrol could be higher than 30°C, resulting in some overheating hours being out of the range. Therefore, it is defined in this study that when $\theta_{rm} \geq 30°C$, $\theta_{max} = 31.7°C$.

## 3. RESULTS AND DISCUSSION

### 3.1 FUTURE CLIMATE CHANGE

Table 3 presents the temperature increase at F1 and F2 when compared with present average values. M2 obtains the largest variation, with the highest increase of average temperature increase in both Climate zone I and II at F1 and F2. In this projection, the average temperature gain by the end of the century reaches to more than 6°C. In the most moderate projection (M1), the average temperature rise is higher than 2.8°C.

**Table 3 Average temperature increase (in °C) in different climate projections (M1,2,3,4) in near future (F1) and far future (F2) compared with present scenario (P)**

| | Climate zone I | | | | Climate zone II | | | |
|---|---|---|---|---|---|---|---|---|
| | *M1* | *M2* | *M3* | *M4* | *M1* | *M2* | *M3* | *M4* |
| *F1* | 0.90 | 2.23 | 1.16 | 0.67 | 0.47 | 2.13 | 1.29 | 0.66 |
| *F2* | 3.04 | 6.16 | 5.13 | 3.22 | 2.84 | 6.11 | 5.61 | 3.31 |

### 3.2 BUILDING ENERGY USE

Figure 3 shows the heating energy use in each future projection of Climate zone I and II. The average heating energy use increases from Climate zone I to II. However, it should be emphasized that the different characteristics of reference buildings (differences in building function, layout, volume, etc.) may prevent the direct comparison of their energy performance. Retrofit solutions reduce the heating energy use significantly in both Climate zones. In Climate zone I, energy retrofit could save 92.2% of the heating energy at the present scenario (P), and 93.4% at near future scenarios (F1) and 95.8% at far future scenarios (F2). In Climate zone II, energy retrofit is slightly less effective compared with Climate zone I in terms of the ratio of energy saving. However, the absolute energy saving of Climate zone II is higher. In summary, retrofit solutions have a substantial impact on heating energy use, and it could achieve better effects in a warmer climate and in Portici houses when comparing the energy-saving ratio.

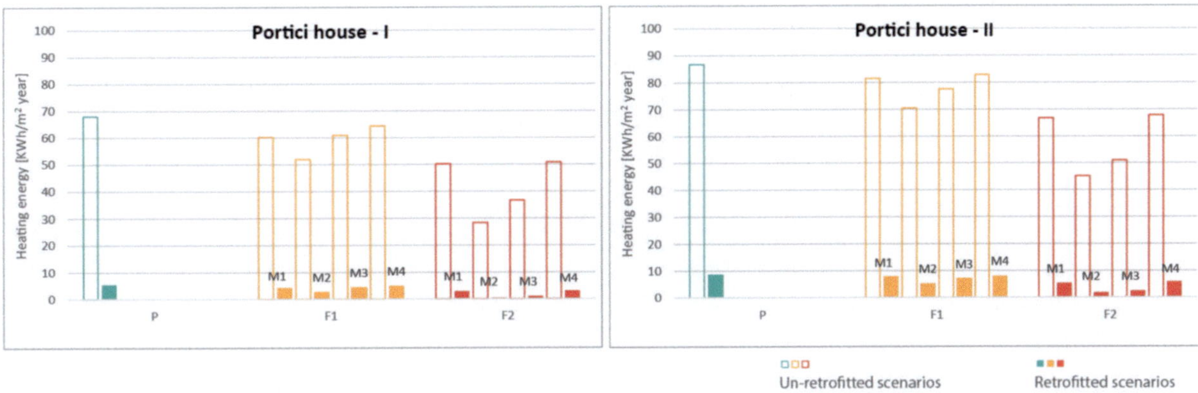


**Figure 3 Average annual heating energy consumption of the whole building in kWh for retrofitted/un-retrofitted and present/future scenarios. left: Portici house-I; right: Portici house-II**

Future climate also affects heating energy use, but its impact is only noticeable at F2 when the building is not retrofitted. In Climate zone I, the average heating energy use drops by 12.7% at F1, and 39% at F2 compared with P. The impact of climate change is less intense in Climate zone II compared with Climate zone I, from the perspective of energy reduction ratio. The temperature rise reduces 10.0% of the average heating energy use at F1, and 33.5% at F2. In the case of retrofitted buildings, climate change causes less reduction in the absolute heating energy use, since it is already low. However, the reduction ratio is high. For instance, there is 67.7% less heating being used due to climate change at F2 in Climate zone I, and 57.0% in Climate zone II. To sum up, the impact of climate change on building energy use is more obvious in a warmer climate and in retrofitted historic buildings when using energy reduction ratio as an indicator. In un-retrofitted buildings, the impact is substantial at F2.

## 3.3 THERMAL COMFORT ASSESSMENT

### *3.3.1 Comfort assessment with adaptive approach*

Adaptive thermal comfort model is used to characterize overheating in the living rooms. As shown in Figure 4, the total applicable hours of the adaptive assessment method increase from present scenario to future scenario meaning that the number of hours where the outdoor running mean temperature is above 15°C increases in the future. In un-retrofitted scenarios of Portici house-I and II, underheating is the main comfort problem in both present and future scenarios. With temperature increase, the underheating problem is slight mitigated. However, climate change also brings some overheating problems in the far future. In retrofitted scenarios, overheating is already a concern in present scenario of Portici house-I, and the situation further deteriorates in future scenarios. While in retrofitted Portici house-II, overheating is not a crucial problem until far future.

Current retrofit solutions change the thermal comfort state of the living rooms in all climate zones. Its impact on overheating risk is more pronounced in Portici house-I (Figure 4). The main source of discomfort changes from under-heating in un-retrofitted scenarios to overheating in retrofitted scenarios not only in future scenarios but also in the present scenario. In Portici house-II, retrofit interventions do not lead to substantial overheating hours at present and F1 (Figure 4).

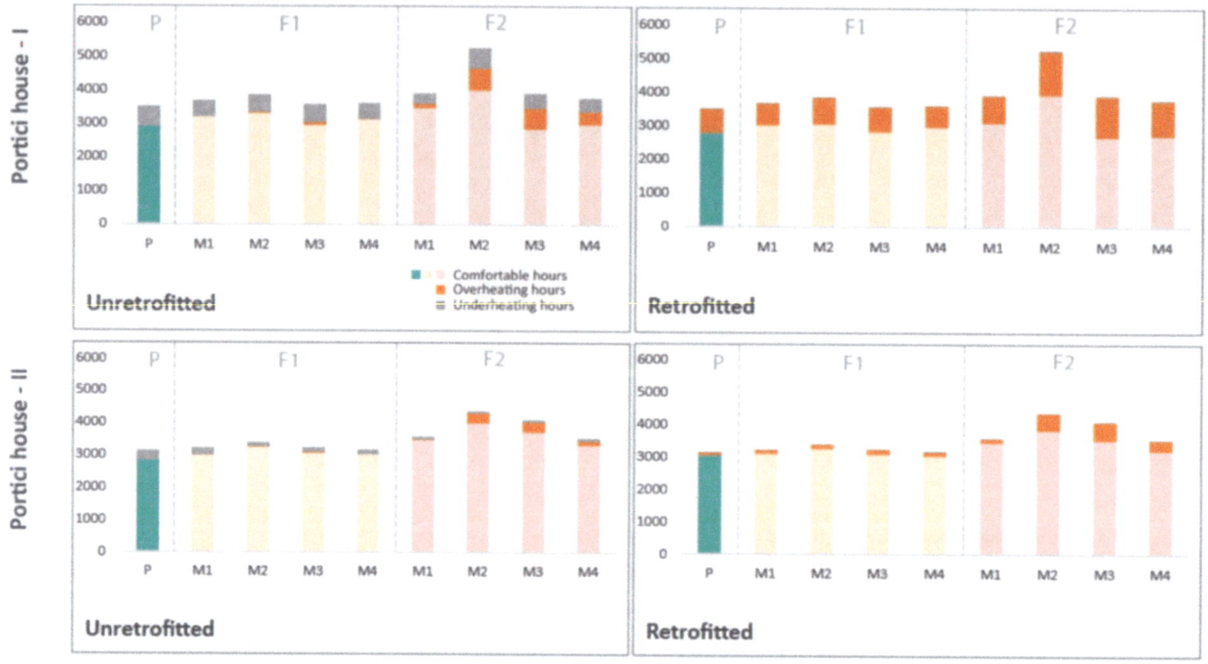


**Figure 4 Thermal comfort state in the Livingroom of Portici house-I (up) and Portici house-II (down) according to adaptive thermal comfort model of EN15251**

### *3.3.2 The impact of climate change and energy retrofit on overheating*

To further analyse the impact of climate change and retrofit interventions on overheating, three parameters are defined:

In un-retrofitted buildings,

Δ1= number of overheating hours in the future scenario – number of overheating hours in the present scenario (i.e. overheating hours caused by climate change in un-retrofitted buildings);

In retrofitted buildings,

Δ2= number of overheating hours in the future scenario – number of overheating hours in the present scenario (i.e. overheating hours due to climate change in retrofitted buildings);

Δ3= number of overheating hours in the retrofitted scenario – number of overheating hours in un-retrofitted scenario at the same time period (impact of retrofit interventions on overheating hours). Table 4 shows the three parameters for Portici house-I and II.

As a result of the significant temperature increase at F2, Δ1 and Δ2 rise notably at F2 scenario in both climate zone I and II. In Portici house-II, Δ2 is higher than Δ1 in all climatic projections, implying that climate change leads to more overheating hours in retrofitted scenarios. However, this phenomenon is not clear in Portici house-I.

When comparing the overheating hours increased by retrofit interventions (Δ3 in Table 4), there are more overheating hours induced by retrofit in Portici house-I than II due to the warmer climate in zone I. In Portici house-I, Δ3 decreases slightly from present to F2. This phenomenon indicates that the negative effect of retrofit interventions falls slightly in future scenarios. However, in Portici house-II, Δ3 rises at F2, meaning that the negative impact of insulation escalates with temperature increases.

Table 4 The number of overheating-hour per year of Livingroom due to climate change/retrofit interventions

| | | P | F1 | | | | F2 | | | |
|---|---|---|---|---|---|---|---|---|---|---|
| | | | M1 | M2 | M3 | M4 | M1 | M2 | M3 | M4 |
| Portici house-I | Δ1 | - | 2 | 34 | 90 | 10 | 138 | 650 | 624 | 384 |
| | Δ2 | - | -59 | 87 | 27 | -77 | 90 | 584 | 519 | 333 |
| | Δ3 | 715 | 654 | 767 | 652 | 627 | 667 | 648 | 690 | 664 |
| Portici house-II | Δ1 | - | 0 | 21 | 29 | 0 | 17 | 318 | 318 | 130 |
| | Δ2 | - | 25 | 39 | 61 | 42 | 26 | 446 | 471 | 234 |
| | Δ3 | 91 | 116 | 109 | 123 | 132 | 99 | 219 | 244 | 194 |

## 4. CONCLUSIONS

Through the analysis of the thermal comfort conditions in historic buildings, the impact of climate change can be quantified in both un-retrofitted and retrofitted scenarios. Climate change alone does not bring any substantial overheating risk in the near future in climate zones I and II, and yet its impacts become critical in the far future. On the other hand, retrofit interventions increase the operative temperature and cause significant overheating risk in the Portici house in Climate zone I at present scenario and future scenarios. It is worth noting that overheating problem exists even without retrofit in the far future. In the Portici house of Climate zone II, retrofit interventions only lead to severe overheating risk in scenarios with climate change in the far future scenario.

The results of this study highlight the importance that a warming climate will have in designing interventions in the built heritage in the near and far future. Any retrofit will have to consider the implications on aspects like thermal mass or ventilation and include measure of adaptation that will ensure an efficient energy performance without the risks of overheating that might lead to discomfort, or increased energy use due to the need of mechanical cooling in summer. Even though this study is limited to residential buildings in South Tyrol (IT), there are reasons to suggest that studying the implications of climate change in summer comfort in historic buildings should be extended to other typologies and climates in future research.

## 5. REFERENCES


[1] M. Zebisch *et al.*, "Klimareport Südtirol 2018," Eurac Research, Bolzano, 2018.

[2] L. Hao, D. Herrera-Avellanosa, C. del Pero, and A. Troi, "What Are the Implications of Climate Change for Retrofitted Historic Buildings? A Literature Review," *Sustainability,* vol. 12, no. 7557, 2020.

[3] E. Shaviv, A. Yezioro, and I. G. Capeluto, "Thermal mass and night ventilation as passive cooling design strategy," *Renewable Energy,* vol. 24, no. 3, pp. 445-452, 2001.

[4] A. Gagliano, F. Patania, F. Nocera, and C. Signorello, "Assessment of the dynamic thermal performance of massive buildings," *Energy and Buildings,* vol. 72, pp. 361-370, 2014, doi: 10.1016/j.enbuild.2013.12.060.

[5] A. D. Peacock, D. P. Jenkins, and D. Kane, "Investigating the potential of overheating in UK dwellings as a consequence of extant climate change," *Energy Policy,* vol. 38, no. 7, pp. 3277-3288, 2010/07/01/ 2010, doi: https://doi.org/10.1016/j.enpol.2010.01.021.

[6] K. M. S. Chvatal and H. Corvacho, "The impact of increasing the building envelope insulation upon the risk of overheating in summer and an increased energy consumption," *Journal of Building Performance Simulation,* Article vol. 2, no. 4, pp. 267-282, 2009, doi: 10.1080/19401490903095865.

[7] V. Tink, S. Porritt, D. Allinson, and D. Loveday, "Measuring and mitigating overheating risk in solid wall dwellings retrofitted with internal wall insulation," *Building and Environment,* vol. 141, pp. 247-261, 2018/08/15/ 2018, doi: https://doi.org/10.1016/j.buildenv.2018.05.062.

[8] S. Cirami, G. Evola, A. Gagliano, and G. Margani, "Thermal and Economic Analysis of Renovation Strategies for a Historic Building in Mediterranean Area," *Buildings,* vol. 7, no. 3, p. 60, 2017, doi: 10.3390/buildings7030060.

[9] L. Hao, D. Herrera-Avellanosa, C. Del Pero, and A. Troi, "Categorization of South Tyrolean Built Heritage with Consideration of the Impact of Climate," *Climate,* vol. 7, no. 12, 2019, doi: 10.3390/cli7120139.

[10] D. Jacob *et al.*, "EURO-CORDEX: New high-resolution climate change projections for European impact research," *Regional Environmental Change,* Article vol. 14, no. 2, pp. 563-578, 2014, doi: 10.1007/s10113-013-0499-2.

[11] D. B. Crawley, C. O. Pedersen, L. K. Lawrie, and F. C. Winkelmann, "EnergyPlus: Energy Simulation Program," *Ashrae Journal,* vol. 42, no. 4, pp. 49-56, 2000.

[12] L. Hao, D. Herrera, A. Troi, M. Petitta, M. Matiu, and C. Del Pero, "Assessing the impact of climate change on energy retrofit of alpine historic buildings: Consequences for the hygrothermal performance," in *IOP Conference Series: Earth and Environmental Science*, 2020, vol. 410, 1 ed., doi: 10.1088/1755-1315/410/1/012050. [Online]. Available: https://www.scopus.com/inward/record.uri?eid=2-s2.0-85079605755&doi=10.1088%2f1755-1315%2f410%2f1%2f012050&partnerID=40&md5=d8f24450b0935020e03d694e033d87ed

[13] CasaClima, "Criteri CasaClima: Per l'esecuzione delle prove di tenuta all'aria," Agenzia per l'Energia Alto Adige CasaClima, 2015.

[14] *EN 15251 Indoor environmental input parameters for design and assessment of energy performance of buildings- addressing indoor air quality, thermal environment, lighting and acoustics*, 2007.

# Launching an experimental label on both energy retrofitting and heritage preservation: early lessons

E. Heberle[1], J. Borderon[1], Y. Dervyn[2], A. Sage[2], H. Arnichand[2]

[1] Strasbourg Laboratory, Center for studies and expertise on risks, environment, mobility and urban and country planning, (Cerema), Strasbourg, (France. elodie.heberle@cerema.fr, julien.borderon@cerema.fr)

[2] Effinergie association, Montpellier, France. (dervyn@effinergie.org, sage@effinergie.org, arnichand@effinergie.org)

**Abstract** – Retrofitting historic buildings is a perilous exercise because they carry huge heritage values and are very sensitive to inappropriate works. The Effinergie association, which operates the French energetic label for new and refurbished low-energy-consumption buildings ("BBC" label), launched in February 2019 an experimental label called "Effinergie Patrimoine". This label addresses all types of heritage buildings, listed or not, and aims at promoting heritage preservation in energy retrofitting projects. To obtain it, two commissions assess the project: the first one decides whether the project is a historic building or not and the second whether it deserves the label or not, based on a set of criteria. The project energy consumption should be as close as possible to the one required for the "BBC" label. The experimentation is running for two years. Cerema, which operates the French centre for responsible retrofit of historic buildings (CREBA), took an active part in it.
**Keywords** – energy label; heritage preservation; energy retrofitting

## 1. INTRODUCTION

Retrofitting existing buildings is the main lever to reduce energy consumption and greenhouse gases emissions generated by the building sector. In 2015, the Act for Energy Transition toward Green Growth (LTECV) [1] was approved in the French law. It sets that in 2050, all buildings will have to be retrofitted to the "BBC" level, which corresponds to a given calculated primary energy consumption for heating, hot water, cooling, ventilation, pumps and lightings [2]. For housing, the consumption should not exceed 80 kWh/m².yr modulated by the location and altitude, whereas for tertiary buildings it should be 40% lower than a calculated reference.

This energy consumption level is still considered as hard to achieve for heritage buildings in France. Moreover, retrofitting heritage buildings is a perilous exercise because they carry huge heritage values and are very sensitive to inappropriate works [3]. It results that very few heritage buildings were retrofitted to the "BBC" level and that some of these retrofits are not satisfying, neither from the heritage conservation nor from the technical point of view [4].

The Effinergie association operates the "BBC-Effinergie Rénovation" certification, which requires the "BBC" energy consumption level and other aspects such as an airtightness testing procedure. Effinergie launched in February 2019 an experimental label called "Effinergie Patrimoine" to address all types of heritage buildings, listed or not, and to promote heritage preservation in energy retrofitting projects.

## 2. BACKGROUND OF THE EXPERIMENTAL LABEL

Effinergie was created in 2006 and is dedicated to energy efficiency and comfort in buildings. Its members are local authorities, various environment agencies and associations involved in the building sector.

In 2012, a first working group was set up on the initiative of the French Ministry of Sustainable Development. Effinergie was invited to take part in it, as well as the French Ministry of Culture, the association Sites et Cités Remarquables de France, that gathers cities that have a protected cultural heritage area and Cerema (Center for studies and expertise on risks, the environment, mobility and planning), that is the engineering office of the Ministry of Sustainable Development. The aim was to think about concrete actions to conciliate energy efficiency and heritage conservation. Rapidly, the idea of a label emerged but was not been carried through.

In 2017, Effinergie set up the working group again. It gathered members of the first working group but also members of the association. Cerema was also involved as it was working with other partners on the creation of the CREBA (centre for responsible retrofit of heritage buildings) since 2016 [5].

The working group was first a place to discuss about energy efficiency of heritage buildings. Several initiatives were presented, like the "Habitat ancien en Alsace" study, carried out by Cerema [6] or the EN 16883 "Guidelines for improving energy efficiency of architecturally, culturally or historically valuable buildings" [7]. This contributed to build a common culture between architects and engineers from the working group. It was also decided that Effinergie would create a section dedicated to retrofitted heritage buildings in its "BBC" observatory [8] and to better document some of them.

Indeed, it has always been possible for heritage buildings to obtain the "BBC-Effinergie Rénovation" label, but some retrofitted heritage buildings did not achieve the required energy consumption because of heritage conservation or technical issues, and renounced to apply to the label, even if they provided good practice examples. On the other hand, there is in France a sharp debate between heritage associations and associations for the defence of the environment about, for example, exterior insulation of heritage buildings [9].

The working group decided to create a specific label addressing heritage buildings, with more flexible energy requirements but with careful considerations about heritage preservation. The label project, called "Effinergie Patrimoine", was presented to the French Agency of the Environment and the Control of Energy (ADEME), which accepted to finance a two years-experimentation and set the target to 40 delivered labels. The label was named "experimental" because such kind of labels does not exist in France and because the pursuing of the project depends on its conclusions.

## 3. FUNCTIONING OF THE LABEL

The label was officially announced in February 2019 at the BePositive show in Lyon (France). Effinergie issued a press release and gave some interviews in specialised press. In parallel, every member of the working group was invited to communicate about the label in their own networks. During the rest of the year, the working group met five times to build the certification procedure.

The procedure was released for the first time in September 2019 on the Effinergie website [10]. It is open to on-going projects and consists in three main steps: a preliminary advice that decides whether the project is a heritage building or not; the signature of a contract with a certifying body in charge of collecting and checking documents on both energy and heritage aspects; and a consolidated advice that decides whether it deserves the label or not. At the end of these three steps, the label is given by the certifying body once it is completed.

The role of the commission and each step are described below.

## 3.1 COMPOSITION AND ROLE OF THE COMMISSION AND EXPERTS

The commission is the decision-making body of the label. It is composed of experts that are also part of the working group. They meet every three months since the beginning of 2020 in order to evaluate projects. They had to sign a charter to ensure their independence. Notably, if one of them is involved in a project presented to the commission, he or she will not participate in the debate and deliberate. The private organisms of the commission are remunerated by Effinergie, if their members participate in at least three commissions per year. In return, the association receives funds from the French Ministry of Sustainable Development and from the ADEME.

In addition, seven members of the commission are designated as experts: three heritage architects on heritage preservation, and four building physics engineers on energy retrofitting. Their role is to assess projects during the second step, the consolidated advice. After having studied the documents sent by the project owner, they prepare a reasoned opinion and present it to the rest of the commission that finally decides whether the project deserves the label or not. Therefore, each expert had to demonstrate his or her skills on their field of expertise. As it takes time to analyse the documents related to one single project, each assessment is remunerated thanks to the financing scheme previously described.

## 3.2 PRELIMINARY ADVICE

### *3.2.1 Procedure*

The preliminary advice aims at deciding if the building can be considered as a heritage building or not, on the basis of documents provided by the project owner. This decision is taken by a vote of the commission after a short presentation of the project by Effinergie. If the building is not considered as a heritage building, the candidate is redirected to the regular "BBC-Effinergie Rénovation" label.

### *3.2.2 Documents required for the preliminary advice*

To assess the heritage value of the building, the candidate has to fill out a form and provide several documents. If the building is listed, this form is merely a formality as the building has already been identified as a heritage building. If the building is not listed, more documents are required:

- a 10-to-20 photographic report,
- site plan of the building,
- work progress,
- 3000-characters presentation of arguments justifying the heritage value of the building, with date of construction and historical context, relevant architectural details, details on anterior retrofitting works, materials and composition of the building envelope,
- mass lay-out of the building and plane of the old façades if they exist.

### 3.3 CONTRACTUALIZATION WITH A CERTIFYING BODY AND COLLECTION OF THE REQUIRED DOCUMENTS

Once a project has received a preliminary advice, candidates have to choose a certifying body and sign a contract with it. Indeed, Effinergie does not interact directly with the project owners, but through one of the four certification-bodies participating in the experimentation. They will support the project owner to collect the required documents that will be analysed by the experts as described in the section above. After checking their completeness, the certifying body forwards these documents to Effinergie. Certifying bodies are indeed the executive body of the label, as they give the label at the end of the procedure. The certification cost goes from less than 1000 € to about 10 000 €, depending on the type (tertiary or residential) and the surface area of the building.

### 3.4 CONSOLIDATED ADVICE

#### *3.4.1 Procedure*

The consolidated advice aims at deciding if the project has preserved the architectural aspect of the building while reaching a sufficiently low energy consumption level. This decision is taken on the basis of the documents provided by the project owner. The members of the commission vote, based on the reasoned opinions of one heritage expert and one energy expert. The commission can then either give a negative advice, request for additional information or give a positive advice. In this case, the certifying body will have to validate in any case the label after the reception of the works.

To improve their chances of passing the consolidated advice, project owners and their contractors are encouraged to follow the guidance promoted in the CREBA charter for a responsible retrofit of heritage building [11].

#### *3.4.2 Documents required for the consolidated advice*

The candidate that has passed the preliminary advice has to provide another set of documents. First of all, the energy consumption of the project should be as close as possible to the one required for the “BBC-Effinergie Rénovation” label described in the introduction. If not, the reasons why have to be explained and will be assessed by the commission, which will decide if they are relevant or not. Among the required documents, it is asked to provide:

- a note justifying the choice of the proposed retrofit solutions regarding technical, thermal and architectural aspects and indicating the prescription of the French government building architect if the building is listed,
- an architectural and technical study including: the actual and future use of the building; a description of the construction materials and existing equipments and of their state of deterioration; a note about the hygrothermal compatibility of the proposed retrofit solutions; a dynamic thermal simulation or a note about comfort in summer after retrofitting (depending of the size of the building); a note about acoustic comfort; and a note about indoor air quality.

## 4. FEEDBACKS

### 4.1 NUMBER OF APPLICATIONS

The first commission took place on February 10, 2020. Since then, they are organised each 3 months. The dates of the two forthcoming commissions are always published on the page dedicated to the experimentation, so that the candidates know when their projects will be assessed. 20 projects were validated so far for the preliminary advice described above and amongst these projects:

- 3 left the experimentation, due to a lack of interest for the label or to the cost-related issues.
- 2 were about to transfer consolidated advice documents to the certifying body they choose.
- 1 single project was assessed by the experts for a consolidated advice and was actually presented to the commission.

### 4.2 A FIRST ASSESSED PROJECT: THE TOWN HALL OF BRIAS

The first assessed project is the town hall of Brias, a town of 300 inhabitants near Lille in the north of France. It is a former presbytery built in 1865 by a Parisian architect, Pierre-Charles Dusillion on behalf of the earl Charles-Marie de Bryas, who gave its name to the town. As shown in Figure 1, the building is made of bricks and chalk in a Victorian style, which was very appreciated by the earl. Since 1999, the town hall occupies the first storey and a municipal housing the second one.

**Figure 1: Exterior façades of the town hall of Brias**

The exterior façade has chalk details at corners, edges and window frames. The interior is tiled with blue stones, typical of that part of France, or have chestnut parquet flooring. The chimney is of red marble and the staircase has turned-wood balusters. A vaulted cellar built of bricks is well preserved. The walls were never insulated but 5 cm of mineral wool was installed on the roof. A 40 kW fuel boiler provides heat to the building. The windows were replaced long before with plastic windows. The building suffers from multiple moisture disorders due to rising damp in the cellar and to the absence of a sufficient ventilation. Besides, the bricks seem to be very porous and the joints are in poor state.

The retrofitting project consists in insulating the attic floor with 26 cm of recycled cotton wool and the ground floor with 10 of polyurethane foam. In order to avoid more moisture disorders, the walls will not be insulated but rendered with a high-insulating plaster made up of aerogel.

Following the assessment for the consolidated advice, the commission was not satisfied with the project and requested some evolutions. Among the propositions made by the commission, one can mention the preservation of the plaster on lath on the first-floor ceiling, removal of plastic roller blind and improvement of the wooden windows with a solution closer to the original one. The appearance of the latter was found in historical pictures shown in figure 2 after the advice of the commission. A dynamic hygrothermal simulation was also advised to make sure that the interior render is appropriate.

**Figure 2: Original appearance of the windows**

## 5. DISCUSSION

The strength of this label and the main difficulties encountered start to emerge. These two aspects are described below.

### 5.1 DIFFICULTIES ENCOUNTERED BY THE EXPERIMENTATION

After 3 of the 8 commissions, 17 projects are currently involved in the procedure, far from the 40 delivered labels. The pandemic and the diversity of the project owners make the communication on the experimentation difficult. A global communication plan has currently started in order to include more projects within the experimentation thanks to various channels (articles, newsletters, podcasts, etc.)

The time scale of the building sector is also a major challenge for this experimentation. Planning of some retrofit projects may be spread over several years, slowing down the progress toward the certification. Some projects may eventually end up later than the deadline of this experimentation. An extension of the experimentation is therefore currently considered.

In addition, the procedure is time consuming for project owners, as the number of required documents is important and as these documents have sometimes to be drafted especially for the certification procedure. It should be understood that the application for the label is a voluntary procedure amongst other mandatory ones, on which project owners may want to focus first. That can explain why some project owners left the experimentation after the preliminary advice.

Another issue is the cost of this experimentation itself. For the moment, experts and private organisms participating in the commission are remunerated by funding's coming from the Ministry of Sustainable Development, the ADEME and Effinergie. A long-term solution could be that the certifying bodies may integrate heritage experts in their teams (in addition to energy experts that are already integrated), which is not the case today. However, this would increase the cost of the certification, which has already been identified as a major issue, especially for small private projects including individual houses (for large projects this cost represents a minor percentage of the whole refurbishment project). An estimation of the certification cost can now be provided to project owners entering the experimentation.

Finally, the first project applying for the consolidated advice showed that even if the number of required documents is important, clarifications are always needed on particular points. It is always difficult to deeply understand a project and a balance has to be found regarding the amount of information required.

### 5.2 STRENGTH OF THE EXPERIMENTATION

First, the notion of heritage building seems to be shared amongst the members of the commission, even if they come from diverse backgrounds. Indeed, the decision related to the preliminary advices was often taken without much debate. Therefore, a simplification of the preliminary advice may be considered for future improvement of the procedure.

One of the concerns that arose before the beginning of the experimentation was a lack of diversity among the projects that could be involved. Indeed, the risk of having a large number of similar buildings, like luxurious Parisian office buildings that are able to support the cost and the time dedicated to the procedure, was high. As shown in figure 3, this was not the case as the 17 projects currently involved come from 10 of the 13 French regions, and one comes even from Tunis (Tunisia).

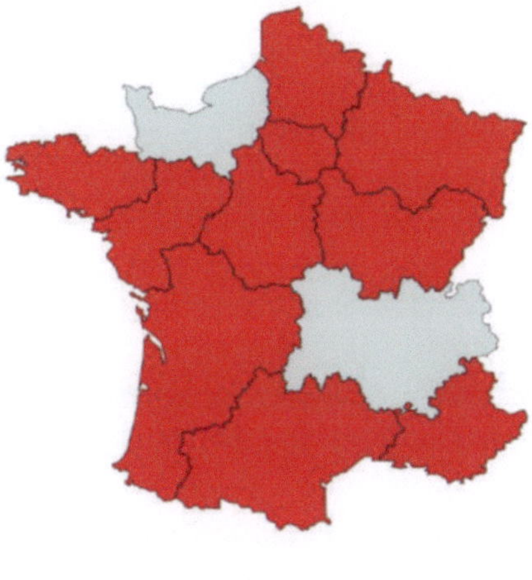

**Figure 3: French regions (shown in red) in which at least one project is involved in the experimentation**

In addition, all types of building are involved in the experimentation: individual houses, apartment buildings and buildings from the tertiary sector. As shown in Figure 4, a broad diversity of projects is also observed regarding the construction period: most of the buildings are from the last three centuries, but two of them are from the 16-17th century and two of them from the middle-age.

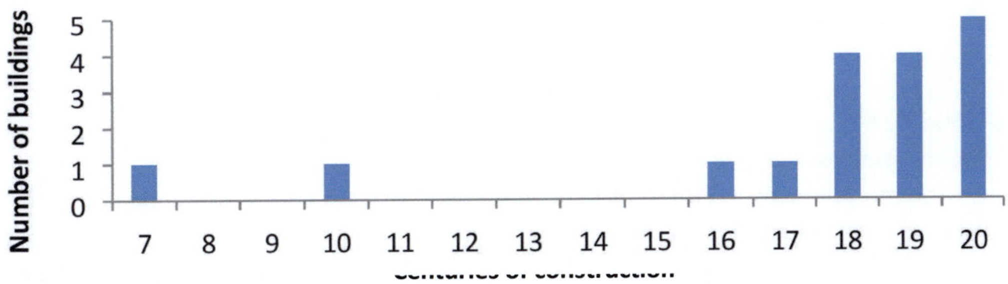


**Figure 4: Distribution of the number of projects as a function of the construction period**

## 6. CONCLUSION

This experimentation is the result of a work started in 2012 and now developed by Effinergie with the support of multiple organisms such as Cerema. Started in early 2020, it now involves 17 projects in a certification procedure and targets a total number of 40 buildings.

A large diversity has been observed among the participating projects, regarding their regions, construction period, and types. However, the time scale of the building sector represents real challenges for this experimentation that has to be conducted in only two years. An efficient communication should help reach the label objectives. Other issues are the cost of the certification and the time required by the project owner to produce the documents for the preliminary and the consolidated advice.

The aim of this experimentation is to demonstrate that heritage preservation is possible when retrofitting a heritage building. To do so, the projects labelled will be referenced in two database providing feedbacks on both energy (the "BBC" observatory) and patrimonial aspects (the CREBA buildings website) and a best practice guide will be published after the end of the experimentation. Finally, this experimentation will allow Effinergie to prepare a lasting label with an improved certification procedure.

## 7. REFERENCES


[1] Ministère de l'Écologie et du Développement durable. "Loi n° 2015-992 du 17 août 2015 relative à la transition énergétique pour la croissance verte." Internet: https://www.legifrance.gouv.fr/loda/id/JORFTEXT000031044385/2020-11-16/, Aug. 18, 2015 [Nov. 16, 2020].

[2] Effinergie. "Le label Effinergie Rénovation." Internet: https://www.effinergie.org/web/les-labelseffinergie/effinergie-renovation, Nov. 8, 2011 [Nov. 16, 2020].

[3] J. Borderon, E. Héberlé, A. Cuny and J. Burgholzer. "Investigation of post-insulated walls with wooden beam ends: Risk analysis for different insulation techniques." Conference report of the 3rd International Conference on Energy Efficiency in Historic Buildings, 2018, 120-126

[4] R. S. Okutan, T. Kershaw, M. Herrera Fernandez, D. Coley. "A socio-mathematical approach to exploring conflicts between energy retrofit and perceived heritage character." Building and Environment, vol 138, 1120, 2018.

[5] CREBA. "Centre pour la réhabilitation responsable du bâti ancien." Internet: www.creba.fr, Oct. 2018 [Nov. 16, 2020].

[6] E. Héberlé, J. Borderon and J. Burgholzer. "Guidance for Finding a Sustainable Balance between Energy Savings and Heritage Preservation When Retrofitting Heritage Buildings." Restoration of Buildings and Monuments 2017.

[7] European Committee for Standardization. "Guidelines for improving energy efficiency of architecturally, culturally or historically valuable buildings." EN 16883:2017.

[8] Effinergie. "Observatoire BBC. Patrimoine." Internet: https://www.observatoirebbc.org/patrimoine, Jan. 8, 2020 [Nov. 16, 2020].

[9] Société pour la Protection des Paysages et de l'Esthétique de la France. "Nouveau décret isolation par l'extérieur: bilan de notre lobbying." Internet: http://www.sppef.fr/2017/05/11/nouveau-decretisolation-par-lexterieur-bilan-de-notre-lobbying/, May 11, 2017 [Nov. 16, 2020].

[10] Effinergie. "Le label Effinergie Patrimoine." Internet: https://www.effinergie.org/web/les-labelseffinergie/le-label-effinergie-patrimoine, Sept. 8, 2020 [Nov. 16, 2020].

[11] CREBA. "Charte de réhabilitation responsable du bâti ancien." Internet: http://www.rehabilitation-batiancien.fr/fr/a-propos/charte-de-rehabilitation-responsable-du-bati-ancien, Sept. 13, 2018 [Nov. 16, 2020].

# Retrofit plan for energy performance improvement of masonry historic buildings on campus: cultural heritage building cases

Hyun Mi Cho[1], Hyeonseong Yuk[2], Ho Hyeon Jo[2], Beom Yeol Yun[2], Sungwoong Yang[2], Jihee Nam[2] and Sumin Kim[2]*

[1]Department of Building Research, Korea Institute of Civil Engineering and Building Technology, Goyang 10223, Republic of Korea.

[2]Department of Architecture and Architectural Engineering, Yonsei University, Seoul 03722, Republic of Korea.

**Abstract –** In recent years, the energy retrofit of historic buildings has been considered as a strategy to reduce carbon dioxide emissions and building energy consumption. Although retrofit projects and studies of historic buildings are being actively conducted in Europe, the study of historic buildings in Asian countries including Korea is insufficient. Therefore, in order to promote the research of historic buildings and to reduce the building energy consumption in Korea, this study analyzed the energy performance of four representative 20th century masonry historic buildings on the campus of Yonsei University using building simulation. In addition, the Energy Efficiency Measure (EEMs) package of each building is applied, and the optimal retrofit solution is proposed. As a result, the energy performance of the building was improved after applying the optimal energy retrofit solution, and the energy consumption of the building was reduced by up to 60%.

**Keywords** – Historic building; Energy performance; Energy retrofits; Building simulation

## 1. INSTRUCTIONS

Over the past several decades, global population growth and rapid economic development has led to significant increases in energy consumption, especially with regard to buildings [1]. To reduce greenhouse gas emissions and building energy consumption, building materials and energy technologies must be optimized [2]. In contrast to building reconstruction, retrofitting is an efficient method of improving the energy performance of a building [3]. Because historical buildings are typically characterized by relatively low-performance building and energy systems, energy retrofits are highly beneficial for older historical buildings [4]. Common energy retrofits include adding walls or insulating roofs, renovating windows, upgrading heating systems, installing ventilation and air conditioning (HVAC) systems, and optimizing system operation schedules. Further energy retrofit actions include lighting improvements such as lamp replacement and the use of lighting control systems, the introduction of solar energy systems, improvements to mechanical equipment, and use of renewable energy. There are many retrofit technologies, but not every modern technology can be applied to historical buildings. Historical buildings must not be damaged, especially in the case of cultural buildings. Thus, an alternative to retrofitting must be considered for historical buildings [5].

This study uses a building simulation program to analyze the energy consumption of a cultural building that is a representative educational historical building in Korea. An energy-saving retrofitting methodology for the continued use and conservation of cultural buildings is presented for historical buildings that have low energy performance. In many previous studies, energy retrofitting was used as a strategy to reduce the energy consumed by historical buildings [6], [7]. Given this background, the purpose of this study is to select an educational historical building from among cultural buildings that consume a great amount of energy and to suggest an energy retrofit method suitable for historical buildings based on in-situ measurements and a simulation analysis. The proposed methodology combines energy modelling and a dynamic simulation of a real historical university building.

## 2. MATERIALS AND METHODS

### 2.1 APPLIED METHODOLOGY

First, energy modelling of the historical building was performed using DesignBuilder, the EnergyPlus-based simulation program [8], [9], by taking into consideration the actual drawing and in-situ measurement data. Before executing the energy retrofit, a detailed energy modelling of the building was performed using the measured thermal transmittance of the external wall and temperature and relative-humidity data to determine the annual heating and cooling energy consumption. The heating and cooling energy consumption derived from the simulation was calibrated based on the actual energy usage bill. The energy technologies that were appropriate for the purpose were then packaged based on an analysis of the in-situ measurement data and calibrated-simulation energy use. Energy technology packages were compared and analysed using an energy simulation program. Finally, the optimal package solution was derived, and the energy cost savings of the optimal package solution were analysed.

### 2.2 STUDIED BUILDING

The selected building was Underwood Hall, a five-story masonry university building situated in Seoul, Korea (Fig. 1). This building was built in 1924 to commemorate Dr. Horace G. Underwood, the founder and first principal of Yonsei University. The building has stone walls as the external façade, reinforced concrete slabs as floors, and a lightweight wood construction as the roof. The external wall is composed of a combination of mica schist stone obtained from the ground and plain concrete and is decorated using granite. The windows have wooden frames with 3-mm clear glazing and internal blinds.

**Figure 1. Underwood Hall of Yonsei University, Seoul, Korea**

The building was designed with the intention of using steam heating at the time of the first plan; however, the steam heating system was removed, and an air conditioning system (electric heat pump: EHP) is now used. The building area is 2,894.84m$^2$, and it currently has offices such as the president's office and meeting rooms. Underwood Hall is at the centre of the Yonsei University campus and is a representative masonry educational building of the early 20th century. This building is valuable because it is an example of western-style architecture and was built in the early 1900s in Korea. Underwood Hall also has great historical and architectural value because it retains its original shape (function and form) and is well-preserved. The main characteristics of the building are summarized in Table 1.

**Table 1. Characteristics of studied building (Underwood Hall)**

| | | |
|---|---|---|
| *Location* | Seoul, Korea N 37.566333, E 126.93875 | |
| *Usage* | Educational facilities | |
| *Completion year* | 1924 | |
| *Construction* | Masonry(envelope), reinforced concrete(slab), wooden structure(roof) | |
| *Window system* | 3-mm clear glazing + wooden frame<br>Window-to-wall ratio = 35% for all walls | |
| *HVAC system* | EHP (Electric heat pump) | |
| | Heating temperature (°C) | Cooling temperature (°C) |
| | 20 | 25 |
| *Building size* | Number of floors: 1 basement level and 4 ground levels | |
| | Building area/ Total floor area: 694.67 m$^2$ / 2,894.84 m$^2$ | |
| *Occupancy (people/ m$^2$)* | 0.110 | |
| *Operating schedule* | Weekdays 9:00–17:00 | |

## 2.3 ASSESSMENT OF BUILDING

An assessment of the heat flow between the building envelope and the outside was performed using infrared (IR) thermography. IR thermography is widely used for quantitatively evaluating building diagnostics such as the evaluation of thermal properties of building envelopes and the heat exchange or the detection of excessive heat-loss zones, air leakages, and missing or damaged thermal insulation in the building elements [10]. As shown in Fig. 2, the surface temperatures of the external wall (0.4 °C) and window (–6.3 °C) are higher than the outdoor temperature (0°C), and as it is an old building without insulation, heat loss occurs through the external walls. The IR thermography image indicates that significant amounts of heat and air are being transferred through the walls and windows, respectively. Air leakages, infiltrations, heat bridges, and heat energy outflow from the external walls and windows increase the heating and cooling loads, and they are a major cause of the thermal discomfort of occupants, especially during the winter. These results were considered in the energy retrofit process.

The indoor air temperature and relative humidity were monitored in the fall (from September 25, 2019 to October 31, 2019) and winter (from November 1, 2019 to December 16, 2019). The indoor air temperature and relative humidity were automatically recorded with a 1-h sampling time using a TESTO 174H data logger. From the monitoring, it was observed that the average temperatures in the offices, meeting room, and staircase were 20.9 ℃, 20.2 ℃, and 17.6 ℃, respectively. As shown in figure 2, O-Office and U-Office, capital letters O and U represent the building use schedule (with or without occupants), and mean Occupied and Unoccupied, respectively. In the case of the air-conditioned space of the offices and meeting rooms, the average temperature was higher than in the non-air-conditioned space of the staircase. Furthermore, considering that the office is used daily and the meeting room is used occasionally, the daily temperature difference in the office is greater. This is because the density of the occupancy in the office is higher than those in the meeting rooms (fig. 2) with respect to the indoor relative humidity, the offices, meeting rooms, and staircases exhibited similar trends, and the indoor relative humidity in winter was lower than that in the fall owing to the cold and dry winter climate. The average relative humidities in the office and meeting room were 40.6% and 43.9%, respectively. In addition, it was seasonally confirmed that the average outdoor temperatures measured in the fall and winter seasons were 10.7 ℃ and 0.4 ℃, respectively. As the outdoor temperature decreases, the daily indoor temperature difference in the office increases; and in the case of the staircase, which is a non-air-conditioned space, the indoor temperature decreases owing to the influence of the outdoor temperature.

The thermal transmittance of the vertical envelope was assessed based on the continuous measurement of the heat flux, surface temperature, and air temperature (according to ISO 9869-1: 2014) [11]. In-situ measurements were performed using a multifunction measuring instrument, TESTO 435, for 72 h, and the average thermal transmittance rate of wall thus obtained was 1.344 W/m²K. The thermal transmittance of the window was measured using the heat flow meter (HFM) method; the measured value was 1.79 W/m²K. Although the glass itself is of a single size, the actual building was renovated like double glass, and the thermal transmittance value showed better performance than single glass.

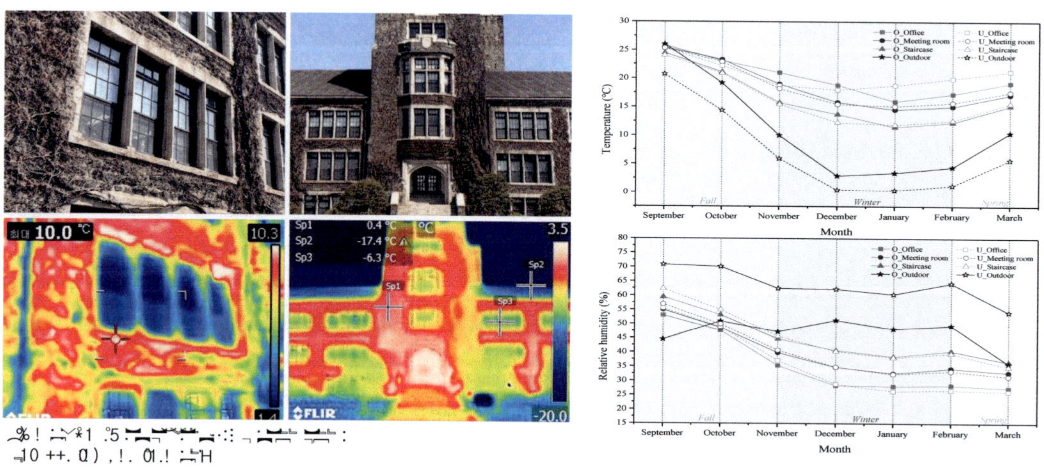


**Figure 2. Assessment of IR thermography images and monitoring data of indoor temperature and relative humidity**

## 2.4 ENERGY MODELING AND CALIBRATION

To analyze the energy performance of Underwood Hall, the DesignBuilder simulation program was used to model the building and determine the monthly energy consumption. The parameters entered into the DesignBuilder simulation program include the occupancy density, heating and cooling setpoint temperature, building facade composition, type and schedule of lighting, and HVAC system. Input parameters were calibrated through building user surveys and field measurements. The building operation schedule, lighting schedule, density of occupancy, and heating and cooling setpoint temperature were calibrated and entered into the simulation. The building is a historical masonry building, and the window-to-wall ratio was set as 35% in the modeling of the windows. All windows have 3-mm clear glass in a wooden frame, and the HVAC system is an EHP system. The external façade has a U-value of 1.344 W/m²K.

To calibrate the simulated energy consumption, the actual energy consumption for 2 y was obtained from the energy usage bill. Because the EHP was installed as the HVAC system, electrical energy was used for the cooling and heating of Underwood Hall. The actual energy consumption in 2018 was less than in 2017, but the cooling energy consumption in summer was higher in 2018 than in 2017. The average of the actual building energy consumption over 2 y was 411,076 kWh, and the simulation data provided an annual building energy consumption of 410,959 kWh. The average error rate between the actual energy usage and simulation value was 10.47%. The actual energy consumed per unit area was 151 kWh/m²y, and the heating energy consumption was higher than the cooling energy consumption. It was determined that the actual heating and cooling energy consumption and the values obtained from the simulation were sufficiently calibrated, and EEM(Energy Efficiency Measures) packages were applied to the simulation data to analyze the energy-consumption reduction effect (fig. 3).

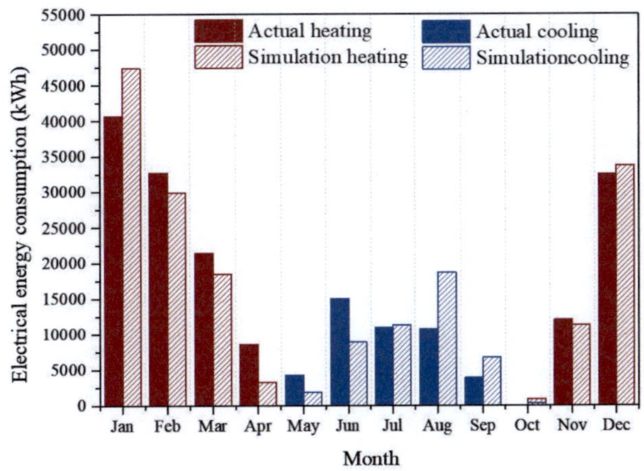


**Figure 3. Heating and cooling energy consumption analysis based on energy consumption obtained from bills**

## 3. ENERGY RETROFITTING OF HISTORICAL BUILDING

The main challenge in retrofitting a historic building is to improve its performance while maintaining its appearance. The insulation of the envelope cannot be added in a simple and easy way like in a normal building, and the window changes are also different. In this work, the previously studied EU project Robust Internal Thermal Insulation of Historic Buildings (RIBuild) was referenced as a method for energy retrofitting. The EU research project RIBuild provides guidelines on how to install internal thermal insulation in historic buildings while maintaining their architectural and cultural heritage. The purpose is to reduce energy consumption in historic buildings in order to meet the EU 2020 climate and energy targets [12].

### 3.1 EEM PACKAGES

The EEM package consists of a combination of passive, active, and renewable energy technologies. Each technology applied in the EEM is presented in Table 2. Commonly used technologies were selected to derive the optimal solution for building energy retrofit. Three analysis methods were used to examine the suitability of each technology. The three analysis methods are: ( i ) Analysis of energy savings of each technology; ( ii ) Impact on the historical value of buildings when each technology is applied; ( iii ) Economics of each technology. The suitability of each technique was examined by combining the results of these three analysis methods. This paper proposes the optimal energy retrofit technology package solution based on the suitability results of each technology. The passive technologies include the application of 200 mm of extruded polystyrene (XPS) insulation to the interior of the roof and walls, changing the window system, improving the airtightness performance, and installing internal blinds and exterior overhangs. The airtightness performance was improved from 2.6 ACH (Air change per hour) to 2.0 ACH to realize improved energy performance. In general, historic buildings have poorer airtightness performance than modern buildings. Improvements were made to the window system by replacing the existing 3-mm clear glazing and wood frames with low-e double glazing and low-e triple glazing with a polyvinyl chloride (PVC) frame to improve the heat transmission rate to 1.499 W/m²K and 0.786 W/m²K. Active technology was used to replace 100% of the LED lights, a high-efficiency HVAC system was installed, and the renewable technology applied uses PV panels with an efficiency of 19%.

**Table 2. EEM packages**

| | *EEM packages* | *EEM1* | *EEM2* | *EEM3* | *EEM4* | *EEM5* | *EEM6* |
|---|---|---|---|---|---|---|---|
| *PASSIVE* | *Roof insulation U-value 0.15* | | *O* | *O* | | | |
| | *Wall insulation U-value 0.17* | | *O* | *O* | | | |
| | *Window system Low-e Double* | | *O* | *O* | | | |
| | *Internal blinds* | *O* | *O* | *O* | *O* | *O* | *O* |
| | *Infiltration 2.6 ACH →2.0 ACH* | *O* | *O* | *O* | *O* | *O* | *O* |
| *ACTIVE* | *LED lights* | *O* | | *O* | | *O* | *O* |
| | *High-efficiency HVAC system* | *O* | | *O* | | *O* | *O* |
| *RENEWABLE* | *Photovoltaic panel (PV)* | | | | | | *O* |
| *IMPROVEMENT* | *Window system Low-e Triple* | | | | *O* | *O* | *O* |
| | *Exterior overhang* | | | | *O* | *O* | *O* |
| | *20% improvement in roof performance* | | | | *O* | *O* | *O* |
| | *20% improvement in wall performance* | | | | *O* | *O* | *O* |

## 4. RESULTS AND DISCUSSION

### 4.1 RESULTS OF EEM PACKAGES

This section presents an analysis of the energy savings for EEM packages. The reason for the difference between passive technology and active technology is to distinguish and analyze the scenario in which the building itself can reduce energy loss (passive technology) and the scenario in which energy can be recycled and effective energy savings can be realized (active technology). In the case of the current condition of the building in use after reinforcement in 2010, internal blinds were installed on the interiors of the windows, and some LED lights were also installed. In the case of EEM1, which shows the current state of the building, 53.5 kWh/m²y per unit area, 17.6 kWh/m²y per unit area, and 47.7 kWh/m²y were consumed as heating energy, cooling energy, and lighting energy, respectively. This resulted in a total energy consumption of 118.9 kWh/m²y per unit area per year. For the EEM2 and EEM4 packages having passive technology only, the total annual energy consumption was 90.2 kWh/m2y and 88.1 kWh/m²y, respectively (Table 3).In the case of EEM4, methods such as those of EEM2 were applied; however, in order to improve the envelope performance by 20%, window glazing was applied (low-e triple glazing instead of low-e double glazing) and exterior overhangs were installed. In addition, the thermal transmittance of the wall insulation was improved by 20%. Between EEM2 and EEM4, the difference in energy savings was insignificant even when the envelope performance was improved by 20%. However, the cooling energy-consumption savings of EEM4 was slightly higher. The energy savings realized with EEM3, EEM5, and EEM6, which combined passive and active technologies, was greater than 60%. The energy-saving effect was much higher than that of packages using only passive technologies, and these packages showed the maximum energy-consumption reduction effect. In the case of active technologies, 100% of the lights were replaced with LED lights, and a high-efficiency HVAC system was installed, with EEM3 and EEM5 providing 61.2% and 62% reductions in the total energy consumption, respectively. The cooling energy consumption in the case of the technology packages with only passive technology (EEM2 and EEM4) was higher than that of EEM1, but the application of active technology (EEM3 and EEM5) could reduce the cooling energy consumption by more than 60%. EEM6 used renewable energy technology and involved the application of a PV panel, and the total energy consumption per unit area was 41.9 kWh/m2y per unit area owing to the effect of the energy production, which was the maximum energy savings. In addition, the EEM package was able to reduce the heating and cooling loads. In a manner similar to the reduction of the energy consumption, the heating and cooling loads were reduced. Using only passive technology, EEM2 and EEM4 reduced the heating and cooling loads by 30% and 34%, respectively. EEM3, EEM5, and EEM6 used active technology and could reduce the heating and cooling loads by more than 45%. Even with active technology, the heating and cooling loads were reduced. A comparative analysis of the energy consumption of the EEM packages suggests that passive and active technologies can be applied together to achieve optimal energy performance retrofitting (fig. 4).

**Table 3. Energy savings results obtained when applying EEM packages**

| *Energy savings (%)* | *Heating* | *Cooling* | *Lighting* | *Total* |
|---|---|---|---|---|
| *EEM 2* | *51.8* | *-6.4* | *4.3* | *24.1* |
| *EEM 3* | *69.0* | *62.0* | *52.2* | *61.2* |
| *EEM 4* | *52.8* | *0.5* | *5.1* | *25.9* |
| *EEM 5* | *69.4* | *65.0* | *52.5* | *62.0* |
| *EEM 6* | *69.4* | *65.0* | *52.5* | *64.8* |

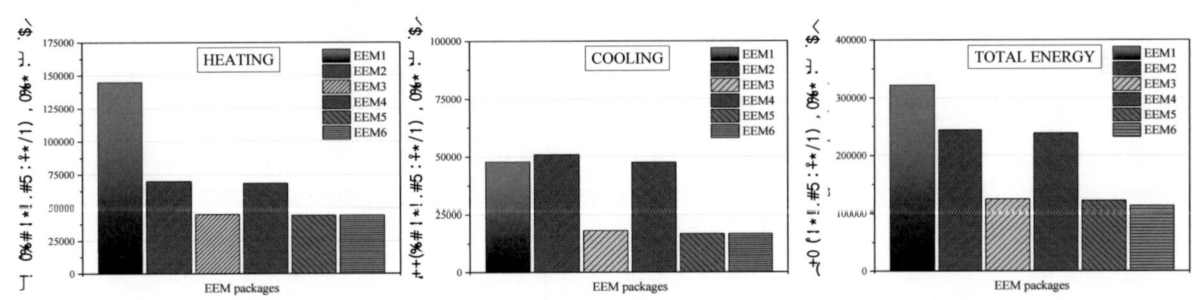


**Figure 4. Heating, cooling, and total energy consumption results obtained after applying EEM packages**

## 4.2 OPTIMAL ALTERNATIVE SOLUTION AND ENERGY COST

When EEM3 was applied, the passive technology and active technology were combined to show the optimal energy saving effect, and EEM3 was selected as the optimal scenario considering the cost used for the technology and the degree of damage to the historical value. In other words, EEM3 was selected as the optimal scenario considering three factors: energy savings, cost applied to technology, and impact on historical value. In consideration of the conservation of historical value, utilization of buildings, and the construction of technology, EEM3 was selected as the best retrofit solution, and an energy consumption analysis was conducted by comparing the results of EEM3 and the actual energy consumption obtained from the energy consumption bill. As for the monthly savings, Korea has a distinct climatic zone with four distinct seasons, so in general, the monthly savings in winter and summer are considered important in Korea. After applying EEM3, the heating energy consumption was reduced by an average of 54.2% from December to February in winter, and the cooling energy consumption was reduced by an average of 42.6% from June to August in summer (fig. 5). The application of EEM3 reduced the total annual energy consumption by 47.9% in current buildings. Figure 14shows the cost of energy consumed per month. Underwood Hall consumed approximately USD$41,280 per year before the application of EEM3, but the energy cost after applying EEM3 was approximately USD$21,460 per year, a savings of approximately USD$19,790 per year.

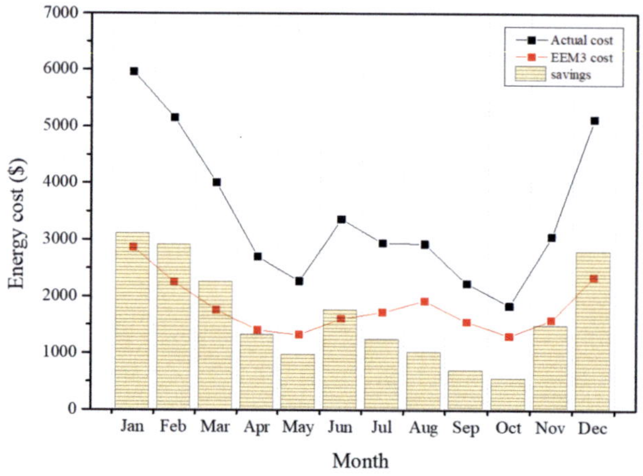


**Figure 5. Actual energy cost reduction realized after application of EEM3**

## 5. CONCLUSIONS

Energy retrofits have the potential to reduce building energy consumption and carbon emissions. There exist difficulties in the implementation of energy retrofits in historical and traditional buildings. Retrofitting these buildings is a complex task where many criteria are balanced against each other when attempting to realize the continued use of the building. Thus, historic cultural buildings must be retrofitted in such a manner that the building remains undamaged. In this study, to improve the energy performance of a historical educational building, the EEM package, which consists of a combination of passive, active, and renewable-energy technologies was applied. The energy consumption analysis showed that the EEM3 package, focusing on the preservation of historical value and building usability and improving the energy savings, reduced the total energy consumption by 61.2% and 45% for the heating and cooling loads, respectively. The application of EEM6 resulted in a 7% reduction in total energy consumption via the application of PV panels, which have high energy productivity. This study demonstrated that EEM3 is the optimal solution for realizing energy saving retrofits of historical buildings while preserving their historical value and usability. The application of EEM3 resulted in a reduced average energy consumption in Underwood Hall by 54.2% in winter and 42.6% in summer. The application of EEM3 also reduced the annual energy costs by 47.9%.

## 6. ACKNOWLEDGEMENT

This work was supported by the National Research Foundation of Korea(NRF) grant funded by the Korea government(MSIT)(No. 2021R1A4A1032306)

## 7. REFERENCES


[1] S. Wi, S. G. Jeong, S. J. Chang, J. Lee, and S. Kim, “Evaluation of energy efficient hybrid hollow plaster panel using phase change material/xGnP composites,” Appl. Energy, vol. 205, no. August, pp. 1548–1559, 2017, doi: 10.1016/j.apenergy.2017.08.156.

[2] G. Bin and P. Parker, “Measuring buildings for sustainability: Comparing the initial and retrofit ecological footprint of a century home - The REEP House,” Appl. Energy, vol. 93, pp. 24–32, 2012, doi: 10.1016/j.apenergy.2011.05.055.

[3] S. Nagpal, J. Hanson, and C. Reinhart, “A framework for using calibrated campus-wide building energy models for continuous planning and greenhouse gas emissions reduction tracking,” Appl. Energy, vol. 241, no. August 2018, pp. 82–97, 2019, doi: 10.1016/j.apenergy.2019.03.010.

[4] F. Ascione, N. Cheche, R. F. De Masi, F. Minichiello, and G. P. Vanoli, “Design the refurbishment of historic buildings with the cost-optimal methodology: The case study of a XV century Italian building,” Energy Build., vol. 99, pp. 162–176, 2015, doi: 10.1016/j.enbuild.2015.04.027.

[5] A. L. Webb, “Energy retrofits in historic and traditional buildings: A review of problems and methods,” Renew. Sustain. Energy Rev., vol. 77, pp. 748–759, Sep. 2017, doi: 10.1016/j.rser.2017.01.145.

[6] L. Bellia, F. R. D’Ambrosio Alfano, J. Giordano, E. Ianniello, and G. Riccio, “Energy requalification of a historical building: A case study,” Energy Build., vol. 95, pp. 184–189, 2015, doi: 10.1016/j.enbuild.2014.10.060.

[7] F. Wang, K. Pichetwattana, R. Hendry, and R. Galbraith, “Thermal performance of a gallery and refurbishment solutions,” Energy Build., vol. 71, pp. 38–52, Mar. 2014, doi: 10.1016/j.enbuild.2013.11.059.

[8] “DesignBuilder EnergyPlus Simulation Documentation for DesignBuilder,” 2019. http://www.designbuilder.co.uk/.

[9] A. Tindale, “DesignBuilder software,” Des. Softw. Ltd, 2005.

[10] A. Kylili, P. A. Fokaides, P. Christou, and S. A. Kalogirou, “Infrared thermography (IRT) applications for building diagnostics: A review,” Appl. Energy, vol. 134, pp. 531–549, 2014, doi: 10.1016/j.apenergy.2014.08.005.

[11] ISO 9869, Thermal insulation — Building elements — In-situ measurement of thermal resistance and thermal transmittance — Part 1: Heat flow meter method. 2014.

[12] E. J. de P. H. Andra Blumberga, Kristaps Kašs, Edīte Kamendere, Gatis Žogla, Agris Kamenders, Dagnija Blumberga, Armands Grāvelsiņš, Reinis Purviņš, Marika Rošā, Lelde Timma, Hans Janssen, Peggy Freudenberg, Fredrik Stahl, Ruut Peuhkuri, Pierryves Padey, Sebastien Lasva, “Robust Internal Thermal Insulation of Historic Buildings_State of the art on historic building insulation materials and retrofit strategies,” 2019. [Online]. Available: https://eur-lex.europa.eu/legal-content/PT/TXT/PDF/?uri=CELEX:32016R0679&from=PT%0Ahttp://eur-lex.europa.eu/LexUriServ/LexUriServ.do?uri=CELEX:52012PC0011:pt:NOT.

# Balancing energy efficiency and conservation aspects of Terragni's Casa del Fascio in Como: thermal analysis, energy modelling and intervention proposals

S. Mauri, V. Pracchi

Dept. of Architecture, Built Environment and Construction Engineering (DABC), Politecnico di Milano, Milan, Italy. (sara.mauri@polimi.it; valeria.pracchi@polimi.it)

**Abstract** – This paper aims to present a methodology for planning energy improvement interventions in modern listed buildings, carried out through a combination of desk research, on-site survey, monitoring campaign and energy modelling. For this purpose, Giuseppe Terragni's Casa del Fascio has been chosen as our case study. The building is characterised by indoor microclimate conditions that are far from the standard requirements: summer is the most critical period because there is no possibility of limiting overheating due to the absence of the original devices designed by Terragni. Based on monitored data and building characteristics, an energy dynamic model was carried out with the aim of simulating the recovery of the original solutions or the addition of low-impact measures. The simulated options were eventually evaluated by taking into account their compatibility with the building features. This resulted in some strategies suitable for Casa del Fascio, that enable to reconcile the conservative goal with that of improving internal comfort.

**Keywords** – Modern Movement Architecture; Energy Efficiency; Restoration; Thermal analysis; Energy modelling.

## 1. INTRODUCTION

In the last few years, the improvement of energy efficiency of historic built heritage has become an ever more important topic within the scientific community. How to balance sustainability and energy efficiency needs, pushed by EU Directives, with the aim of preserving our heritage is still an open question. Given the complexity of this issue, the regulations in force allow the exclusion of listed building from performance requirements [1]. Hence, the focus of the question shifts from the achievement of standard values to a performance improvement which shall be "as much as possible". However, the lack of a shared decision-making method may lead to the risk of not having a complete overview on both preservation and energy efficiency aspects when approaching the project [2]. That being said, the study aims to present a methodology for planning energy improvement measures based on a combination of different knowledge in a whole building approach. Coherent, coordinated and planned activity of study, technical survey and energy modelling are combined here to improve the building efficiency and the users comfort level, with the least impact on its historical material consistency.

The occasion for the research on Casa del Fascio in Como came from a request from the Italian Superintendency of Archaeology, Fine Arts and Landscape of Milan (that promoted a restoration plan for wooden windows in 2005), after becoming aware that a general building microclimate analysis was necessary to draw up an energy-efficiency project.

## 2. METHODOLOGY

The proposed methodology is divided into some distinct and defined phases. It starts from a knowledge phase in which all bibliographical and archival information is collected and the building transformations from its construction to nowadays are identified. This is supported by an on-site survey: as a matter of fact, not all the building changes are documented or traceable, especially the least relevant ones linked to ordinary maintenance or to the users comfort needs (for example replacing construction elements with others that are different in terms of mode of operation, adding new services, etc.). After this phase, the investigation focuses on the building thermo-hygrometric characteristics and on the indoor microclimate conditions, examined through suitable diagnostic equipment. In this way it is possible to define the building energy profile. Finally, starting from microclimatic data and building characteristics it is possible to develop an energy dynamic model. This allows to simulate the building thermal behaviour and, therefore, evaluate the improvement made to the current situation by some low-impact intervention proposals. Choosing the best possible solutions depends on the above-mentioned analysis and the evaluation of their feasibility must take into account the compatibility with the building typological and material features. This method results in strategies able to reconcile the conservative goal with that of improving internal comfort.

## 3. THE CASE STUDY: CASA DEL FASCIO IN COMO

The proposed methodology is applied to Giuseppe Terragni's Casa del Fascio in Como (Italy), one of the historic masterpieces of the Italian Modern Movement[1] [3].

**Figure 1. (a) Outside view of the Casa del Fascio. (b, c) Inside view of the *Salone delle Adunate* and courtyard.**

The building has four floors with a square plan, and its crux is a large central double height hall called *Salone delle Adunate* (Gatherings room). The gallery on the first floor overlooks this huge space and connects the offices. The second floor follows the distribution system of the first floor: the only difference is that the gallery opens onto the roof of the *Salone delle Adunate*, which is the floor of the inner courtyard, characterised by a skylight in concrete-frame glass blocks crossed in the middle by a walkway. On the top floor the building is divided in two parts by two open galleries: reaching this last floor is only possible by using the secondary staircase, while the other two floors

[1] The construction works of Casa del Fascio began in July 1933 and ended in 1936, when it was inaugurated as the local branch of the National Fascist Party. It held this function for a relatively short period, from 1936 to 1945, the year in which the building was forcibly occupied by the Provincial Federations of the National Liberation Committee Parties, after the liberation of the city of Como from the Fascist Regime. Since 1957, Casa del Fascio houses the Command of the VI Legion of the Italian Finance Police, but in February 2017 a petition was launched proposing its re-use for cultural purposes, namely as a museum of rationalism.

are reachable also through the main staircase [4]. The great peculiarity of this building is its construction system: a mix of autarkic and traditional materials in a reinforced concrete structure [5]. The lack of immediate recognition of the monumental value of Casa del Fascio, listed in 1986 only, led to some maintenance actions carried out over the years by unskilled casual labour with the result that the original design intents of Terragni was altered.

### 3.1 DOCUMENTARY RESEARCH AND ON-SITE SURVEY

Detailed desk research and field survey are essential to develop coherent building intervention proposals. First of all, a documentary (bibliographical and archival) research was carried out. Numerous written works were consulted to reconstruct the building history and Terragni's design intents: for this purpose, the special issue of the monthly magazine *Quadrante*, published in 1936 and completely dedicated to the Casa del Fascio, is of particular interest. It includes diagrams of the orientation and exposure to solar radiation of the building drawn by Terragni on the basis of Ernst Neufert's work. This study influenced the building final position[2] and led to the creation of the open galleries of the South-West façade to cope with the outdoor climatic conditions: in the warm season they help prevent direct sunlight exposure of the façade, while in the cold season they favour the entrance of the solar radiation [6]. The other investigation reports on the building only covers the construction and architectural composition of the Casa del Fascio, not taking into account the events that took place in the course of eighty years of its life which have radically changed its material consistency. Fundamental in this regard was the investigation of two archives, whose study was very complex and extended due to the large number of documents stored there: accounting and notarial documents, unpublished photographic material, local newspapers, private and administrative memories etc.[3] [7]. The documentary research was then accompanied by an on-site survey: this has allowed to identify undocumented changes as well as building characteristics that negatively affect the internal comfort. What has emerged is that the main cause of the lack of summer and winter comfort is the high percentage of transparent surfaces (that account for about 40% of the total area) like concrete-frame glass blocks and wooden and metal windows. Wooden windows, in particular, are in a poor state of conservation due to their high technical complexity, dimensions (the most common window has a length of 4.7 m and a height of 2.63 m), weight (about 263 kg) and to the action of atmospheric agents and variations in temperature and humidity [8]. This has led to the deformation of the window frames and, as a consequence, to problems in the opening/closing system: if the window doesn't close completely there is a high loss of heat in winter, while if window doesn't open properly, the correct air exchange in summer is not allowed. Wooden windows

---

[2] The building is oriented along the orthogonal axes "North-West/South-East" and "North-East/South-West" with a rotation of 25 degrees with respect to the north geographic. This orientation leads to the fact that all four façades are affected by direct solar radiation in summer: the two most stressed ones are those in Via dei Partigiani and Piazza del Popolo, but in the early morning and late afternoon the other two façades are also exposed.

[3] Particularly rich and never systematically explored is the *Municipal Archives in Como* which preserves all the documentation issued by the municipal officers on post-war maintenance interventions. The archive of the Italian Superintendency of Archaeology, Fine Arts and Landscape of Milan has also proved to be very useful: it preserves documents, projects, cost estimate and photographs of the latest restoration works, linked to the institutional activity of historical-artistic protection carried out by the institution on the building from 1986 to the present day.

restoration was partially carried out in 2005, when the Superintendency obtained specific funding: the restoration work on a sample window ended in 2006 and concerned the frames' structural stiffening through the insertion of steel tubulars and the replacement of the single glazing with a safety one [8]. In 2016, two other windows were restored, for a total of three refurbishment projects that were completed. Generally speaking, windows restoration would allow not only to recover their movement system, based on a counterweight mechanism, but also to implement the glass U-value with its replacement.

In addition to the replacement of some building components (e.g. all the metal windows, most of the glass blocks, etc.) [9, 10] and changes in materials and colours [11], the documentary research and the on-site analysis revealed that Terragni (in addition to open galleries) designed some experimental solutions to mitigate the indoor microclimate conditions during summer: the roller blinds on the main façade (Piazza del Popolo)[4], the awnings on via Pessina and on the perimeter walls of the inner courtyard[5] and the air conditioning system[6]. However, these devices were prematurely abandoned due to some problems deriving from their technological inadequacy as a result of their high experimental character. All this led to a drastic worsening of the indoor comfort conditions of the Casa del Fascio.

### 3.2 DIAGNOSTIC ANALYSIS AND MICROCLIMATIC DATA

The field survey is followed by a diagnostic analysis that involved the building's internal microclimate and the properties of its building components. The analysis of the indoor thermo-hygrometric conditions of the Casa del Fascio was carried out using six microclimatic probes[7], located inside and outside the building according to the results of the preliminary psychrometric analysis. This was performed inside the building twice during the day, in the morning and in the afternoon, thus allowing to identify the points in which the greatest daily thermal range occurred, i.e. the most critical issues [12]. In the next phase, the microclimatic probes were positioned in these points: in total five internal points and one external point were identified[8]. The monitoring campaign began on

---

[4] They consisted of a piece of fabric connected to a roller (inserted in a special cavity of the open galleries ceiling) through four cables that formed two inverted V. The surface of the fabric did not entirely cover the span but, with the curtain unrolled, a portion of empty space remained on the top. The particular characteristics of these curtains probably made them very sensitive to the wind action and for this reason they lasted for a very short time. The rollers are still in place.

[5] The structure of the awnings is composed of a roller supported by two side arms. Currently only one window of the inner courtyard still has the piece of fabric.

[6] The air conditioning system consists of air vents located on the floor of *Salone delle Adunate*. During the summer period, the air vents used to introduce fresh air into this huge space while the natural action of the stack effect extracted warm air. The offices could also benefit from this system through the door transom opening, that brought about air circulation. Unfortunately, the system worked for a very short time: most of the air vents were sealed and the door transoms remained blocked in the closed position due to their weight.

[7] EL-USB-2, temperature and humidity datalogger with USB provided by Lascar Electronics.

[8] Probe n°1 was placed at the entrance; Probe n°2 inside the *Salone delle Adunate*; Probe n°3 in a room facing North-West with a damaged window, which remains open throughout the day and night; Probe n°4 in

13th December 2014 and ended on 5th May 2016, for a total of approximately one-and-a-half-year. The monitoring phase showed a strongly oscillating behaviour of both temperature and relative humidity which are the result of the influence of the external climate. By looking at the graph of the daily average temperatures it can be seen that the days in which the values fall into the winter and summer comfort zones (provided by the current standard) are very limited [13]. In winter, the heating system[9] is only partially able to guarantee comfort conditions: the air temperature inside the rooms seldom reaches 20°C. The summer period turns out to be the most critical because there is no possibility of limiting overheating. The microclimatic probes measured internal temperatures very close to those external: the highest recorded temperature is around 35°C (in July). The graph of daily average relative humidity confirms a strong sensitivity to external variations during the year, which corresponds to a significant fluctuation of the indoor conditions [7]. Generally speaking, the indoor microclimate conditions of Casa del Fascio are far from complying with the standard requirements and the main cause is the use of lightweight construction techniques, typical of the 20th century architecture: as a matter of fact, large windows and thin walls favour a strong heat loss in winter and a considerable solar gain in summer when the external temperature rises.

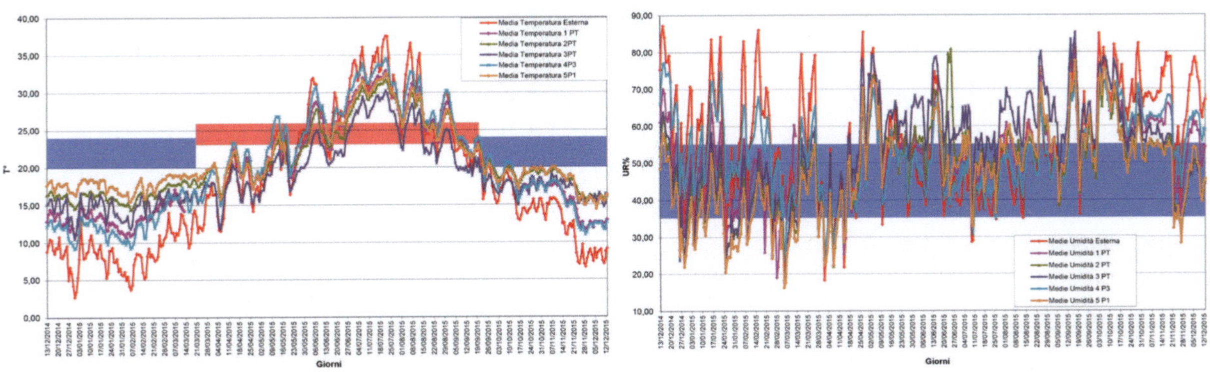


**Figure 2. (a) Graph of daily average temperature. (b) Graph of daily average relative humidity.**

The diagnostic investigation ended with the measurement of the U-value of the North-West and North-East external walls through a heat flow meter[10]. They were carried out according to the procedures provided by the international standard ISO 9869. The final thermal transmittance was then calculated by entering the collected data into a dedicated software (U calc): the U-value of the North-West wall is 1.4 W/m$^2$K while the one of the North-East wall is 1.5 W/m$^2$K [12]. These values were useful for the following energy modelling phase.

### 3.3 DYNAMIC ENERGY MODEL OF THE BUILDING

Starting from the building characteristics and the data monitored, a dynamic energy model was created, in order to simulate the building thermal behaviour. For this purpose, the EnergyPlus simulation engine software was used [14]. The first step was the definition of the model geometry

---

the gallery of the top floor; Probe n°5 in the *Sala del Direttorio* on the first floor and, finally, Probe n°6 was placed externally, on the open gallery of Piazza del Popolo, protected from direct sunlight.

[9] The heating system is characterised by the presence of a centralised boiler, located in the basement. The terminal units are radiators (of different types) positioned in each room. In winter these are accompanied by electric heaters, as reported by the building users.

[10] ThermoZig system, wireless heat flow meter provided by Optivelox.

and the reconstruction of the surrounding context with its shading surfaces. Then, the thermal zones were defined: to avoid any kind of problem, the model was kept as simple as possible with a total of 31 thermal zones, considering the *Salone delle Adunate* as a single unheated area (stairwells and galleries included). After that, a material characterization of the opaque and transparent envelope was done: the comprehension of the exact building components stratigraphy was possible by combining the data from documentary research, direct observation and diagnostic analysis. Two different types of brick masonry, a type of concrete and hollow tiles mixed floor and all the recognised transparent surfaces were included in the energy model. Once it is geometrically completed, it becomes a reliable tool by means of the use of specific parameters: in addition to the on-site measurements, the parameters available are the boiler fuel consumption and the electricity bills. These data were processed to calculate the real primary energy, used as a comparison value in the final phase. But first of all, the energy model must be calibrated and for this reason a schedule for each thermal zone was created by indicating the electrical equipment, the outdoor airflow rate and the heating set-point. In general, the entered data vary over time according to the real use of the rooms. First of all, the internal loads of electrical equipment (number of lights, computers, people, telephones, printers, radio, etc.) were added. Then, the outdoor airflow rate value was calculated based on the volume of the rooms and the air infiltration factor through the windows (as mentioned above, windows are in a poor state of conservation). Finally, two different schedules were created for the heating set-point according to the daily use and type of space, public areas or offices. The temperature values (18°C for public areas and 23°C for offices) were obtained by cross-referencing the data from the microclimatic probes and the psychrometer. The simulation was run with these settings in order to validate the model: what we have obtained is the energy demand, converted into primary energy. In conclusion, the real primary energy (239,088 kWh) was compared with the energy calculated through the model (219,176 kWh) and a deviation of 9% was detected [12].

### 3.4 INTERVENTION PROPOSALS

Thanks to the analyses that were previously described as well as the energy model, it was possible to simulate and then verify the technical compatibility, feasibility and effectiveness of various low impact retrofit interventions, able to combine conservation aspects with those on energy. Interventions have been defined considering both winter and summer periods.

The microclimatic study showed that winter is the least critical period, even if the heating system is not completely able to guarantee acceptable internal comfort conditions according to the current standard. Excluding some very impactful interventions, the restoration of all wooden windows was proposed. Although this intervention is expensive and complex, this would make it possible to safeguard the integrity of a unique and unrepeatable system, while obtaining a 15 percent reduction in heating consumption [12].

As we have seen, summer is the most critical period since there is no possibility of limiting overheating except with the use of existing wooden roller shutters which, however, obstruct the passage of natural light. One of the aims of this study was to insert *ex-novo* internal curtains and to recovery and assess the effectiveness of the external solar shading systems designed by Terragni (roller blinds and awnings), currently no longer in use. From a conservation point of view, their impact on the building is very low because they are removable devices already partially included in the original project. According to these solutions, the addition of interior and exterior high-performance fabrics was proposed: for this purpose, a market survey was carried out. Different types of fabrics were studied by analysing their performance in terms of sun protection, visual permeability, privacy, passage of natural light and ventilation, thickness, weight and fire and wind resistance. The fabrics were then grouped into four categories: coated fabrics and dyed PET fibre fabrics, suitable for both indoor and outdoor use; mixed metal-plastic fabrics for outdoor uses; honeycomb or cellular shades for indoor uses. Finally, the fabrics were compared with each other in order to determine which ones best suit the needs of Casa del Fascio[11] [7]. With regard to the roller blinds, a further fundamental step was the development of a system that enables to recall the compositional features of Terragni's solutions. The most convincing one is a roller, with two side rails, connected with the fabric by means of three stiffening belts. However, this system, being very sensitive to the wind action would require very frequent, complicated and costly maintenance. Our final proposal is therefore a classic roller blind with two side rails and a piece of fabric that, in its open configuration, occupies the entire span up to the banister. In conclusion, the comparison between the dynamic simulation of the external shading systems with the insertion of new internal curtains and the external shading systems with the correct management of the wooden roller shutters has shown that this latter is the most effective scenario as it could lead to a reduction in the internal temperature of min 0.4°C and max 3°C [7]. The microclimate improvement is not enough to ensure acceptable comfort conditions for workers.

### 3.5 TESTING OTHER SOLUTIONS

In view of what emerged in the previous paragraph, the pros and cons of some more invasive interventions were evaluated [7]. A first design solution concerns the shielding of the roof of *Salone delle Adunate* with the insertion of a textile shade sail above the inner courtyard, capable of preventing solar radiation from passing through the skylight. The results show a temperature reduction of 2.5°C in the *Salone delle Adunate* (in July), while the temperature reduction in the other rooms is insignificant. This is therefore an expensive and impactful intervention not justified by the small improvements in terms of thermal comfort. A second solution regards the insulation of the flat roof, i.e. the installation of a reflective insulation material (approximately 4 cm thick) under the original roof finishing layer made of white cement grit tiles. The findings show a benefit in the internal microclimate conditions in winter and a worsening in summer. During the hot season the low thermal inertia of the envelope allows heat to enter during the hottest hours and to come out in the evening and night. The installation of an insulating layer prevents this process causing a greater accumulation of heat during the summer. This is the reason why this intervention was not considered advantageous. The last intervention we took into account is the replacement of the glass blocks with insulating ones. This would make it possible to level out the building appearance while improving the energy performance of the transparent envelope. However, the results of the simulation indicate a situation similar to the flat roof insulation. As a consequence, even this strategy is not feasible. That said, some alternative solutions that could be considered for future assessment are the recovery of the air conditioning systems and the insertion of new high-performance systems.

[11] "Soltis 92" and "Soltis 99" by Serge Ferrari (coated fabrics) were selected respectively for the external roller blinds of the main façade and the internal curtains, while "Tempotest Star FR" by Parà (dyed PET fibre fabric with flame retardant properties) was chosen for the awnings of via Pessina and the inner courtyard.

## 4. CONCLUSIONS

Within the restoration project, the definition of energy efficiency retrofit interventions starts from the recognition of the architectural composition and geometrical studies, archival and bibliographic research combined with on-site survey. All this made it possible to define a picture of the historical changes and stratifications. These results, together with the microclimatic and diagnostic analysis, allowed to create an optimized energy model capable of identifying the thermal performance data and assessing the energy impact of few feasible retrofit options, already verified from the conservation point of view. Hence, a prudent and careful choice of techniques and materials was made, also giving priority to the recovery and enhancement of the existing building elements designed by Terragni. So, a first conclusion regards the importance of considering this aspect inside the typical methodology of a conservation design process. Usually, the improvement of energy efficiency is considered a specialized sector and it is treated separately. A second conclusion, more referred to this experience, shows the difficulties in reaching the current parameters of internal temperature and relative humidity without inserting a new impacting air conditioning system. Should the building turn into a new museum, this part will be necessary, taking into consideration the interaction between systems and interior design. The Superintendency considers this experience and the acquired data as the basis for a dialogue between conservation and energy needs.

## 5. REFERENCES


[1] Directive 2010/31/EU of the European Parliament and of the Council of 19 May 2010 on the energy performance of buildings (recast), 19th May 2010.

[2] A. Buda, V. Pracchi. "Potentialities and criticalities of different retrofit guidelines in their application on different case studies", in The 3rd International Conference on Energy Efficiency in Historic Buildings, 2018, pp. 283-293.

[3] S. Poretti. La Casa del Fascio di Como. Rome: Carocci, 1998.

[4] A. Artioli, Giuseppe Terragni la Casa del Fascio di Como. Guida critica all'edificio: descrizione, vicende storiche, polemiche, recenti restauri. Rome: BetaGamma, 1989.

[5] F. Dal Falco, Stili del Razionalismo. Anatomia di quattordici opere di architettura. Rome: Gangemi, 2002.

[6] G. Terragni, "Tavola ricavata dagli studi di Ernst Neufert sull'insolazione degli edifici", Quadrante. Documentario sulla Casa del Fascio di Como, no.35-36, p. 43, 1936.

[7] F. Busnelli, S. Mauri. "Fragilità di un'«Architettura perfetta». Conoscenza degli aspetti materici e microclimatici della Casa del Fascio di Giuseppe Terragni a Como", M.S. thesis, Dept. ABC, Politecnico di Milano, Milan, Italy, 2016.

[8] A. Artioli et al. "Palazzo Terragni ovvero ex Casa del Fascio di Como (1932-1936) di G. Terragni. Esperienze di lavoro verso una conservazione programmata", in Conservazione programmata per il patrimonio architettonico del XX secolo. A. Canziani, Ed. Milan: Electa, 2009, pp. 161-169.

[9] V. Pracchi et al. Como: ottant'anni di vita della Casa del Fascio di Terragni. Ananke, no.80, pp. 75-81, 2017.

[10] A. Buda, S. Mauri. "Building survey and energy modelling: an innovative restoration project for Casa del Fascio in Como", in International Society for Photogrammetry and Remote Sensing (ISPR), 2nd International Conference of Geomatics and Restoration, XLII-2/W11, May 2019, pp. 331-338.

[11] S. Mauri, V. Pracchi. "The role of colour in Modern Movement architecture. The case of Terragni's Casa del Fascio in Como", in Colour and Colorimetry. Multidisciplinary Contributions, XVIB, 2020, pp. 99-106.

[12] A. Farinella, D. Fedeli, M. Stellini. "Il fascismo è una casa di vetro in cui tutti possono e devono guardare. L'analisi microclimatica della Casa del Fascio di Giuseppe Terragni a Como", M.S. thesis, Dept. ABC, Politecnico di Milano, Milan, Italy, 2015.

[13] Ergonomics of the thermal environment. Analytical determination and interpretation of thermal comfort using calculation of the PMV and PPD indices and local thermal comfort criteria, UNI EN ISO 7730:2006, Feb. 2006.

[14] G.G. Akkurt, N. Aste, J. Borderon et al. (2020, Feb.). "Dynamic thermal and hygrometric simulation of historical buildings: Critical factors and possible solutions". Renewable and Sustainable Energy Reviews. [Online]. 118.

# Traditional and innovative materials and solutions to improve the energy efficiency of historic windows: a literature review

S. Mauri, V. Pracchi

Dept. of Architecture, Built Environment and Construction Engineering (DABC), Politecnico di Milano, Milan, Italy. (sara.mauri@polimi.it; valeria.pracchi@polimi.it)

**Abstract –** During the last decades the improvement of the energy efficiency of historic built heritage has taken on increasing importance: this has led to the production of a great amount of research works within the scientific community. Among the building components, windows are commonly considered the weakest element of the envelope and, therefore, the first to be replaced in historic buildings. Contrary to what one may think, more "sustainable" solutions are possible: there are several strategies that can be applied to enhance windows thermal performance, sustainability and conservation without substituting them. Our goal is to outline the research state-of-the-art in this field through a literature review: to this purpose we collected many publications for a total of 126 documents. The result is a as complete as possible view of the research status on window interventions, with particular attention on problems and future perspectives of the high-performance materials integration in the historical context.

**Keywords** – Historic windows; Energy efficiency; Literature review; High-performance materials; New technological solutions

## 1. INTRODUCTION

Over the last decades, the enhancement of energy efficiency for historic heritage has been a widely discussed topic among the scientific community: this has led to the production of an ever-increasing amount of research works. Considering that approximately 30% of the European building stock consists of historic buildings [1], whether listed or not, any energy management and performance improvement in those buildings may lead to a significant reduction in the global energy consumption and greenhouse gas emission. As a consequence, the historic buildings' energy requalification is undergoing a strong acceleration. Unfortunately, it is widely believed that historic buildings are not energy-efficient and therefore need to be "radically upgraded". Actually, the energy performance of most historic buildings can be improved, but it is essential to find alternative and compatible energy retrofit approaches that harmonize energy efficiency needs, sustainability and conservation principles. Hence, improving the thermal performance of historic buildings is something that must be done with great care. Among the interventions on the building envelope, the solutions move from recovering or replacing parts of the building to adding high-performance elements to existing components. In this research, in particular, we want to focus the attention on windows. Historically windows configuration is linked – as for shape, size and type – to climate factors, compositional requirements and building constructive structure. Recognizing the importance of these architectural elements and the contribution they provide to the building is the first step toward deciding the proper line of action. Windows are an irreplaceable resource and should be preserved and repaired as much as possible. However, windows are commonly considered the weakest element of the envelope and, therefore, the first to be replaced in historic buildings in the name of a significant energy saving which is actually minimal compared to other interventions. As a matter of fact, in the document published by CRESME (Centre for Economic, Sociological and Market Research for Construction Industry and Environment) entitled *"Analysis of the socio-economic impact of 55% tax deductions for upgrading the energy efficiency of existing buildings"*, the average annual savings achieved by type of intervention shows that the replacement of windows has the lowest saving equal to 2.6 MWh, quantifiable between 80 and 125€ with payback achieved in about 12-25 years [2]. Unfortunately, windows replacement is a widespread practice, also promoted by tax incentives, and has caused the loss of a large number of traditional windows, especially in minor historical centres. For this reason, the Italian Ministry of Economic Development and ENEA (National Agency for New Technologies, Energy and Sustainable Economic Development) have proposed a revision of the incentive's mechanism because *"it is not appropriate to demand further performance of U-values at our latitudes with risk of false or useless benefits, without paying attention to walls, floors and roofs as well"* [3]. Moreover, from the environmental point of view, the replacement of the original windows by new ones leads to a 7-fold increase in $CO_2$ emissions into the atmosphere, due to the whole production cycle of the new building components and to the disposal of those removed [4]. Contrary to what one may think, more "sustainable" solutions are possible: there are several strategies that can be applied to enhance windows thermal performance, sustainability and conservation without having to replace them or negatively affecting the building. This paper aims to provide a as complete as possible overview on energy retrofit interventions of historic windows, with a careful assessment of the wide range of possible solutions and a look at the problems and future prospects of the integration of new technological solutions and high-performance materials in the historical context.

## 2. LITERATURE REVIEW METHODOLOGY

In order to outline the research status on window interventions, a literature review was carried out: studies on windows energy retrofit strategies and on the development and application of new technological solutions, also with the use of high-performance materials, have been collected. The search for articles published in scientific journals and conference proceedings took place through three electronic databases of peer-reviewed literature (Scopus, Web of Science and Google Scholar). After analysing all these publications, 63 were picked. To these should be added guidance instruments such as handbooks, guidance, booklets and so on published on the websites of associations as well as governmental and non-profit organizations from all over the world. They are addressed to architects, building contractors, owners and users with the intention to explain the most appropriate conservation practices, heritage management and energy improvement solutions for historic buildings. Finally, there are many Research Institutes and Centres that have developed long-term research programs and projects which significantly contributed to the production of study reports. In addition to the previous documents, other 43 handbooks, booklets etc. and 20 study reports were taken into consideration, for a total of 126 documents. The literature review process was then summarized in three research fields: retrofit solutions; new technological experimentation and perspectives; high-performance materials and solutions applied to existing buildings.

### 2.1 RETROFIT SOLUTIONS

The starting point of the literature review is a research carried out a few years ago in which 21 study reports, published in Europe and America, were examined in order to understand how the interventions on windows in historic buildings are dealt with in the current debate [5]. These reports take into account different parameters to evaluate the performance of historical windows: transmittance is the most recurrent value (it is present in 13 reports), but parameters such as air leakage, saved energy, costs, payback and LCA are also reported. What has emerged from the comparison of results is that although single-glazed windows have a U-value of about 5,18 W/m$^2$K, significant performance improvements can be achieved through a series of measures that gradually balance energy performance with preservation of historical material consistency, without having to replace the original window.

| | 2 English Heritage | 3 Historic Scotland | 4 Ireland | 5 Norwegian Institute | 6 Grand Poitiers | 7 Berner Fachhochschule | 8 ReHab | 13 US Department | 15 Iowa | 16 Vermont | 17 Victoria | 18 Winsconsin | 20 NCPTT | Average [W/m2K] |
|---|---|---|---|---|---|---|---|---|---|---|---|---|---|---|
| Unimproved single glazing | 4,3 | 4,5 | 5,4 | 4,7 | 4,4 | 4,5 | 4,9 | 4,9 | 6,2 | 5,2 | 6,8 | 5,6 | 5,9 | 5,18 |
| Heavy curtains | 2,5 | 3,2 | 3,2 | | | | | | | | | | 3,2 | 3,03 |
| Closing shutters | 1,8 | 2,2 | 2,2 | | | | | | | | | | | 2,07 |
| Insulated shutters | 1,7 | | | | | | | | | | | | | 1,70 |
| Internal window film | 1,9-2,3 | | | | | | | | | | 3,2 | 2,8 | 3,1 | 2,66 |
| Secondary glazing system | 1,8 | 1,7 | 1,7 | | 2,5 | 2,6 | | | 2,8 | | 3,6 | | 2,3 | 2,38 |
| Secondary glazing and shutters | 1,6 | 1,1 | 1,1 | | | | | | | | | | | 1,27 |
| Insulating glazing | | | | | | | 1,4 (vi) | | | | | | | 1,4 (vi) |
| Double glazing | | 1,9 | 1,9 | 2,8 | 1,4 | 1,9 | | 2,8 | | 2,8 | | 1,4 | | 2,11 |
| Polycarbonate panel | | 2,4 | | | | | | | | | | | | 2,4 |
| Internal storm window | | | | 2,6 (vs)<br>2,0 (vd) | 1,1 (vi) | 2,3 (vs)<br>1,3 (vd)<br>1,0 (vi) | 2,3 (vs)<br>1,1 (vi) | | | | | | | 2,40(vs)<br>1,65 (vd)<br>1,06 (vi) |
| Internal and external storm windows | | | | 2,7 (vs) | | 2,6 (vs) | 2,3 (vs)<br>1,1 (vi) | | | | | | | 2,5 (vs)<br>1,1 (vi) |
| External storm window | | | | | | 2,2 (vs) | | 2,7 (vs)<br>2,2 (vi) | 3,3 (vs)<br>1,9 (vi) | 2,9 (vs) | | 2,8 (vs)<br>2,1 (vi)<br>1,8 (vd) | 2,1 (vi) | 2,78 (vs)<br>2,07 (vi)<br>1,8 (vd) |

**Figure 1. U-value results for window retrofit interventions: vs means single glazing, vi insulating glazing and vd double glazing. All the options have previously undergone draught-proofing operations. Source: [5]**

Thanks to the above-mentioned study reports and the analysis of further documents for a total of 93, it was therefore possible to provide a complete overview of the window retrofit options. These were then grouped into three levels of increasing impact on the heritage, in the perspective of a step-by-step approach to energy efficiency improvements. The first level concerns low-impact interventions, i.e. conservative options potentially applicable to any building, such as repairing windows, improving their airtightness, recovering shutters and adding curtains. The second level regards medium-impact interventions: in this case the sustainability of the intervention depends on the characteristics of the building and the climatic zone in which it is located. It includes strategies such as installing a secondary glazing, inserting an external or internal storm window, adding window film and replacing existing glass. Finally, the third level has the greatest impact and correspond to the window replacement: this needs careful reflection because it is aimed at achieving high performances while taking little account of the values of the historic buildings. These strategies cannot be applied automatically to the building, but a case-by-case evaluation is needed. Moreover, the building components are connected to each other and in the assessment of the window retrofit interventions it is necessary to adopt a global vision of the building. As a matter of fact, the performance of historical windows can be improved by working not only on glass, frames and shading systems, but also on the connections with the walls. This is a weak point because thermal bridges and air infiltrations are concentrated here. To cope with these problems, a very common solution in current practice is to insulate the interior side of the wall by turning the insulation over the reveals and windowsill. This kind of intervention can be greatly simplified by the use of high-performance materials such as aerogel which, thanks to its low thickness, limit the narrowing of the opening compartment allowing the original window to be maintained [6]. Another aspect of utmost importance is the climate context in which the building is placed. This is proved in a study published by the National Trust for Historic Preservation Green Lab [7] that examines multiple window improvement options, comparing their energy, carbon and cost saving in five cities representing various climate types of the continental U.S. The results of this analysis demonstrate that the best retrofit options for a heating-dominated climates may not be right for a cooling-dominated climates. However, for all cities at least one and often two of the selected retrofit options can achieve energy savings within the range of savings expected from new high-performance windows (at a fraction of the cost). Finally, a study carried out in "La Specola" museum of Florence shows that even in the same climate context, windows with different solar exposures need tailor-made solutions [8]. Generally speaking, it is better to follow the historic buildings characteristics and take inspiration from the solutions used in the past. Strategies such as inserting or recovering indoor and outdoor curtains and shutters are widespread in hot climates where the biggest problem is preventing the entrance of solar radiation. On the contrary, in cold climate measures that limit air infiltration and heat loss through the glass, like the installation of a secondary glazing, are more commonly used.

**Figure 2. Example of installation of a secondary external glazing in Villa Piazzi in Nerviano (MI) by the restorer Ercole Livio Rini. (a, b) View of the external side. (c) Inside view of the closed window.**

As we have seen, in the assessment of possible interventions we must not only consider quantitative criteria, referred to comfort and thermal properties, but also qualitative criteria, linked to the restoration's principles (compatibility, reversibility, invasiveness, etc.). Only a clear methodology based on a multidisciplinary and integrated approach allows to make informed decisions directed towards balancing all the aspects at stake.

### 2.2 NEW TECHNOLOGICAL EXPERIMENTATION AND PERSPECTIVES

An interesting field of research concerns the future prospects of the technological evolution of windows and their shading systems. The literature review highlights 33 documents that focus on building components, devices and products made with high-performance materials and technologies resulting from universities experimental research and never applied to new or existing buildings because they are not yet available on the market. Aerogel, phase-change materials (PCMs) and advanced and smart glazing systems are the materials and technologies most frequently applied to experimental solutions for the high performance they can provide. Studies on PCMs account for 79% of the analysed documents. PCMs are characterized by a melting point near the comfort temperature and they accumulate and release latent heat during the transition phase from the solid to the liquid state, without changing their surface temperature. During the day, when the temperature increases, PCMs absorb heat and as a consequence liquefy, cooling the room; when the temperature decreases, the material returns solid and releases heat that can be dissipated by ventilation [9]. Thanks to the possibility of integration in limited thickness components, they are often used for shutters and glass prototypes. In this regard, following research works provide valuable insight: the studies carried out by the University of Aveiro in Portugal on a shutters system with aluminium blades filled with PCM[1] [10] and the studies carried out by the Politecnico di Torino in Italy on a cutting-edge system with PCM inserted between the glass panes[2] [11]. Glass is also the element on which some research works on aerogel are focused. They account for 12% of the analysed studies. Aerogel is a light, highly porous material produced from silicon dioxide and is composed of approximately 96% of air and the remaining 4% of open-pores structure of silica which gives great lightness to the system thanks to its small specific weight. Aerogel has the lowest thermal conductivity among solid materials, even lower than that of the air, thus making it an excellent thermal insulation product. The introduction of aerogel in glazing systems occurs both using monolithic and granular aerogels in the glazing interspace[3] [12]. The University of Perugia is particularly active in aerogel experimentation and boasts a series of studies conducted in laboratory on glass and polycarbonate systems with granular aerogel [13]. Finally, 9% of the analysed documents present a complete overview of the latest developments in the world of high-performance glass [14]: intelligent, vacuum, photovoltaic and photochromic glazing are just some examples of the products designed for new buildings that can have interesting applications also in the historical context. In this regard, it should be pointed out that in the last few years a real breakthrough has taken place in the field of photovoltaic glazing: in 2018 the University of Milano-Bicocca developed a patent of a transparent photovoltaic window that uses LSC (Luminescent Solar Concentrator) technology, integrated in the transparent component, to convert sunlight into infrared rays. These are then reflected inside the panel up to the edge where a strip of silicon photovoltaic cells converts it into electrical current. Unfortunately, at the moment there are no experimentations in historic buildings, but it would be interesting in the near future to test the prototypes in traditional windows to verify their actual efficiency [15].

Despite the strong limitations due to the experimental character of these technologies, some of them are quite interesting and could lead to the development of new products for historic buildings, obviously to be tailored, in terms of compatibility and sustainability, to the specific needs of the built cultural heritage. On the other hand, others are still too "embryonic" to be used in the historical context (this is the case of the PCMs).

---

[1] The research presents the results of an experimental campaign of a full-scale outdoor test cell, composed by two side-by-side compartments with and without PCMs. The results reveal the PCM potential for the thermal regulation of indoor spaces during winter and summer periods. In the cold season, the compartment equipped with PCM shutters reached a maximum internal temperature of 37.2°C compared to 53.8°C detected in the sector taken as reference (16.6°C less). During the summer season, the compartment with PCM shutters reduced the indoor temperature from roughly 22% to 18% and decreased the maximum and minimum temperature peaks by 6% and 11% respectively.

[2] The performances of the prototype were monitored in an experimental campaign and compared with those of a conventional double glass unit. It was found that PCM glazing is able to contribute to a better indoor thermal environment for most of the time during the various season compared to the reference glazing.

[3] The monolithic aerogel is more suitable for use in glazing systems because it offers the best compromise between light transmission and thermal insulation. Due to its fragility, excessively high cost and difficulties linked to production processes, it is not yet widely marketed. The use of granular aerogel in glazing therefore offers an alternative solution to the monolithic version because it is cheaper, more robust and easier to produce on a commercial scale. However, it is characterised by poor transparency and a translucent aspect, which strongly limits the view towards outside with hazily deformation of optical images.

## 2.3 HIGH-PERFORMANCE MATERIALS AND SOLUTIONS APPLIED TO HISTORIC BUILDINGS

The last field of research investigates the use of retrofit solutions with high-performance materials and technologies in historic buildings: in particular, 8 documents out of 93 (referred to "Retrofit solutions" paragraph) are related to the built cultural heritage. The results of the literature review show that two are the most practiced research lines: working on the glazing systems or on shading systems like shutters.

With regard to glazing, the Italian guideline includes a collection of sheets on available materials and types of intervention in which high-performance technologies play a leading role: insulating glazing with TIM (Transparent Insulating Materials) and aerogels, chromogenic glazing and photovoltaic glazing are just some examples of the systems considered [16]. However, in the section dedicated to the case studies, these types of glazing are never present: on the contrary, there is a considerable preference for the replacement of the whole window. In general, a wide range of high-performance glazing is available today, but the choice depends on the window state of conservation, its material and dimensional characteristics, the strength of the existing frames and the weight of the new glazing system. A remarkable study entirely dedicated to the replacement of the existing single glass unit with a variety of high-performance double-glazing products (retaining their original frame in 6 cases out of 10) is that carried out by Changeworks in some listed buildings located in Edinburgh [17]. Conventional double glazing consists of two layers of glass up to 25 mm apart with dry air or inert gas in the cavity. Most traditional windows, however, have a glazing bar with a shallow rebate, designed to take a single sheet of glass of about 3 mm: this means it is usually impracticable to replace old glass with standard double-glazed units. For this reason, the focus of this study is the installation and on-site monitoring of slim-profile double glazing, characterised by a significantly smaller cavity and lower weight compared to conventional double glazing (from 8.2 to 16 mm), but with a similar thermal transmittance (ranging between 1 $W/m^2K$ to 2.8 $W/m^2K$) [17]. Different types of this kind of glazing were investigated by the authors, but the most innovative is certainly the vacuum glazing, consisting of two glass panes with a vacuum-filled space between them. Although the cavity thickness of the vacuum glazing was only 0.2 mm, it proved to be the best from the U-value point of view, reaching a value of 1 $W/m^2K$ compared to the others that reached values always amounting to 2 $W/m^2K$ or more [17]. Even though vacuum glazing is a valuable resource for the thermal performance improvement of historic windows, it is necessary to take into account some drawbacks related to the choice of these products: they employ metal pillars between glass panes, to prevent glass breakage (due to the pressure gradient), that are visible from a close distance; they can be produced only in limited sizes and their cost is still very high [18]. Despite these limitations, in recent years, the installation of vacuum glazing has been fostered by the institutions involved in preservation. Another interesting innovation in the field of glazing concerns the use of aerogel: 3 are the studies, collected during the literature review process, that simulate the insertion of a new glazing system with monolithic aerogel in the cavity of a double-glazing units of two historic buildings (one of which is listed). Both projects assume to upgrade only a glazed portion of the window, alternating aerogel-enhanced glazing and traditional transparent ones: several alternative configurations were therefore considered (40, 60, 80 and 100% of aerogel). The results of the simulation showed that heating energy consumption decreased by increasing the aerogel proportion in the windows and that cooling energy consumption kept stable with percentages of aerogel above 60%: the greater the quantity of aerogel the lower the window SHGC and U-value (it ranges from 1.2 $W/m^2K$ for 40% of aerogel to 0.6 $W/m^2K$ for 100% of aerogel) [18]. In the review, the only real application example of panels filled with aerogel in a listed building is the Alte Börse in Zürich

(Switzerland), where the existing roof was substituted with aerogel elements with an improvement in the U-value from 2 $W/m^2K$ to 0.6 $W/m^2K$ [19]. Aerogel glazing show significant energy savings which, however, is still burdened with high costs of materials and long payback times: on top of that, they are characterised by a translucent aspect that is an unacceptable alteration for a historic building. Moreover, given the lack of case studies in built cultural heritage, the application conditions and compatibility of the aerogel glazing have not yet defined comprehensively.

**Figure 3 (a, b). Vacuum glazing installed in a historic building located in Edinburgh. Source: [17]**
**Figure 4. The roof of the Alte Börse in Zürich after the renovation with the aerogel glazing. Source: [19]**

Finally, regarding shutters interventions, 2 are the studies that must be mentioned: an in-lab testing carried out by the Historic Scotland, followed by a practical application in one of their case studies. The first study measured in laboratory the performance of a traditional shutter and a modified shutter with a 9 mm thick aerogel insulation blanket inserted into panels and covered with plywood (the insulated area was 55%). The traditional shutter showed a U-value of 2.2 $W/m^2K$, while the insulated shutter led to a U-value of 1.6 $W/m^2$, equivalent to low-E double glazing [20]. A similar type of shutter, upgraded using a 10 mm aerogel quilt, was used in a tenement flat in Edinburgh a few years later. Here, starting from a U-value of 2.2 $W/m^2K$ the shutters reached a value down to 0.4 $W/m^2K$, with an 82% improvement [21]. Although promising, this type of intervention cannot be applied to all types of shutters: they must have enough internal space to house the insulation, so if they consist of a single piece of wood, alternative tailor-made measures are required.

## 3. CONCLUSIONS

In many countries, historical windows are disappearing at an alarming rate, replaced by highly efficient windows with a great environmental impact. Three are the main reasons: disregarding their importance in representing the material culture of the craftsmanship, tax incentives and misunderstanding about sustainability. Despite this, nowadays it is possible to prove that many alternative retrofit solutions are available for improving the performances of ancient windows. In this paper we tried to define a picture of the research state-of-the-art on window interventions though a literature review. This showed that managing several quantitative and qualitative criteria is crucial in selecting the best intervention and, therefore, a multidisciplinary approach is required in order to balance energy efficiency and sustainability needs with conservation aspects. High-performance materials and technologies can help in this regard: in recent years, the growing interest in this field has led to a strong acceleration in the technological development of new efficient products. However, as pointed out by Milone et al. [23] *"the Best Available Technologies for building components characterized by high level of thermal performances show, not rarely, a limited compatibility with the architectural integrity of the building [...] to which a certain artistic, historic and/or architectural merit is recognized"*. Hence, the delicate relationship between building protection and energy efficiency cannot be solved through the uncritical application of the best technologies from a performance profile, but through a good balance between advanced technologies and the conservation of the identity of the historic buildings. Unfortunately, at the moment, although the first results from the thermal point of view are encouraging, there is not enough research works investigating in detail the application conditions (including climatic aspects) and compatibility of these technologies in heritage buildings. Our purpose is therefore to implement our knowledge and experience in this field through the assessment of a wide range of case studies.

## 4. REFERENCES


[1] European Commission. "Energy-efficient buildings PPP. Multi-annual roadmap and longer term strategy". Brussels: Publications Office of the European Union, 2010, p. 18.

[2] CRESME. "Analisi sull'impatto socio-economico delle detrazioni fiscali del 55% per la riqualificazione energetica del patrimonio edilizio esistente". Milan, 2010, pp. 6-7.

[3] Ministro dello Sviluppo Economico and ENEA. "Impatto delle detrazioni fiscali del 55% sul mercato dei prodotti e dei servizi incentivati ed effetti macroeconomici indotti". 2010.

[4] S. Fossdal. "Windows in Existing Buildings – Maintenance, Upgrading or Replacement?". Norwegian Building Research Institute, Oslo, Norway, Project report 192, 1996, p. 17.

[5] V. Pracchi, N. Rat, A. Verzeroli. "Historic windows. Conservation or replacement. What's the most sustainable intervention? Legislative situation, case studies and current researches", in Proceedings of the International Conference on Energy Efficiency and Historic Buildings, 2014, pp. 432-449.

[6] E. Genova. Edifici storici ed efficienza energetica. Palermo: 40due, 2017, p. 98.

[7] National Trust for Historic Preservation Green. "Saving windows, saving money: evaluating the energy performance of window retrofit and replacement". 2012.

[8] F. Sciurpi et al. (2015, May). "Environmental monitoring and microclimatic control strategies in "La Specola" museum of Florence". Energy and Buildings. [Online]. 95, pp. 190-201. [Dec. 9, 2020].

[9] R. Parameshwaran. "Nanomaterial-Based PCM Composites for Thermal Energy Storage in Buildings", in Nano and Biotech Based Materials for Energy Building Efficiency, P.F. Torgal et al., Ed. Cham: Springer, 2016, pp. 215-243.

[10] T. Silva et al. (2015, Feb.). "Development of a window shutter with phase change materials: Full scale outdoor experimental approach". Energy and Buildings. [Online]. 88, pp. 110-121. [Nov. 12, 2020].

[11] F. Goia, M. Perino, V. Serra. (2013, Feb.). "Improving thermal comfort conditions by means of PCM glazing systems". Energy and Buildings. [Online]. 60, pp. 442-452. [Nov. 12, 2020].

[12] E. Cuce, P.M. Cuce, T. Guclu, A.B. Besir. "Thermal Properties of Aerogels: from Theory to Practice", in Translucent Silica Aerogel: Properties, Preparation and Applications, C. Buratti, Ed. New York, Nova science publishers, 2019, pp. 103-130.

[13] C. Buratti and E. Moretti. "Misura delle proprietà energetiche e luminose di sistemi vetrati con aerogel granulare", in Atti del 62° Congresso Nazionale Ati, 2007.

[14] S.D. Rezaei, S. Shannigrahi, S. Ramakrishna. (2017, Jan.). "A review of conventional, advanced and smart glazing technologies and materials for improving indoor environment". Solar Energy Materials and Solar Cells. [Online]. 159, pp. 26-51. [Nov. 12, 2020].

[15] F. Meinardi and S. Brovelli. "Vetrocamera con concentratore solare luminescente per produzione di energia elettrica". Internet: https://www.glasstopower.com/pages/brevetti/, 2018 [Dec. 9, 2020].

[16] MiBACT. "Linee di indirizzo per il miglioramento dell'efficienza energetica nel patrimonio culturale. Architettura, centri e nuclei storici ed urbani". 2015.

[17] Historic Environmental Scotland. "Technical paper 9: slim-profile double glazing. Thermal performance and embodied energy". 2010.

[18] U. Berardi. (2015, Sept.). "The development of a monolithic aerogel glazed window for an energy retrofitting project". Applied Energy. [Online]. 154, pp. 603-615. [Nov. 13, 2020].

[19] M. Ganobjak, S. Brunner, J. Wernery. (2020, March). "Aerogel materials for heritage buildings: Materials, properties and case studies". Journal of Cultural Heritage. [Online]. 42, pp. 81,98. [Nov. 13, 2020].

[20] Historic Environmental Scotland. "Technical Paper 1: Thermal Performance of Traditional Windows".2010.

[21] Historic Environmental Scotland. "Five Tenement Flats, Edinburgh". 2012, p. 41.

[22] D. Milone et al. (2015, May). "Are the Best Available Technologies the only viable for energy interventions in historical buildings?". Energy and Buildings. [Online]. 95, pp. 190-201. [Dec. 9, 2020].

# A Life Cycle Assessment Methodology to Compare Carbon Emissions from Building Refurbishments to Demolition and New Build

A. Duffy[1], A. Nerguti[2], C. Engel Purcell[3]

[1] School of Civil and Structural Engineering, Technological University Dublin, Dublin, Ireland.

[2] Researcher, Carrig Conservation International Ltd., Dublin, Ireland.

[3] Head of Research, Carrig Conservation International Ltd., Dublin, Ireland.

**Abstract** – To meet the national commitments made through the Paris Agreement, the UK building regulations and planning system have primarily focused on limiting operational carbon emissions. This inadvertently promotes the demolition of older and unprotected historic buildings in favour of new, more energy efficient buildings. However, the whole carbon footprint of such a decision is not usually considered and embodied emissions are often ignored. To better understand the potential carbon emissions that could be saved through a policy of retain and retrofit rather than demolish and build new, Carrig Conservation International Ltd in collaboration with Prof. Aidan Duffy of Technological University Dublin were commissioned by Historic England to conduct a Life Cycle Assessment (LCA) of four dwelling retrofits (1890s end-of-terrace, early 1900s mid-terrace, 1970s semi-detached and 1990s detached) and one conversion (ca. 1850s).

Over a reference study period of 60 years, the study estimated that the retrofit of the dwellings saved 458 tonnes of carbon in total compared to how the buildings were previously operating, that embodied emissions of the retrofits are lower than new builds, and that in some cases, the new builds' total life cycle emissions may never be as low as the retrofit options even with lower operational emissions. Additionally, if embodied emissions are ignored, then up to a third of total life cycle emission are not accounted for.

**Keywords** – Life Cycle Assessment; refurbishment; traditional buildings; embodied carbon; operational carbon

## 1. INTRODUCTION

The publication of the Intergovernmental Panel on Climate Change (IPCC) Special Report in 2018 inspired many national governments to commit to reducing their carbon emissions in an effort to keep global temperatures below 1.5°C above the pre-industrial levels. The UK is the first of the G7 nations to make a legally binding commitment to becoming carbon neutral by 2050 but it is off track to achieve the emissions reduction targets for 2030 [1, 2]. With growing concern over the impacts of climate change, each sector within the UK will have to make a concerted effort to reduce their share of emissions.

The built environment sector is currently responsible for 35–40% of total greenhouse gas (GHG) emissions in the UK [3] and for over 30% of global GHG emissions [4, 5]. These emissions arise from three stages of a building life cycle – construction, operation and end-of-life. Under the current UK Part L building regulations, the focus is primarily on reducing the operational emissions of buildings. However, this ignores embodied emissions (emissions associated with the manufacture, transport and demolition of buildings), which for new residential buildings can account for more than 50% of a building's total GHG emissions over its lifetime [6].

By accounting for operational emissions only, a false impression can be given that only new buildings are a viable solution to reduce emissions from the building sector. This has inadvertently given preference to the demolition of existing dwellings in favour of the construction more operationally energy efficient new buildings. By doing this, the full environmental cost of the demolition of existing buildings and the manufacture, transport and use of new materials in new construction is not considered. Therefore, while it may appear that the building sector is reducing its emissions, the embodied emissions associated with the construction of the new buildings may be increasing. This strategy also does not account for the historic and cultural value of existing buildings, which although not protected may be of local interest and contribute to a sense of place [7, 8].

To combat this, the total life cycle emissions of a building project must be accounted for to reveal the true environmental cost of current building practice. There are currently little data available to policymakers on the total potential energy and carbon savings from the energy efficient refurbishment of existing buildings. More data and research are therefore required to assess the extent to which the refurbishment of existing buildings can reduce embodied carbon and operational emissions over a certain length of time when compared to other options for carbon mitigation in the building stock. Life Cycle Assessments (LCA) of refurbishment projects is one way to produce the necessary supporting data, which this study demonstrates.

## 2. METHODOLOGY

Life cycle assessment (LCA) is an assessment of the environmental performance of products and services, from raw material extraction and manufacturing to disposal and possible recycling. It is these 'cradle-to-grave' or 'cradle-to-cradle' approaches to the environmental assessment of building materials, which include embodied as well as operational emissions.

For each case study, this research considered the full life cycle (embodied and operational emissions) for three different options: Base-Case where there was no intervention to the existing building; Refurbishment which involved energy efficiency upgrades to the existing building to 2019 standards; and New-Build involving a new construction after the demolition of the existing building. The emissions associated with each option were calculated for all five case studies (see Section 3 for details).

The embodied emissions of each material used in the Refurbishment and New-Build options were estimated as the product of the quantity of material and its carbon emissions' intensity; these were summed for all materials to give the total embodied emissions for each option. Emissions' intensity data were found either in Environmental Product Declarations (EPDs) or in the Inventory of Carbon and Energy (database). These emissions include the emissions associated with the replacement of materials throughout a reference study period (RSP), which was set at 60 years for this study In the case of the New-Build option, the embodied emissions associated with the demolition of the existing building were included in addition to those resulting from the materials used in its construction. As no material changes are being made to the Base-Case, this options has no embodied emissions.

Operational energy requirements were estimated using an hourly energy balance model complying with BS EN ISO 52016-1:2017 Energy Performance of Buildings [9]. These were converted to emissions using fuel and electricity emissions factors from Hill, 2008 [10]. An important input to this model is the internal temperature in the building. There is evidence that the occupants of older, less well insulated buildings choose lower operating temperatures [11]. For this reason, the sensitivities of the LCA results to internal temperatures were assessed by modelling different internal temperature scenarios.

We also undertook an analysis of the marginal abatement costs (MAC) with the Refurbishment versus New-Build options of each case study to estimate which option was the most attractive to policymakers and homeowners respectively. More details of the methodology and calculations used can be found in the original report [12].

## 3. CASE STUDIES

To support this study, data from five completed energy refurbishment projects were obtained for analysis. The case study details are summarised in Table 1. More detailed information on the case studies can be found in the original report [12].

**Table 1. Description of the case studies analysed in this study and summary of the refurbishment measures introduced to the case studies**

| Case Study | Description | Refurbishment measures |
|---|---|---|
| **Chapel (c.1850s)** | Derelict Gothic Revival chapel | Internal wall insulation, underfloor heating |
| **Victorian Terrace (1890s)** | Vacant end-of-terrace house, with solid brick walls | Internal wall insulation |
| **Mid-Terrace (1900s)** | Mid-terrace solid brick building, constructed between 1900 and 1910 | Internal wall insulation, MVHR unit installed |
| **Semi-detached (1970s)** | Solid brick masonry walls and cavity walls | Internal wall insulation, heat pump installed |
| **Detached (1990s)** | Cavity and solid concrete walls | Internal wall insulation, MVHR unit installed |

The life cycle carbon emissions of the Refurbishment option for the five case studies were assessed against the Base-Case and New-Build.

The Base-Case option represents the case study building before any energy efficiency upgrades or material changes were made. The operation of the building therefore continues as normal, and by modelling the current energy use and sources, we can see how much carbon the building can be expected to emit over the reference study period if no improvements are made to the building fabric.

The New-Build option is based on data from a newly constructed residential building that has been designed to meet current building regulations and standards. The dwelling is representative of new residential construction in the UK, with concrete block cavity walls and high levels of insulation in the walls, roof and ground floor slab. The life cycle carbon emissions include the embodied emissions of the new structure, the embodied emissions of any structure that was demolished on that site (if applicable) and the operational emissions over the reference study period.

## 4. RESULTS

### 4.1 EMISSION RESULTS

In all case studies, the Base-Case option had the highest total carbon emissions at the end of the 60 years. For the Semi-Detached (1970s) and the Detached (1990s) case studies, the emissions of the Refurbishment option were lower than the emissions of the New Build option. The Victorian Terrace (1890s) and Mid-Terrace (1900s) Refurbishment options showed slightly higher life cycle emissions than the New Build, whereas the Chapel Conversion (1850s) has much higher life cycle emissions than its New-Build counterpart.

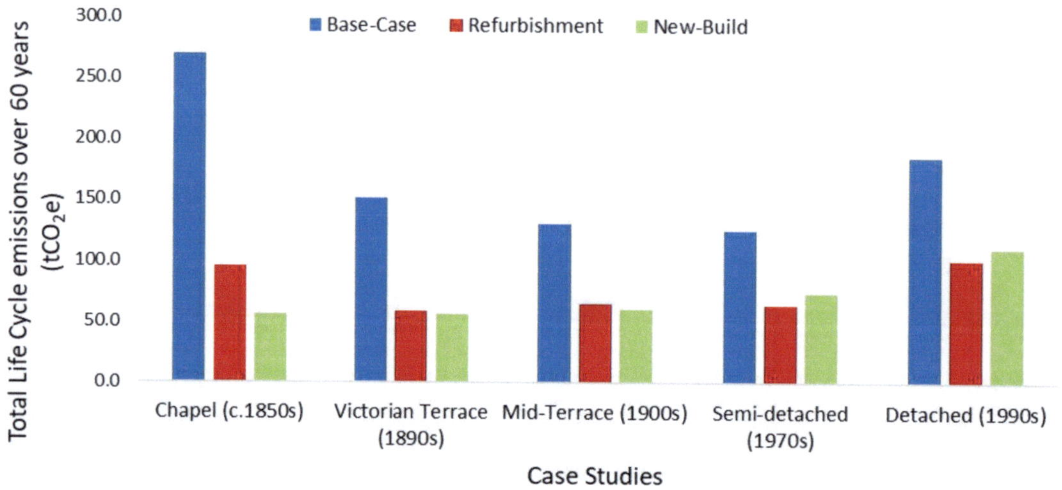


**Figure 1. The total life cycle emissions of the case studies within the Reference Study Period of 60 years.**

On average, the embodied emissions made up a higher percentage of total life cycle emissions of New-Builds (23.6%) than of the Refurbishment case studies (4.8%) (see Table 2).

**Table 2. The breakdown of embodied and operational carbon emissions of each case study, for the Refurbishment and New-Build options.**

| | Refurbishment (% of total life cycle carbon | | New-Build (% of total life cycle carbon) | |
|---|---|---|---|---|
| **Case Study** | **Embodied Carbon** | **Operational Carbon** | **Embodied Carbon** | **Operational Carbon** |
| **Chapel (c.1850s)** | 10.3 | 89.7 | 26.5 | 73.5 |
| **Victorian Terrace (1890s)** | 2.1 | 97.9 | 24.8 | 75.2 |
| **Mid-Terrace (1900s)** | 6.4 | 93.6 | 27.9 | 72.1 |
| **Semi-detached (1970s)** | 4 | 96 | 21.6 | 78.4 |
| **Detached (1990s)** | 1.4 | 98.6 | 17 | 83 |
| ***Average*** | *4.8* | *95.2* | *23.6* | *76.4* |

### 4.2 MARGINAL ABATEMENT COSTS

The marginal abatement cost (MAC) calculations showed that in almost all cases, the Refurbishment options were a more cost-effective way of introducing carbon-reducing measures. The Chapel Conversion, however, has a much higher MAC than its New-Build option due to the amount and expense of the new materials used during the unique conversion process required for that building (see Figure 2).

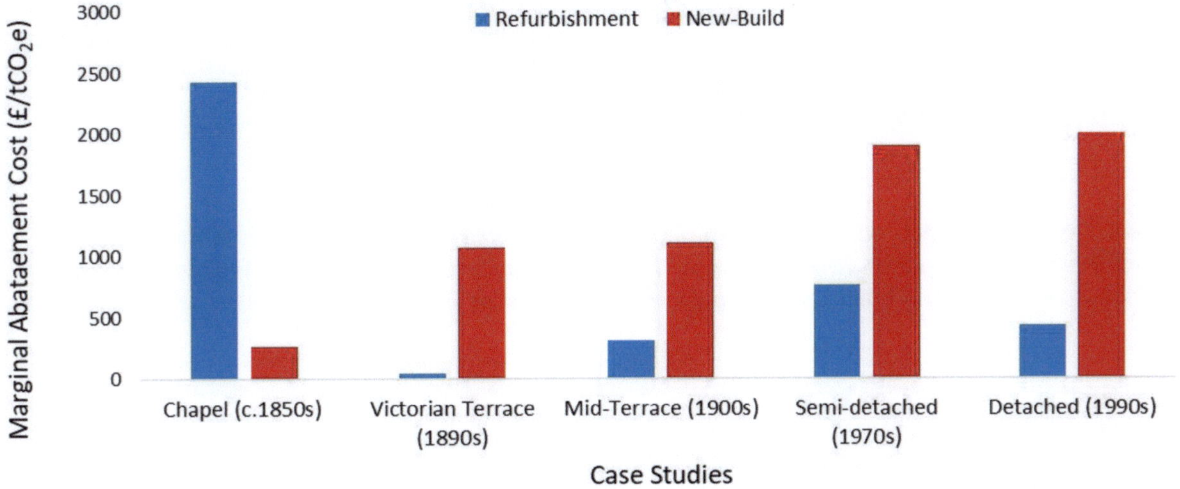


**Figure 2. The Marginal Abatement Cost of each case study for Refurbishment and New-Build cases.**

### 4.3 INTERNAL TEMPERATURE SCENARIO

The emissions results for the case studies were estimated under a specific temperature scenario. The scenario was developed to take account of the fact that poorly-insulated buildings (e.g. the Base-Case scenario) tend to be operated at lower ambient temperatures and that highly-insulated buildings (e.g. Refurbished or New-Build) may be operated at higher temperatures (often referred to as 'comfort taking' or 'the rebound effect') [11]. For these reasons, the sensitivity of the model to the following internal temperature scenario was investigated:

- Base-Case: 18°C
- Refurbishment: 19°C
- New-Build: 21°C

These temperatures are not meant to reflect a level of perceived thermal comfort, but merely possible representative temperatures at which refurbished and new buildings may be operated.

The results show that in all but one case study, the New-Build does not outperform the Refurbishment within the 60-year RSP, i.e. the emissions of the New-Build are consistently higher than the Refurbishment during that period. For two case studies, the Semi-detached and Detached, the New-Build never outperforms the Refurbishment.

In all case studies, the New-Build outperforms the Base-Case within the 60-year RSP.

**Figure 3. Time periods after which the New-Build outperforms the Base-Case and Refurbishment for the internal temperature scenario 18-19-21, in terms of total life cycle emissions (i.e. cumulative New-build emissions are lower). New-build never outperforms Refurbish for Semi-detached and Detached case studies.**

## 5. DISCUSSION

Although this is a small sample of five buildings, they are typical of many residential building types in the English housing stock. Assuming some degree of representativeness, the results show that in order to reduce carbon emissions across the whole building sector, it is not an option to allow existing buildings to continue to operate as they are. Instead, both energy efficient refurbishment and new-build can play a significant role in reducing carbon emissions from the built environment.

The Semi-Detached and Detached Refurbishment case studies had lower cumulative emissions than New-Build within the 60-year reference study period, which, coupled with the results from the MAC calculations show that in terms of reducing carbon emissions refurbishment rather than new construction is capable of providing value for money to policymakers.

While the Victorian Terrace and Mid-Terrace Refurbishment emissions may be slightly higher than the New-Build emissions over the 60-year reference study period, the MAC results show that the cost of achieving the lower emissions associated with the New-Build is much higher than the costs associated with the Refurbishments. It is worth noting that less money was

invested in the Refurbishments (except for the Chapel Conversion) to achieve either a similar level of life cycle emissions or lower life cycle emissions than the New-Builds.

Of all the case studies analysed, the Chapel Conversion was the least representative case study. The building had been derelict, was architecturally important and therefore required significant intervention to bring it up to modern standards and convert it into a different use while satisfying stringent conservation requirements. The budget for this project was thus higher than others, and therefore more money was invested in improving the building. More new materials were used in this project than would normally be used in a typical refurbishment, resulting in the highest embodied carbon percentage out of all of the case studies (10.3% of total life cycle emissions, Table 2). The high life cycle emissions can be explained by the fact that if more materials are introduced, the embodied carbon emissions naturally increase. This case study therefore shows the importance of considering the embodied carbon and quantity of new materials used during the early design stages of refurbishment projects.

The results also show that the consideration of embodied carbon is equally important for New-Builds. Based on these five case studies, the average embodied emissions of a New-Build equalled 23.6% of its total life cycle emissions, which is 18.8% higher than the average Refurbishment equivalent. This means that if legislation does not require the building industry to take account of the embodied emissions of building projects, they could be underestimating the true life cycle carbon emissions by nearly a quarter.

## 6. CONCLUSION

This study set out to compare the life cycle emissions of three building options, Base-Case, Refurbishment and New-Build, for five case studies over a reference study period of 60 years. The study found relatively little differences between the estimated life cycle carbon emissions of New-build and Refurbishment approaches. However, Refurbish had higher operational emissions than New-build, but these were offset by lower embodied emissions. When taken in combination with the additional historic value of existing buildings, these findings lend weight to a reassessment of the policy of considering operational-only emissions which favours demolish and rebuild. A second important finding for policymakers is that the marginal abatement costs of energy efficient retrofitted buildings may be lower than new buildings, so that greater emissions reductions could be achieved at lower cost using this approach.

## 7. REFERENCES


[1] The Committee on Climate Change. (2019, 17 September). Carbon Budgets: How We Monitor Emissions Targets. Available: https://www.theccc.org.uk/tackling-climate-change/reducing-carbon-emissions/carbon-budgets-and-targets/

[2] C. Stark, A. Gault, and D. Joffe, "Reducing UK Emissions: 2018 Progress Report for Parliament", London June 2018, Available: https://www.theccc.org.uk/wp-content/uploads/2018/06/CCC-2018-Progress-Report-to-Parliament.pdf.

[3] S. Sturgis, "Embodied and Whole Life Carbon Assessment for Architects", 2017, Available: https://www.architecture.com/-/media/gathercontent/whole-life-carbon-assessment-for-architects/additional-documents/11241wholelifecarbonguidancev7pdf.pdf.

[4] P. Huovila, M. Ala-Juusela, L. Melchert, and S. Pouffary, "Buildings and climate change: Summary for Decision-Makers", Paris 2009, Available: https://europa.eu/capacity4dev/unep/document/buildings-and-climate-change-summary-decision-makers.

[5] IPCC, "Climate Change 2014: Synthesis Report", in "Contribution of Working Groups I, II and III to the Fifth Assessment Report of the Intergovernmental Panel on Climate Change," Geneva, Switzerland 2014, Available: https://www.ipcc.ch/report/ar5/wg3/.

[6] S. Sturgis and A. Papakosta, "Whole Life Carbon for the Built Environment", London 2017, Available: https://www.rics.org/uk/upholding-professional-standards/sector-standards/building-surveying/whole-life-carbon-assessment-for-the-built-environment/.

[7] P. Drury and A. McPherson, "Conservation Principles, Policies and Guidance for the Sustainable Management of the Historic Environment", London, April 2008, Available: https://historicengland.org.uk/images-books/publications/conservation-principles-sustainable-management-historic-environment/.

[8] Ministry of Housing, Communities and Local Government (2019). ISBN 978-1-5286-1033-9, National Planning Policy Framework. Available: https://www.gov.uk/government/publications/national-planning-policy-framework--2

[9] "BS EN ISO 13790:2008 – Energy performance of buildings – Calculation of energy use for space heating and cooling", ed, 2008.

[10] N. Hill, E. Bonifazi, R. Bramwell, E. Karagianni, and B. Harris, “Greenhouse Gas Reporting: Conversion factors 2018“, E. Department for Business and S. Industrial, Eds., ed, 2018.
[11] BRE Group, “Report 2: Mean Household Temperature“, in “Energy Follow-up Survey 2011“, London 2013, Available: http://doc.ukdataservice.ac.uk/doc/7471/mrdoc/pdf/7471_2_mean_household_temperatures.pdf.
[12] A. Duffy, A. Nerguti, C. Engel Purcell, and P. Cox, “Understanding Carbon in the Historic Environment“, Historic England Research Report Series Historic England, p78, 2020.

# Historic buildings and energy efficiency in New Zealand: A theoretical study of a holistic refurbishment of a 1930s heritage building's windows

R. Paschoalin, N. Isaacs

Faculty of Architecture, Victoria University of Wellington, Wellington, New Zealand (rachel.paschoalin@vuw.ac.nz; nigel.isaacs@vuw.ac.nz).

**Abstract –** Taking account of cultural and environmental importance in the renovation of historic buildings has become a focus in many countries in the last decade, particularly within the European Framework of guidelines for energy renovations. Despite the international interest, New Zealand does not have any guidelines balancing cultural significance and energy performance, what is critical within the context of the recently passed Zero Carbon Act 2019 aiming for zero carbon emissions by 2050. This paper presents a theoretical study of a holistic renovation of the windows of a 1930s heritage building in Wellington, New Zealand. It evaluates three different window refurbishments, coupled with stakeholder interviews to discuss possible trade-offs among multiple benefits and constraints of alternatives. The results present the stakeholders' perspective and barriers within the holistic refurbishment, and provide an analysis of the costs and benefits of the different options, along with recommendations for future action.

**Keywords** – Holistic refurbishment; historic building; trade-off; energy efficiency; multiple criteria; stakeholders.

## 1. INTRODUCTION

Many countries have adopted new policies, projects and standards to mitigate effects on climate change from greenhouse gas (GHG) emissions [1]. This includes advocating energy efficiency improvements of historic and heritage buildings [1], [2]. ICOMOS recognition and the campaign of cultural heritage supporting the sustainable development of cities [3],[4], has encouraged discussions on sensible energy retrofits of historic buildings, and the existing challenges and constraints of this process [5], [6]. In the past decade there have been an increasing number of publications on energy efficiency measures in historic buildings [1], [7]. Guidelines for the renovation of historic buildings, including European 16883:2017 [8] and American ASHRAE 34P:2019 [9], are now available to support balanced decision-making across multiple-criteria, including historic conservation and energy performance, and IEA-TASK 59 "Towards Zero Energy in Historic Buildings" is developing a best practice database (Hiber Atlas) [10]. Internationally, there is a wide range of existing documents and renovation tools, each with their own assessment criteria and methods, but all with a similar systematic multi-criteria decision-making process [11].

In New Zealand (NZ), despite the recently approved Climate Change Response (Zero Carbon) Amendment Act 2019 providing a framework to "develop and implement clear and stable climate change policies" [12] there are no available country-specific tools or guidelines. Besen and Boarin suggest that this is due to the lack of incentives and policies for energy upgrades of historic buildings [13]. This might change in a near future as there is a proposal to revise the New Zealand Building Code (NZBC) to take account of lifecycle environmental impacts, although at the time of writing only new building requirements are being proposed [14]. As this Act [12] is a key driver for this revision, it is possible that existing, including heritage or historic, buildings could be included in future action.

Therefore, while there is an urgency to develop appropriately tailored guidelines there is limited research on how to holistically refurbish historic or heritage buildings considering conservation and cost constraints. This paper presents a case study of a holistic, but theoretical, refurbishment of the windows in a 1930s heritage building. It explores possible trade-offs of window retrofits under different holistic criteria assessments, such as energy savings compared to heritage impact and costs. For the different proposed options it analyses the costs, benefits and barriers, and concludes with recommendations for future action.

## 2. CASE STUDY AND PROPOSED RETROFIT OPTIONS

The Chevening Flats (Figure 1) were built in 1929 in the suburb of Kelburn, Wellington, NZ. They were designed by the architect Llewellyn Williams for an unusual client, a single woman and senior teacher at Wellington Girls' College, Miss Emma Rainforth. It is one of the first examples of self-contained luxury flats built in the city, with many original interior components still preserved. Built in concrete and brick, the elegant four-storey building was designed in a classical style [15]. Recently gifted by Susan Price to Heritage New Zealand Pouhere Taonga (HNZPT), the national historic heritage agency, the building is included in the HNZPT list as a Category II historic place "of historical or cultural significance or value" and is protected under the Wellington City District Plan [16]. The building has been always used as rented apartments, under the previous ownership to academics and students from the close-by Victoria University of Wellington and this is continued under HNZPT.

**Figure 1. Chevening Flats. Source: Studio Pacific Architecture**

After many years of neglect, it was purchased by the Price family in 1979. The rusting steel window frames led to cracks many glass panes [15]. While some panes were soon replaced, in 1991 due to ongoing failures they decided the best cost-effective option was to replace the original steel with new aluminium-frame windows. Original leadlight glazing was transferred to the new framing preserving the Art Deco design [15]. They tried to find the original window manufacturer, but this proved impossible. The building was not listed or protected under any heritage agency at the time.

In 2011, the last major refurbishment occurred to seismically strengthen and restore the interior to its 1920s elegance. Some energy upgrade measures were adopted, including replacement of interior brick partition walls with insulated lightweight timber-frame walls and adding new roof insulation. The single glazed aluminium windows were not changed. In 2016 due to the cold indoor temperatures experienced by tenants, new heat pumps were installed in each flat's living room [17]. This is not an uncommon experience in NZ where houses are often described as cold, damp and difficult to heat due to poor insulation and inefficient heating. Research has found many existing NZ houses fail to meet the World Health Organization (WHO) recommendations of indoor temperatures between 18°C to 24°C [18].

Before, during and after the refurbishment reports were generated by the working team. Non-destructive tests, such as IR-thermography and other visual inspections were carried out to identify possible plaster cracking and leaks. These highlighted the fact the post-refurbishment concrete walls still had lower thermal performance than required by NZBC in new houses, and that moisture damage was due to external moisture rather than internal condensation [19]. The low R-value, single-glazed, non-thermally broken frame windows were still the weakest point in the thermal envelope. The report recommended as a long-term solution changing to double-glazed Insulated Glazing Units (IGU) with a thermally broken frame, coupled with improved ventilation to eliminate the formation of condensation on the internal surface of the glass or frame [19]. The architect recommended window replacement within 10 years.

Enhancing the window performance can benefit occupants' thermal comfort and health while reducing damage to the construction from dampness and mould [20]. As a result of the poor window frame performance, the recommendation for future replacement, and the inadequate thermal comfort, this case study explored three different window retrofits against the base (2011) model detailed on Table 1.

**Table 1. U-values and solar heat gains coefficients of different windows options.**

| Window type | U value *($W/m^2.K$)* | Solar heat gain coefficient |
|---|---|---|
| *(1) Base model – 2011 existing aluminium windows* | 6.7 | 0.84 |
| *(2) double-glazed IGU thermally broken frame* | 3.28 | 0.69 |
| *(3) secondary glazing using an internal removable acrylic sheet* | 2.9 | 0.7 |
| *(4) secondary glazing using a non-removable low-E glass* | 1.8 | 0.69 |

## 3. METHODOLOGY

Historic documents and technical reports from the 2011 refurbishment were assessed to understand the building's history and physical conditions. The qualitative and quantitative data collection, assessment and analysis is by a two-part mixed method: firstly, quantitative assessments of different window refurbishment options and secondly, qualitative expert interviews based on these results.

### 3.1 QUANTITATIVE WINDOWS REFURBISHMENT ASSESSMENTS

The quantitative assessments are based on multiple-criteria and methods collected through a literature review [11]. The holistic concept for historic building renovation with a goal of reducing environmental impact can include assessment criteria such as heritage impact, energy efficiency, carbon footprint, indoor environmental quality (IEQ), cost and other environmental related aspects. For this case study, quantitative assessments were carried out in terms of thermal comfort (% of comfort hours during the whole year), energy life cycle analysis, carbon footprint, and cost analysis (Net Present Value (NPV)). Each retrofit measure was assessed considering an energy and carbon life cycle analysis (LCA) of 90 years (as the historic building has already 90 years), and a life cycle cost (LCC) for a period of 20 years.

Different tools and methods were chosen for each assessment. The thermal comfort hours and operational energy (OE) were generated using the energy simulation software EnergyPlus (U.S. Department of Energy) coupled with a SketchUp/Open Studio generated geometry, using local weather data to generate the indoor temperatures of different zones (e.g. living and bedrooms) for a year of 8760 hours. The free running thermal comfort simulation results (indoor air temperature) were graphed in terms of % of annual hours of comfort with a thermal comfort range from 18°C to 25°C [18]. For operational energy, the heating set point was 20°C with standard dwelling occupancy, schedules and people loads, lighting and equipment from the NZS 4218:2009 [21].

For the life cycle energy (LCE) and carbon analysis (LCCA), the software LCA-Quick Version 3.4 from Building Research Association of New Zealand (BRANZ) was used as the more reliable database of material coefficients for product and construction stages. The LCE/LCCA in the LCA-Quick is calculated in A, B, C and D phases, following EN 15978:2011 [22]. Phase A comprises the embodied energy of materials in the product and construction stages; Phase B comprises maintenance and replacement of components, and operational energy within the use stage of 90 years; Phase C considers end-of-life stage – demolition; and Phase D the benefits and loads beyond system boundaries, such as reuse, recovery, recycling and energy generation/export potential after onsite consumption [22]. For this research, Phase C, end of life, was excluded as the main purpose of upgrading historic buildings is to protect them from demolition while increasing useful life. In an average NZ house, the space heating represents the largest single energy use, about one-third of the total energy use [18] so in this study, the operational energy inputted in LCA-Quick for the final results is only the EnergyPlus heating energy (in kWh per year). The life-cycle energy and carbon footprint are presented respectively as Total Primary Energy (TPE) in units of TJ and potential impact in global warming by kg of $CO_2$ equivalent.

The Net Present Value (NPV) of the total investment, comprising construction, maintenance, operational and replacement costs was calculated over a period of 20 years building's life cycle. NPV is the present value of a total investment and it covers all costs incurred during a specified period of time or life-cycle. The future costs are discounted from the date on which they occur back to the present date and then added to the whole life cycle cost [23]. The maintenance costs were not included in this analysis as they are assumed to be equal in each refurbished scenario. To calculate the construction and replacement costs, the Rawlinson's New Zealand Handbook was used as a national database for purchasing, construction and labour rates of different materials providing reliable cost information [24]. The operational cost considers the cost of electricity for residential space heating, and it was taken as the March 2019 rate of 29.08 cents/kWh [25]. An interest rate of 8% averaged from 1985 until 2019 [26] was used to calculate the NPV for a period of 20 years. The final graphed results of hours of comfort (%), LCE, LCCA and LCC were presented and discussed in the qualitative expert interviews in order to understand levels of heritage impact of proposed options and possible gaps and barriers.

### 3.2 QUALITATIVE INTERVIEWS

Individual interviews were undertaken with 10 experts from private and public organizations – 5 conservation architects, 1 assets manager, 1 City Council heritage advisor, 1 policy advisor, 1 urban planner, and 1 building science and services engineer. A convenience and snow ball sampling resulted in this range of professionals representing the stakeholders involved in holistic renovations of historic buildings. Their selection was also based on their different disciplines and background, level of professional experience, and involvement in the heritage conservation sector. The 10 one-hour recorded interviews used an in-depth semi-structured survey that was later transcribed. The questions asked during interviews aimed to understand 'which' and 'why' retrofit options would beconsidered the best and the worst, 'how' the ICOMOS conservation principles would align with their choices, 'which' assessment criteria and tools would be considered relevant for a sustainable retrofit, and ultimately possible barriers including costs and gaps in need of improvement for future action.

The terms 'renovation', 'retrofit', and 'refurbishment' used in this paper are considered synonyms, representing the historic building performance upgrade or improvement, while 'historic building' refers to any building with a historic value, not just heritage buildings.

## 4. RESULTS AND DISCUSSION

### 4.1 QUANTITATIVE WINDOWS REFURBISHMENT RESULTS AND DISCUSSION

Two occupancy zones with different orientations – bedrooms (NE/SE) and living (NW/NE) – were assessed. Figure 2 shows the comfort results (zone indoor air temperature) of the existing building (Option 1 or base model) and the 3 refurbished solutions (Options 2, 3 & 4). There is little influence from the new windows on comfort levels. Options 2 and 3 have same results with a slight increase of 2% for comfort hours in both zones. Option 4 has the best result with 3% increased comfort hours due to the higher R-value of 0.57±0.08 $m^2$ K/W [27] from adding low-E glass. No option significantly contributes to increased indoor temperatures due to the building's low window to wall ratio (WWR).

The same trend of a small decrease in heating energy, annual energy costs and all life-cycle analysis (energy, carbon and costs) is seen in Table 2, highlighting that merely retrofitting windows is not the optimal solution for this building. All window refurbishment analyses show a decrease of less than 15% in energy consumption, carbon footprint and costs. The similar pattern in results is due to OE being the main contributor, so over 90 years Phase B2 (operational energy for heating) has the greatest influence on the final results compared to embodied energy and maintenance

From all retrofit options, the best in terms of savings is the secondary low-E glazing – not only due to the best OE savings, but also because it has the lowest embodied energy (506.8MJ). The life cycle analysis is important to understand not only which phase(s) affect the final result, but also whether it is possible to identify which materials have lower embodied energy and carbon, or even which have benefits and loads beyond system boundaries (phase D). The LCC reveals that even in a shorter period of 20 years the heating energy is still contributing more than construction, maintenance and replacement costs. This may be due to the lifespan of materials being more than 20 years. For a longer analysis period these results may change, but this depends on future energy prices.

The act of undertaking this analysis found there is a need for more environmental product declarations (EPD) to provide reliable data. In this case study, LCA-Quick material library [22] was used for material coefficients; but it is still incomplete. For example, data on acrylic sheet was not available, so PVC with 35 years lifespan was used as the closest available option. The impact on analysis is unknown. However, it is important to highlight that even with such uncertainties, the LCA results are an important tool for retrofit measure comparisons. The finding of the high influence of operational energy and the low benefits from window improvements are valuable to assist in the refurbishment decision-making process.

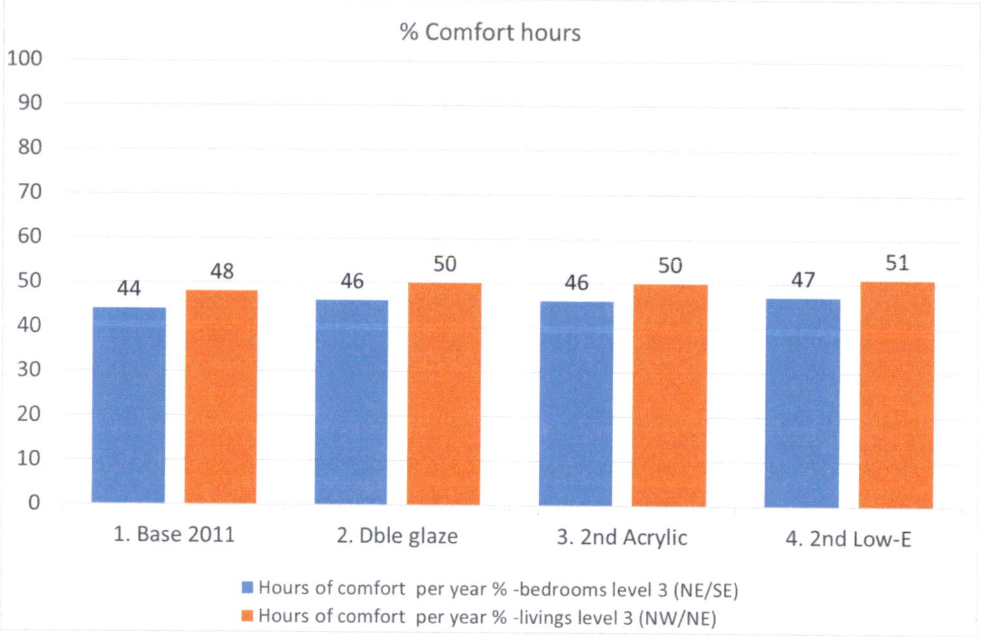


**Figure 2. Comfort hours (%)**

**Table 2. Results of different retrofit options compared to base model**

| **Window option** | **Heating energy per conditioned area** *(kWh/m² year)* | **Annual energy costs** *(NZ$/m² year)* | **LCE** *(TJ) over 90 years* | **LCCA** *(million kg CO2 eq.) for 90 years* | **LCC** *(NZ$ million) for 20 years* |
|---|---|---|---|---|---|
| *(1) Base model – 2011 existing windows* | 56.0 | 1,628 | 7.2 | 51 | 6.4 |
| *(2) double-glazed IGU thermally broken frame* | 51.3 | 1,493 | 6.6 | 46.8 | 5.9 |
| *(3) secondary glazing using an internal, removable acrylic sheet* | 52.1 | 1,515 | 6.7 | 47.5 | 6 |
| *(4) secondary glazing using a non-removable, low-E glass* | 48.6 | 1,413 | 6.3 | 44.3 | 5.6 |

### 4.2 QUALITATIVE INTERVIEW RESULTS AND DISCUSSION

The ten experts were individually presented with the case study results and questioned on their view of the 3 retrofit measures and how the conservation principles aligned with their chosen refurbishment solution. The multiple assessment criteria were also analysed to understand the current practice, and challenges or gaps to implement these criteria.

Regarding the heritage impact, some professionals considered that the options were similar. Interviewee 5 said: *"since the steel windows have gone, I don't feel like there is really that much difference from thinking about heritage, whether you go double glaze or put in a secondary glazing"*. However, the secondary glazing was not chosen by any expert. Interviewee 3 responded: *"I'd look at those secondary ones, for instance if the windows would still remain the same"*. As the windows had already been changed, most accepted double glazing; however, 5 experts recommended *"double glazed steel frames to see if it was possible to going back to something that was closer to the original design"*. Secondary glazing was often associated with issues on correct installation, revealing a lack of knowledge and technical retrofit solution guidance. Interviewee 1 said *"condensation is always a problem"* suggesting other measures are also needed to avoid dampness or mould.

Heritage conservation experts were also asked about how the recommended ICOMOS NZ Charter [28] conservation principles would relate to their choice of retrofit solution. As the windows didn't show significant energy savings or comfort improvements, some professionals considered that if windows were still original they would leave them, following the minimum intervention principle. Interviewee 1 said: *"I would tend to if it's still original single glazed steel frame for example, to keep that and not change it, but then add pretty good insulating curtains"*. Beyond minimum conservation, three important principles were also often highlighted by experts – visual appearance, compatibility and reversibility, although conservation architects, *"try to achieve all of those (ICOMOS NZ) principles"* said Interviewee 1. Compatibility is also key as it avoids future problems or original fabric damage. Interviewee 6 said: *"There's no point in doing an alteration to a building or introducing new materials, if it's going to be detrimental to the existing"*. This statement aligns with the concept of whole-building approach that advocates that every changed component has an impact, and if not looked holistically it may have a bad effect for the building.

For assessment criteria and methods, thermal comfort (often associated with health) and $CO_2$ emissions were the two most important criteria, following after heritage conservation as the leading choice by 9 of the 10 experts. Interviewee 3 mentioned *"the quality of the environment for the people who are living there is the most important"* and Interviewee 8: *"...that's human health really...we spend millions every year in hospital stays...people staying in houses that are out to not good for human health"*. Energy efficiency was not always chosen as a leading refurbishment criterion, but they recognized that it is related to $CO_2$ emissions and savings. Overall, cost was considered a huge barrier. Interviewee 1 said: *"I think cost would be a major issue"*. Moreover, experts also expressed the importance of having durable and sustainable sourced materials and agreed the LCA analysis was a helpful tool.

Professional skills, guidance and expertise are essential when working on historic buildings. However, lack of knowledge which in turn suggested miscommunication among different stakeholders, was found to be one of the main challenges. This is an important issue that needs to be addressed before retrofit policies are implemented. Interviewee 4 mentioned that *"there's not that knowledge of how you can retrofit windows"*, and the *"limited number of conservation architects in NZ, and that there's no accreditation process for them"*. Internationally there is a wealth of technical guidance and given most of the conservation experts interviewed have an overseas diploma or experience, this possibly reveals restricted local training resources for the small number of local heritage professionals. The experts also confirmed a lack of local guidelines and proper advice for historic building retrofits, and acknowledged that there is important guidance material overseas but not that much tailored for the country. All professionals described an overall lack of experience and a poor level of understanding in energy upgrades in historic buildings from different stakeholders.

## 5. CONCLUSION

The case study assessment presented here coupled with the interview findings confirmed a NZ knowledge gap of technical solutions, methods, processes, experience and even practice in renovating historic buildings to reduce their environmental impact. This research was received with interest from interviewees, but with concern from the conservation professionals' sub-group, highlighting the lack of retrofit practice and hence novelty of the subject.

In this specific case, the assessment results for thermal comfort, LCE, LCCA and LCC revealed that only refurbishing windows does not have significant improvements from either energy or conservation experts' perspective. This results from few areas of existing windows compared to the area of uninsulated 'cold' walls, therefore little energy savings coming from improved windows. From a conservation view, it would also not justify changing the windows if they were still original. As the windows had already been modified, replacement with new frame double-glazed windows was acceptable, but the cost of implementing would be a barrier, especially as the long-term benefits are not high. It should be noted that this might not be the case for another building. However, the improved performance of windows, coupled with improved ventilation, could contribute to a better indoor environment with reduced condensation on windows and lower risks of mould. Thus, a whole-building approach should be recommended instead of single measure retrofits [29], [30]. "Whole-building" analysis and measures, such as envelope insulation and ventilation systems should help avoid future (for example) moisture problems even with compatible materials.

The interviews with experts found the use of suitable criteria and methods were important for decision-making process; however, there was a general lack of knowledge as to the use of these tools. LCA was confirmed to be useful to ensure the choice of appropriate materials and to reduce GHG emissions, although the limitations that both thermal and LCA simulations have due to the assumptions. LCA tools are still new to the building sector and have room for improvement, especially as much material data is unknown. User behaviour may differ from simulated data and this will interfere in the results. For this reason, in any renovation work it is important to have previous monitoring coupled with post occupancy evaluation on completion so that necessary adjustments can be made. Health and comfort criteria often seemed to be of more interest to the conservation professionals than energy efficiency. Interestingly, the research suggests that if policies were driven in this direction both IEQ and energy have potential to benefit.

The main gap identified as needing future actions in NZ is an urgent need for education in both the energy and conservation fields. Beyond investment in training, an accreditation process for professionals working in conservation would be of longer benefit along with providing feedback to continue to improve the performance of this sector and the long-term sustainability of historic and heritage buildings.

## 6. REFERENCES


[1] A. Martínez-Molina, I. Tort-Ausina, S. Cho, and J.-L. Vivancos, "Energy efficiency and thermal comfort in historic buildings: A review," *Renew. Sustain. Energy Rev.*, vol. 61, pp. 70–85, Aug. 2016, doi: 10.1016/j.rser.2016.03.018.

[2] A. L. Webb, "Energy retrofits in historic and traditional buildings: A review of problems and methods," *Renew. Sustain. Energy Rev.*, vol. 77, pp. 748–759, Sep. 2017, doi: 10.1016/j.rser.2017.01.145.

[3] J. Hosagrahar, J. Soule, A. Potts, and L. Girard, "Cultural Heritage, the UN Sustainable Development Goals, and the New Urban Agenda." Sep. 2015.

[4] UN, "The Sustainable Development Goals Report 2017," United Nations, 2017. [Online]. Available: https://sustainabledevelopment.un.org
[5] F. Haas, D. Herrera, W. Hüttler, D. Exner, and A. Troi, "Historic Building Atlas Sharing best practices to close the gap between research & practice," presented at the EEHB2018, Visby Sweden, Sep. 2018.
[6] S. Bond and D. Worthing, *Managing Built Heritage: The Role of Cultural Values and Significance.*, Second edition. Chichester, West Sussex, United Kingdom: Wiley Blackwel, 2016.
[7] S. Lidelöw, T. Örn, A. Luciani, and A. Rizzo, "Energy-efficiency measures for heritage buildings: A literature review," *Sustain. Cities Soc.*, vol. 45, pp. 231–242, Feb. 2019, doi: 10.1016/j.scs.2018.09.029.
[8] CEN, "EN 16883:2017, Conservation of cultural heritage — guidelines for improving the energy performance of historic buildings," CEN, Brussels, 2017.
[9] ASHRAE, "ASHRAE guideline 34P, Energy guideline for historical buildings," ASHRAE, 2019.
[10] "IEA task 59." http://task59.iea-shc.org/
[11] R. Paschoalin and N. Isaacs, "Holistic renovation of historic and heritage buildings: comparing New Zealand and international scenarios.," *International Journal of Building Pathology and Adaptation*, Nov. 2020.
[12] New Zealand Government, *Climate Change Response (Zero Carbon) Amendment Act 2019*. 2019. [Online]. Available: http://www.legislation.govt.nz/act/public/2019/0061/latest/LMS183736.html
[13] P. Besen and P. Boarin, "Energy retrofit of historic buildings in New Zealand: current practice, challenges and opportunities," presented at the 54th International Conference of the Architectural Science Association 2020, Auckland, N.Z, Nov. 2020.
[14] Ministry of Business Innovation & Employment, "Building for Climate Change: Transforming operational efficiency and reducing whole-of-life embodied carbon." Aug. 2020. [Online]. Available: https://www.mbie.govt.nz/have-your-say/building-for-climate-change-transforming-operational-efficiency-and-reducing-whole-of-life-embodied-carbon/
[15] S. Price, "Chevening, 90 Salamanca Road, Kelburn, Wellington.," Wellington, N.Z., 2019.
[16] Wellington City Council., "Planning our suburbs: Wellington City Council.," 2016.
[17] S. Price, *Maintenance and Housekeeping Manual, 90 Salamanca Road*. Unpublished document prepared by the previous owner, 2019.
[18] Isaacs, N. P., Camilleri, M., Burrough, L., Pollard, A., Saville-Smith, K., Fraser, R., Rossouw, P., & Jowett, J., *Energy use in New Zealand households final report on the household energy end-use project (HEEP)*. Porirua, N.Z: BRANZ, 2010.
[19] P. Arnold, "Chevening Apartments -condensation / waterproofing," eCubed Building Workshop Ltd., Wellington, N.Z., Jul. 2012.
[20] A. Troi and Z. Bastian, *Energy Efficiency Solutions for Historic Buildings: A Handbook.* Berlin, München, Boston: DE GRUYTER, 2015.
[21] Standards New Zealand, *Thermal insulation: housing and small buildings.* 2015.
[22] CEN European Committee for Standardization, "EN 15978:2011 Sustainability of construction works – Assessment of environmental performance of buildings – Calculation method.," CEN, Brussels, 2011.
[23] N. Mithraratne, B. Vale, and R. (Robert J. D. Vale, *Sustainable living: the role of whole life costs and values*. Abingdon: Routledge, 2011.
[24] Rawlinsons Media Limited, "Rawlinsons New Zealand Construction Handbook." Rawlinsons Media Limited, 2012.
[25] Ministry of Business Innovation & Employment, "Real and nominal price data relating to New Zealand's energy prices — petrol, diesel, fuel oil, natural gas and electricity." 2020. [Online]. Available: https://www.mbie.govt.nz/building-and-energy/energy-and-natural-resources/energy-statistics-and-modelling/energy-statistics/energy-prices/
[26] Reserve Bank of New Zealand, "Monetary Policy Statement." 2019. [Online]. Available: https://www.rbnz.govt.nz/monetary-policy/monetary-policy-statement
[27] N. Smith, N. Isaacs, J. Burgess, and I. Cox-Smith, "Thermal performance of secondary glazing as a retrofit alternative for single-glazed windows," *Energy Build.*, vol. 54, pp. 47–51, Nov. 2012, doi: 10.1016/j.enbuild.2012.07.038.
[28] ICOMOS NZ, "ICOMOS New Zealand Charter for the Conservation of Places of Cultural Heritage Value." 2010. [Online]. Available: https://icomos.org.nz/charters
[29] N. May and N. Griffiths, "Planning responsible retrofit of traditional buildings," *Sustain. Tradit. Build. Alliance STBA*, p. 24, 2015.
[30] English Heritage, "Energy Efficiency and Historic Buildings - Application of Part L of the Building Regulations to historic and traditionally constructed buildings." 2012.

# The Carbon Value of the U.K.'s Historic Housing Stock

Dr James Ritson

UCEM, Reading, United Kingdom. j.ritson@ucem.ac.uk

**Abstract** – The need to reduce carbon emissions and lower the energy consumption of the historic built environment is now recognized as a critical factor in helping the U.K.'s government's aim to reach net-zero carbon emissions by the year 2050. This paper proposes rather than encourage historic homeowners to sustainably refurbish their properties, it proposes that the most sustainable option is to adopt a building conservation-focused strategy to maintain and apply small benign changes to the property. The primary data is from testing a range of different sustainable improvement interventions on 20 different historic houses using computer modelling and live data. The paper will show that significant energy and carbon savings can be made without affecting the visual or fabric heritage of the property. The study will go on to show that this strategy is also the most economically effective method for sustainably refurbishing historic dwellings. The paper concludes by defining the balance of the competing priorities of economic capacity, the preservation of the heritage of the historic housing stock and environmental performance improvements happens at a key 'tipping point' which is used to define the 'carbon value' of our historic housing stock.

**Keywords** – Sustainable Refurbishment; Carbon Value; Economic value; Historic Dwellings; Sustainable strategy.

## 1. INTRODUCTION

The reduction of carbon emissions is now regarded as one of societies' most important challenges in the 21st century. With the UK's Existing housing stock contributes 27% of national $CO_2$ emissions,[1] and it is predicted that two-thirds of the dwellings that will be standing in 2050 are already in existence [2]. Improving the performance of existing dwellings is therefore vital in helping to reduce the ecological footprint of the UK as a whole. This paper sets out to define the heritage difficulties and the economic barriers that need to be overcome with the historic housing stock in England if these sustainable refurbishment targets are to be met. The UK has one of the oldest building stocks in the developed world, and there are currently around 4.7 million historic dwellings [3]. So, with the challenge to refurbish this large number of properties to reduce their carbon emissions while at the same time, preserve their heritage and value requires a different approach. This paper proposes that the most suitable method of sustainable refurbishing a historic UK home is to focus on small benign changes and maintenance methodologies rather than an invasive environmentally-focused refurbishment strategy for each dwelling. This large number of refurbishments needed also has enormous economic implications and therefore for the consideration has to be taken to balance the different priorities of the reduction of carbon emissions, the preservation of the historic housing stocks inherent heritage and economic capacity has to be considered if such a refurbishment methodology is to be successful.

## 2. CONTEXT

The U.K. has one of the oldest building stocks in the developed world. This group of buildings are defined as hard to treat, and strategies for their sustainable refurbishment remain ambiguous at best and at worst damaging to the fabric of the building. The U.K.'s carbon reduction target is net-zero carbon by the year 2050. As part of this strategy, the target of all dwellings in England and Wales to have an equivalent environmental performance of an energy performance certificate (EPC) grade C or higher by 2035. The pre-1919 housing stock in the U.K. has, on average, the worst SAP score and the highest carbon emission of any house age group, and typically, over twice the maintenance costs compared with modern housing for basic repairs [4]. There are over 4.7 million of these dwellings in England alone [5] which equates to over 420 home refurbishments every single day from now until 2050 if the net-zero carbon emissions goal is to be met. More drastic is if the target of refurbishing all dwellings by 2035 to be reached this would mean that 850 refurbishments every day need to be completed between now and 2035.

## 3. PROJECT AIMS

The project hypothesis is 'The most sustainable strategy for owners of historic dwelling does not lie in sustainable focused refurbishment of their dwellings but in historic building maintenance and benign improvements.' The overall aim of the project is to show that building maintenance and carefully selected interventions, could significantly improve the environmental performance of historic dwellings and at the same time be economically viable and culturally beneficial to the preservation of the historic asset.

### 3.1 METHODOLOGY

The primary data for this study comes from analysis of 20 different historic dwellings in England. The dwellings came from a range of sources; 4 dwellings from The Reading case-study project [6], 12 dwellings from the Redbridge project [7][8], A further 4 dwelling were tested to complete a range of historic urban/suburban dwelling typologies found in England. The dwellings were tested using a range of techniques; the computer modelled buildings were tested using the Government's Standard Assessment Procedure (SAP) calculation for domestic energy consumption and carbon emissions. The NHER Plan Assessor [9] was used to simulate existing environmental performance of the dwellings and then range of improvements. Data from other dwellings included live and actual energy consumption and carbon emission results collected from the dwelling following refurbishment. In each case, a variety of environmental performance improvement interventions were tested. These included conservation-based maintenance and benign environmental improvements which have little or no effect on either; the visual heritage of the dwelling or damage to the historic fabric of the dwelling or impacted the building's physical properties (such as moisture transfer) (see section 3.3). For comparison also tested were other common sustainable interventions such as replacing the single glazed windows with double glazing (to compare these changes with the benign conservation changes). Each intervention was tested against the following criteria: cost of the intervention, the amount of reduction is $CO_2$ (eqiv.) emissions and reduction in energy consumption. From these results cost to benefit calculations were derived.

### 3.2 HISTORIC BUILDING MAINTENANCE AND BENIGN CHANGES

It is important to understand that the fabric and the appearance of a historic dwelling have cultural significance - the building itself is an artefact and historical asset. The idea of approaching work from a minimum intervention methodology is best summarised by the Burra Charter [10] "as much as necessary, as little as possible". The methodology for this study is the improvement in energy saving and carbon emissions reduction with as little damage or change to the inherent heritage of the historic dwelling. The Historic Town Forum [11] supports this methodology stating that "One of the most energy efficient ways to preserve historic buildings is to ensure that continued, regular maintenance is carried out to safeguard its historic fabric." Both the Historic Town Forum and English Heritage encourage the use of small/benign changes to improve the environmental performance of a historic dwelling. Benign changes are defined as changes to the building that either have little or no effect on the heritage of the dwelling or do not damage the dwelling fabric either to the fabric itself or the way it needs to perform or react. Typical benign interventions include installing of loft insulation, draught proofing the building, insulating the hot water cylinder (if applicable), replacing a non-condensing boiler with a high efficiency condensing boiler, improving the heating controls, installing energy-saving lightbulbs & installing floor insulation in raised timber floors. Maintenance tasks such and servicing of heating systems were also included as well as Periodical renewal of elements with a set lifespan, be they sacrificial elements such as paint or appliances as long as their replacements meet the requirements of a benign intervention were also included in the study.

## 4. THE RESULTS OF THE STUDY AND TIPPING POINT & CARBON VALUE

The aim of this study was to investigate if maintenance and benign changes could be seen as the most sustainable approach for the refurbishment of historic suburban dwellings.

**Table 1 overall energy savings of the benign interventions from the study**

| Action | Percentage Energy Saved % | | | Capital Cost Used in Study (£) | Impact on Fabric Heritage | Impact on Visual Heritage |
|---|---|---|---|---|---|---|
| Upgrading the loft insulation to 300mm | 4.0% | to | 31.1% | £273.00 | LOW | LOW |
| Draft proofing and window repair | 2.0% | to | 10.0% | £50–£2000 | LOW | LOW |
| Hot water cylinder insulation to >75mm | 3.6% | to | 8.7% | £20.00 | LOW | LOW |
| Fitting of a condensing boiler | 16.0% | to | 46.0% | £1,750.00 | LOW | LOW |
| Improved heating controls | 12.0% | to | 14.1% | £250.00 | LOW | LOW |
| Energy saving light bulbs | 0.1% | to | 0.2% | £200.00 | LOW | LOW |
| Floor insulation fitted in raised timber floor | 8.3% | to | 14.0% | £1,000.00 | LOW | LOW |

First overall finding is that this study found that benign maintenance (conservation focus) refurbishment as the potential to save between 30% and 50% of carbon emissions along with this up to around 40% savings in energy consumption. To be considered sustainable cultural, economic and environmental factors have to be considered and ideally in balance. There is a point at which these factors become in equilibrium. This point is defined as the tipping point.

All of the primary results when mapped against carbon reduction and cost of intervention from the primary research showed the same pattern showed in figure 1. Figure 1 compares the cost savings of the building intervention (set of interventions) against the $CO_2$ Saving Incurred.

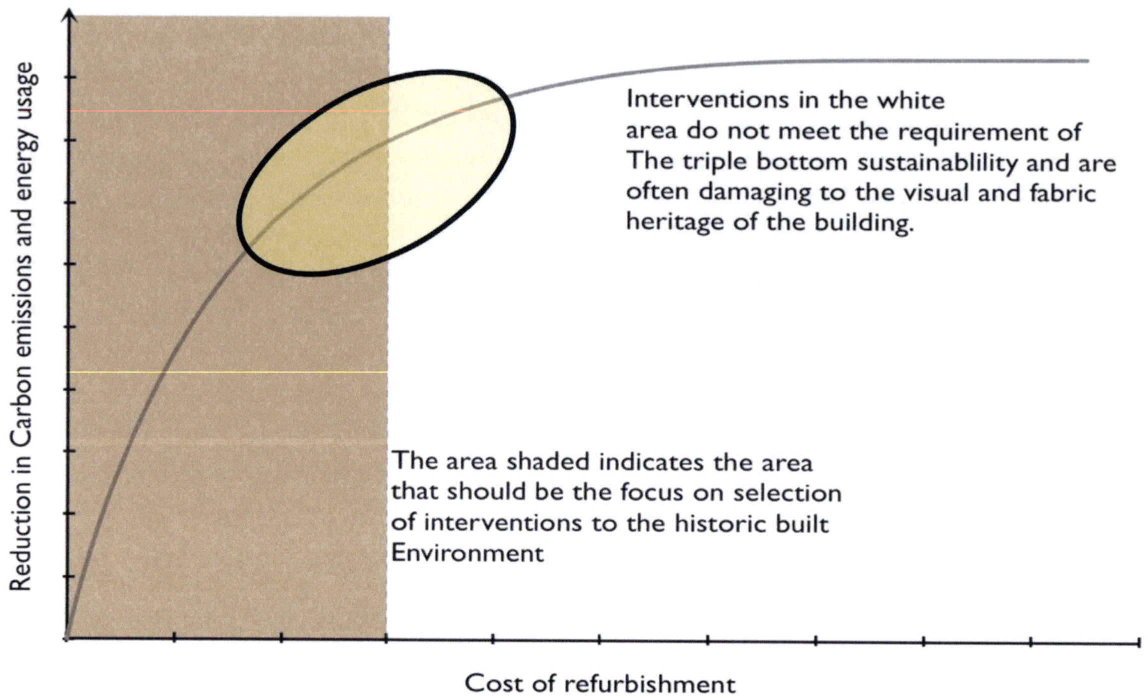


**Figure 1 Comparing the cost savings of the building intervention against the $CO_2$ Saving Incurred tipping point highted in yellow**

The point circled on figure 1 is the point in which the gradient changes significantly this can be seen as a tipping point or the turning point in which the rate of the cost to benefit (carbon saving/energy savings) changes in relation to the amount of financial costs of the sustainable interventions applied to the dwelling. This cost benefit analysis (CBD) starts to put real-world numbers on the findings and the hypothesis of this study. In the cost benefit analysis, the value unit is the financial cost of the intervention and the benefit is the reduction in $CO_2$ emissions from that intervention. The units are defined as: £ per KG $CO_2$ reduced or £ per %$CO_2$ reduced. Steeper the gradient shown in the chart the more $CO_2$ is saved per pound spent, in other words, greater the cost benefit ratio. Smaller the cost benefit ratio and lesser the amount of $CO_2$ saved per pounds spent on the intervention. For the best balance between the economic and environmental values of an intervention should be calculated. All of the primary results follow a similar trend. While the individual buildings follow slightly different result gradients, the trend remains constant. In the Redbridge study the ratio for the benign changes £6.71 per kg $CO_2$ which then rose to £42.90 per kg $CO_2$ past the tipping point. In the Reading study ratio for the benign changes £2.54 per kg $CO_2$ which then rose to £27.02 per kg $CO_2$ past the tipping point. This ratio changes but the tipping point remain within a consistent tipping point. This Tipping point occurs around the £3000-£7000 mark and show a carbon emission savings of between 30% and 50%.

If the triple bottom-line criteria are taken into account this is the point where environmental and economic values could be seen to be in balance or at least to be in the most efficient. This would be the case if all of the interventions are seen as benign, little to no damage to either the visual or fabric heritage. This is the point in which the graph gradient turns from a steep slope to a gradual incline, this tipping point is a key part of the discovery of this study as it shows the balance point between the economic and the environmental values.

### 4.1 CARBON VALUE

From the tipping point it is possible to begin to define what is possible in reducing the carbon emissions from historic dwelling within the financial capabilities of the owner and without damaging the heritage (visual and fabric) of the dwelling. This number when compared to the overall target of reduction in $CO_2$ emissions is labelled the carbon value of the heritage of the dwelling. This is the differential between what is economically and culturally possible calculated against the perceived target of the reduction in $CO_2$ emissions. It is possible to define the carbon value in a simple equation and this could be translated as

| | | | | |
|---|---|---|---|---|
| *Carbon value of heritage* | = | Target Carbon emission Reduction | – | Total Carbon emission saving that can be achieved without damaging the heritage of the building |

As defined earlier in the study the benign changes are interventions that do not have a negative impact on either visual or fabric heritage so therefore the question can be written as

| | | | | |
|---|---|---|---|---|
| *Carbon value of heritage* | = | Target Carbon emission reduction | – | Total Carbon emission saving from the benign changes |

This can be further rewritten as to bring in the third value of economic limitations equation has to be further defined to bring in the economic limitations

| | | | | |
|---|---|---|---|---|
| *The triple bottom line carbon value of heritage* | = | Target Carbon emission Reduction | – | Total Carbon emission reduction of the benign changes which are financially viable |

At this point, the tipping point, results can be used to provide the owner with the best value, this is the best cost to benefit in the case of historic suburban housing. The results show that the tip point occurs in the range of £2000 to £7000 which provides a carbon emission saving of between 30% and 50%. When these numbers are put into the final equation the results show that the triple bottom-line carbon value of Historic dwellings is between 30% and 50% of carbon emissions target.

### 4.2 COST DIFFERENTIAL

While the carbon value shows the gap between the proposed target and what the study finds is optimal in terms of carbon emissions reduction in Historic suburban dwellings (in England). This could be seen as a failure to achieve the desired target. However, the tipping point and the cost benefit analysis highlights the economic reality of trying to reduce the historic housing stocks carbon emissions by the UK government's target of 80% to 100% reduction in carbon emissions by using refurbishment methodology.

**Table 2 Overall Refurbishment costs compared for total Pre-1919 housing stock**

| Type of interventions | cost | number of dwellings | total cost |
|---|---|---|---|
| **Upper full refurbishment** | £80,000.00 | 4,700,000.00 | £376,000,000,000.00 |
| **Mid full refurbishment** | £40,000.00 | 4,700,000.00 | £188,000,000,000.00 |
| **Lower full refurbishment** | £20,000.00 | 4,700,000.00 | £94,000,000,000.00 |
| | | | |
| **Upper benign changes** | £7,000.00 | 4,700,000.00 | £32,900,000,000.00 |
| **Mid benign changes** | £3,000.00 | 4,700,000.00 | £14,100,000,000.00 |
| **Lower Benign changes** | £2,000.00 | 4,700,000.00 | £9,400,000,000.00 |

It would cost between £9 billion and £32 billion to reduce their historic housing stock in the U.K.'s carbon emissions to around 30% to 50%. To reduce the same stock by an additional 30% to 50% (to meet the government target) and additional £61 billion-£373 billion will be needed. This additional cost needs to be seen within the context of the overall cost to benefit for the country as a whole. There is a tenfold increase in financial cost to increase the saving from 30%-50% $CO_2$ emissions to 80%-100% reduction in $CO_2$ emissions. This large jump in cost raises the question whether the cost of the further intervention (above the benign intervention) can be better spent elsewhere in policy such as greening the electricity grid which would benefit the whole of the built environment rather than a small group of buildings.

The building is industry capacity also needs to be taken into account with 4.7 million pre-1919 dwellings in England this would equate to over 420 refurbishments to be completed every day from now until 2050. If the target of refurbishing old dwellings by 2035 is to be reached this would mean that 850 dwellings every day need to be completed between now and 2035.

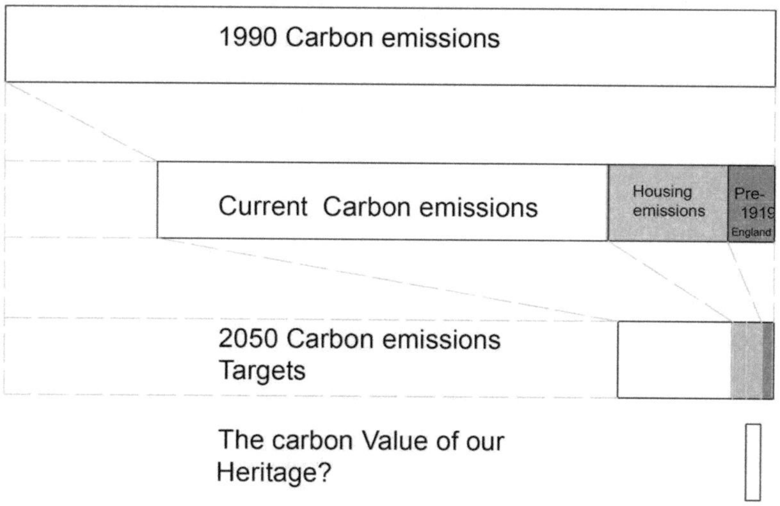


**Figure 2 Carbon Value of heritage 80% emission target**

The figure 2 show the carbon value of the heritage. That show the size of the Triple Bottom line Carbon Value of our historic housing. The small size of the carbon value can be seen against the other savings required.

### 4.3 OTHER USES FOR THE CARBON VALUE

As much as the carbon value was intended to be used as a decision-making tool it could however be used for other purposes. The carbon value could be used in further scenarios: it could be used as a measure for carbon taxation applied to historic dwellings or for the use of offsetting to help meet the target of net reduction in carbon emissions. It is the net target that is important as part of the government's carbon reduction strategy.

## 5. CONCLUSION

While it is accepted that the historic built environment must reduce its energy consumption and lower their carbon emissions but at the same time the need to preserve inherent heritage. This defines the need for change. If we take the definition of building conservation as the management of change [12] and the basis of sustainability is the balance of the triple bottom line. The study has shown that benign changes and maintenance do offer a triple bottom line sustainable strategy for lowering carbon emissions and increased energy efficiency of historic suburban homes.

The study has shown that benign changes and maintenance are cost-effective with a high cost to benefit ratio benign changes help preserve the cultural value and historic fabric of the dwelling and finally the study shows that the Tipping point occurs around the £3000-£7000 mark and show a carbon emission savings of between 30% and 50%. The study clearly shows that the optimal balance between economic measures and environmental improvements can be found at the tipping point. After the tipping point the cost benefit ratio decreases and becomes increasingly more expensive to lower carbon emissions and reduce energy consumption. It must also be noted that as the interventions to the property are either benign or maintenance based there is little or no impact to the visual heritage, furthermore, maintenance is critical to the survival of the historic fabric of the building. Historic building maintenance, periodical renewal and benign improvements methodology can be expected to get most pre-1919 dwellings up to an EPC level C rating. Real-life limitations have to be taken into account such as the amount of financial capital, the capacity of the built environment to be able to refurbish a large number of properties *et cetera*. The vast scale of the number of interventions and refurbishments needs to be understood. While it has been shown that it is technically possible to refurbish a dwelling beyond the tipping point, the time and resources needed to do such a refurbishment provide their own limitations: with 4.5 million of these dwellings in England alone, this equates to at least 425 refurbishment every single day from now until 2050 so any policy/strategy for encompassing all of the historic built environment dwellings needs to able to be scaled up simply to meet the huge number of refurbishments that have to be completed. Another key point to support the hypothesis is that benign changes and maintenance is **not** a set, restrictive strategy. Benign changes and maintenance do **not** restrict other sustainable improvements to take place on the dwelling, if correctly applied, actually they should support them. The strategy does not rely on a single large refurbishment completed at a single point, but in a collection of small interventions done over a period of time.

In final conclusion then, if all historic dwelling sustainably refurbished to their tipping point and those interventions are benign, then the balance between the need to lower carbon emissions and energy consumption, then need to preserve the heritage of the building and the need for the intervention to offer the best values and be affordable will be met and be at the optimal balance of the triple bottom line sustainability requirements. After the tipping point, the question is which of the triple bottom line has to give way to the other criteria or does it require the use of other models such as carbon off-setting or carbon taxation.

## 6. REFERENCES


[1] Francis, A. & Wheeler, J. "One Planet Living in the suburbs", Surrey: WWF. 2006

[2] Boardman, B., Darby, S., Killip, G., Hinnells, M., Jardine, C., Palmer, J. & Sinden, G. "40% House". Oxford: Environmental Change Institute, University of Oxford. 2005

[3] EHCS, "English House Condition Survey 2007, Annual Report" London: Communities and local government. 2009.

[4] EHCS, "English House Condition Survey 2007, Annual Report" London: Communities and local government. 2009.

[5] EHCS, "English House Condition Survey 2007, Annual Report" London: Communities and local government. 2009.

[6] Newman, C. "Reducing Energy Use in Traditional Dwellings". Historic England: Swindon East. 2017.

[7] Thames Group (2012) "Greener Homes for Redbridge", East London Group: London. Available at https://issuu.com/eastthamesgroup/docs/greener_homes_property_portfolio accessed 29-04-14.

[8] Hartless, R. & Standen, M. "Greener homes for Redbridge", BRE: Watford. 2013.

[9] NES. What is an NHER rating? [Online] Available at http://www.nesltd.co.uk/blog/what-nher-rating 2012 [accessed on 20-02-12]

[10] ICOMOS. 'Australia ICOMOS Charter for Places of cultural significance (The Burra Charter 1999)". Paris: International Council on monuments and sites. 1999.

[11] HTF. 'Saving Energy: A Guide for owners of Historic Homes', Bristol: Historic Town Forum. 2011.

[12] Drury, P. & McPherson, A. "Principles, Policies and Guidance for the sustainable management of the historic environment." London: English Heritage. 2008.

# Energy Efficiency Intervention in Brutalist buildings: New Challenges in preservation of 20th century built heritage

E.J. Uranga, L. Azcona, L. Etxepare, I. Lizundia, M. Sagarna

Dept. of Architecture, University of the Basque Country, (UPV/EHU), Basque Country (Spain).
E-mail: enekojokin.uranga@ehu.eus; leire.azcona@ehu.eus; lauren.etxepare@ehu.eus; inigo.lizundia@ehu.eus; maialen.sagarna@ehu.eus

**Abstract –** In order to give continuity to the EU policy, the buildings that make up the building stock should be energetically more and more efficient and emit less $CO_2$ into the atmosphere. To this end it should be analysed if it is possible to improve the energy performance of those existing buildings. The built heritage represents a part of that building stock. A fundamental architectural historic current that occurred during the 20th century is part of this built heritage: Brutalism. This communication presents an assessment of some cases of brutalism around the city of San Sebastian. So as to propose some energy intervention solution, its architecture and constructive solution is analysed and its current energy performance is considered. The challenge of this research is to investigate how an energy efficiency intervention can be made in this type of buildings scattered throughout Europe while preserving their historical and architectural values.

**Keywords** – Energy Efficiency; Built heritage; Intervention; Brutalism

## 1. INTRODUCTION

The problem of excessive energy consumption and $CO_2$ emissions has become one of the most important problems for humanity in the beginning of this 21st century. The EU has identified the construction sector as one of the key areas where action should be taken to improve this situation. Consequently, improving the energy efficiency of buildings has been one of the EU's main objectives for more than two decades. The current building stock must be renewed as it is the sector that consumes the most energy. Thus, it is established by some European directives such as Directive 2002/91/EC [1], Directive 2010/31/EU [2], or Directive 2012/27/EU [3]. But there is still a problem that has not been solved by European legislation, namely what to do with existing buildings that have a heritage value. As the current regulations state, there is a double option when refurbishing energetically existing buildings: if a building is not protected, the refurbishment to improve its energy performance must be the optimum one; on the contrary, if the building has some kind of official protection, it is exempt from any energy improvement. This generates a dichotomy that spans between the intervention that risks the heritage values of historic buildings, and the nonintervention that may lead to an excess of energy consumption and eventually to the abandonment of this type of architecture. A series of valuable buildings from the 60s and 70s are precisely in this situation; they have not yet been officially protected and an indiscriminate energy refurbishment may ruin their initial historical value. These are the buildings of the brutalist style.

## 2. THE BRUTALISM AND ENERGY EFFICIENCY IN BUILDINGS

In the mid-1950s the architect Le Corbusier pioneered a way of conceiving architecture. This new movement subsequently called "Brutalism", or "New Brutalism" was developed in the 1960s and 1970s all around the world [4]. The main idea to conceive these buildings was based on simplicity and honesty in the use of construction materials. Many architects experimented throughout the planet with this new architectural style [5]. In the same way, some of the architects who built in the Basque Country in the 1960s and 1970s adopted this way of conceiving architecture. That includes several cases of brutalist architecture around San Sebastian.

After 50 years since its construction, this architecture style has begun to be valued and some of those buildings are considered as a part of the built heritage. Others, nevertheless, do not yet have such recognition and are not under any official protection. This means that at the moment these buildings can be modified in their architectural configuration, intervened on without any conservation criteria or even demolished. For these reasons, it would be important to recognize officially the value of these buildings as part of built heritage.

On the other hand, we realise that these buildings have poor energy performance. That is due to the fact that, within other reasons, at the time when buildings were designed energy efficiency was not a priority. However, whether they are protected or not, it is considered that an energy refurbishment should be conducted in order to improve their original energy performance. The case of the brutalist buildings is a clear example of the dichotomy that arises in the current energy efficiency legislation in Europe: if the buildings are officially protected it is not mandatory to improve their energy performance; while if they remain unprotected, as some of them are at the moment, they are exposed to an energy intervention where the original architectural and constructive values could be lost. In this type of buildings, there is a clear need to break this dichotomy by ensuring the protection of buildings, but also by improving their energy performance.

Furthermore, the preservation and energy improvement of these buildings is complex particularly due to their construction design. The use of concrete as a unique material for structure and façade enclosure makes it more complicated to intervene in it from an energy point of view. Finding the solution to insulate the building thermally without losing its personality and materiality is a difficult equation to solve.

## 3. BRUTALISM CASES IN SAN SEBASTIAN

In the mid-1960s, when Spain began to open up politically to the world, the international architectural style of the time, brutalism, was beginning to influence Spanish architects. In less than ten years, between 1965 and 1974, three significant brutalist buildings were designed and constructed in the area around San Sebastian by different young architects (Fig. 1).

### 3.1 THREE BRUTALIST BUILDINGS

**– Infant Jesus of Prague School: M. Oriol and G. Lafuente (1965-1967)**

The first building in the brutalist style to be built in San Sebastian was the Infant Jesus of Prague School [6]. The former project that was planned in the 1950s had nothing to do with what it would end up being. On a plot in the new urban development of the Amara district, the architect Luis Jesus Arizmendi designed a first building in a much more traditional style, in accordance with the political ideology of the time. Due to different technical reasons, the commission was finally awarded to two young architects from Madrid: Miguel de Oriol and Gregorio Lafuente. These young architects, more influenced by the growing architectural trend, designed a completely different and impressive building. The project designed in 1965 and built in the following years, used the "béton brut" throughout the different volumes and was the first totally brutalist building to emerge in the city.

**– Official Maritime School in Pasaia: J.L. Zanón and L. Laorga (1966-68)**

Between 1963 and 1968 the architects José Luis Zanón and Luis Laorga designed and built seven Maritime Schools in different cities in Spain. One of them was the Official Maritime School in Pasaia. Although in terms of the programme this school is very similar to the other six, from an architectural and constructive point of view, it is remarkably different from the rest. While the others were based more on the Rationalist style, the school in Pasaia adopted the new international style already known by then as the Brutalist Style [*7]*. The characteristics of this sixth Maritime School are based on the use of the "betón brut" along the whole building, highlighting among other volumetric elements the auditorium and the pyramidal tower. The materiality of most of the building is reflected in the use of concrete as a structural and finishing element. Special mention should be made of the composition of the auditorium in the form of a folded sheet, in which the stalls stand out over a flight of 8.50 metres.

**– Carmelo Balda Fronton: L.J. Arizmendi (1969-1974)**

In 1969, the San Sebastian city council decided that it was necessary to build a new pelota court called Carmelo Balda as a tribute to one of the greatest promoters of Basque pelota in the 20th century. On the occasion of the 6th World Pelota Championships in 1970, as the city of San Sebastian did not have a long pelota court due to the fact that the previous one had been demolished a few years earlier, the design and construction of this new court was undertaken. The architect in charge of this task was the city's municipal architect, Luis Jesús Arizmendi, the same architect who had designed the former project Infant Jesus of Prague School years before. Due to a series of economic and technical problems, its construction was delayed until it was completed in 1974, which meant that the Championship could not be held there. For the design of this building, the architect chose to use the brutalist style as a local response to the incipient international movement [8]. It is true that the two previous cases were already built and that they probably influenced the architect's choice. As in the two previous cases, concrete is used as a structural and finishing element, resulting in a great basic and compact volumetry that responds to the use of the building.

**Figure 1. Infant Jesus of Prague School, Official Maritime School in Pasaia and Carmelo Balda Fronton.**

### 3.2 CONSTRUCTIVE SOLUTIONS

Although the three mentioned buildings present different volumetric and spatial solutions, they all share a single constructional language: skin and structure are basically made of reinforced concrete. Their functional programmes vary greatly, as they are public buildings created for different purposes, but the extensive use of bared concrete is common to all the cases (Fig. 2). Thus, as Brutalism advocated, concrete is used for both structural and enclosure uses. The typical construction procedure is the following: the fresh concrete is poured over a formwork that is removed once the piece has hardened, resulting in a surface finish that copies the roughness of the mold. Normally, wooden formwork is used, but in some cases metal formwork is employed too. The thickness of concrete vary depending on which element it is used in; if it is used in bearing walls, its section width range from 30 cm to 40 cm; in contrast, when it is used in non-bearing ones, it is reduced to 20 cm or 15 cm. We often find that the wall of concrete is backed by a cavity and an inner hollow brick leaf. Thanks to this second leaf, in badly ventilated rooms, condensation was avoided. Finally, in addition to the concrete walls, the joinery is generally made of wooden frames and simple single-pane glass.

It must be taken into account that the use of "béton brut" was motivated by economic reasons, but above all, by aesthetic aims. It is for this reason that the construction is greatly simplified in terms of the use of materials, but also, that this will cause problems in terms of durability.

**Figure 2. Detail of concrete of the three buildings.**

### 3.3 ENERGY PERFORMANCES

Regardless of the building use and construction date, poor thermal response is common to all the projects. When they were designed the current concern about excessive energy consumptions and $CO_2$ emissions into the atmosphere did not exist. For this reason, the lack of insulation is usual and therefore, the energy loss is significant; this is a constant in most brutalist buildings.

Although the problems in the thermal envelope are shared by the three buildings, the need to maintain interior heat differs from one to the other. The variety of uses should be considered as it conditions the user's comfort needs. Thus, the need to heat or cool the spaces in a fronton will not be the same of those of a school. Or it may even happen that in different spaces of the same building the comfort needs differ, as in the case of the Maritime School, whose architectural program includes spaces with such diverse uses as classrooms, workshops or offices. Therefore, it is not only essential to differentiate the energy needs of each building, but also those of each of the spaces within it.

Focusing the attention on the thermal envelope, Brutalist Style's enclosure's performance has proven to be particularly poor as it is shown in Table 1. The exposed concrete walls' thermal transmittance values - U values - are high even when a cavity wall has been implemented. The problem is even greater in the windows, where neither the frame nor the glazing area do respond to the current thermal resistance requirements. Nonetheless, this can be easily solved by replacing the old windows by others with components that incorporate the latest technologies.

In order to verify the thermal response of the cited buildings, the transmittance values of the most common types of walls have been calculated. To this end, the methodology followed has been that of the document DB-HE/1 of the Spanish Technical Building Code (CTE) [9]. For this, three types of opaque enclosures have been selected. On the one hand, two solid exposed concrete walls of 30 cm (Type A) and 20 cm thick (Type B), and on the other, an identical example to the first one but backed by a 5cm cavity and a plastered brick leaf (Type C). The values obtained have been compared to the maximum transmittance value accepted by the regulations for this particular climatic zone, in this case, 0.41 $W/m^2K$. Overall, the results are far from being satisfactory.

**Table 1. Thermal response of different types of walls**

| **Thermal response** | **Façade types** | | | |
|---|---|---|---|---|
| | *Type A* | *Type B* | *Type C* | *D Climatic zone requirements* |
| *Transmittance ($W/m^2 \cdot K$)* | 3.45 | 4 | 1.66 | 0.41 |

## 4. CASES OF INTERVENTION

The cases shown below try to address the improvement of the building's energy performance by intervening in the façade, as it plays a key role in the energy performance of the whole. The action will focus on the element that better represents the image of Brutalism: the concrete wall.

As it has been tested, non-insulated concrete walls are unable to achieve thermal resistance requirements set by current regulations. It is also a fact that in general adding insulation may bring positive benefits (Fig. 3). But on the other hand, any action that includes external cladding may destroy much of the architectural quality of the façade. Concrete repair industry has rapidly grown over the last three decades. When it comes to restoring the integrity of the material, nowadays there are several well-established techniques that are extensively used with success, but still, it is unusual to find approaches that meet the specific needs of each conservation case. Despite the difficulties, some of the most significant brutalist buildings of the 20th century have already been repaired. However, interventions that address the improvement of the building's energy performance are scarce [10].

### 4.1 OPTION 1: PRESERVING THE ORIGINAL

The option that preserves the heritage significance of concrete the most is the one that limits the intervention to the repair of the deteriorated element. Nevertheless, this option does not tackle the problem of energy efficiency. Conventional concrete reparation includes removal of carbonated concrete, cleaning of corroded steel reinforcement, replacement of loss rebar when required and concrete cover reconstruction with cement-based mortars. Nowadays, this is the most widely used method in brutalist buildings.

### 4.2 OPTION 2: ETICS

In the case of choosing this option, application guidelines agree to recommend a previous inspection and the subsequent reparation of the substrate to grant the stability and flatness of the wall. Reparations of concrete such as the ones we mentioned above may be necessary before starting with the installation of an External Thermal Insulation Composite System (ETICS). Once the substrate is prepared, the rigid insulation boards are applied to it by using adhesive and supplementary mechanical fixings, and afterwards, the whole is rendered with several acrylic-based coats. To prevent finishing render from cracking, ETICS should include reinforcing mesh embedded in the basecoat over the whole front face of the solid wall and as stress patches on window edges.

The popularity of these systems versus technically more complex options such as rainscreen cladding systems is due to its comparatively cheaper cost. From an energy point of view, a priori, both systems make it possible to reduce the thermal transmission values to those required by current regulations. Furthermore, applying the appropriate thickness of insulating material from the outside, not only reduces heat losses but also avoids the negative impact of thermal bridges. In addition, although it might be difficult to replicate its texture, the surface of cast on site concrete resembles more to the continuous finishing of ETICS than to the discontinuous cladding of a backvented façade. On the other hand, the use of any system based on external insulation entails assuming the partial loss of the original geometry. This is a major loss in the case of brutalist buildings, whose expressionist profiles are inherent to the architecture style itself.

### 4.3 OPTION 3: INTERNAL INSULATION

The constructional solutions for internal insulation are like external ones in terms of the transmittance values achieved: a 10 cm mineral wool insulation panel, regardless of its coating, will provide the same U values installed both outside and inside. When it comes to preserving the architectural values of the envelope, the internal option would be more respectful, but it also poses other disadvantages. For instance, if the insulation is placed in existing wall cavities, the final thermal transmittance of the enclosure is conditioned by its thickness. When that space is not sufficient, the benefits provided by insulation do not reach those required by regulation, as it happens in the case in 3.b (Fig.4). Another recurring problem with very harmful effects is the appearance of thermal bridges. The surface and interstitial condensations derived from these require complex construction solutions in practice that make the project more expensive [11].

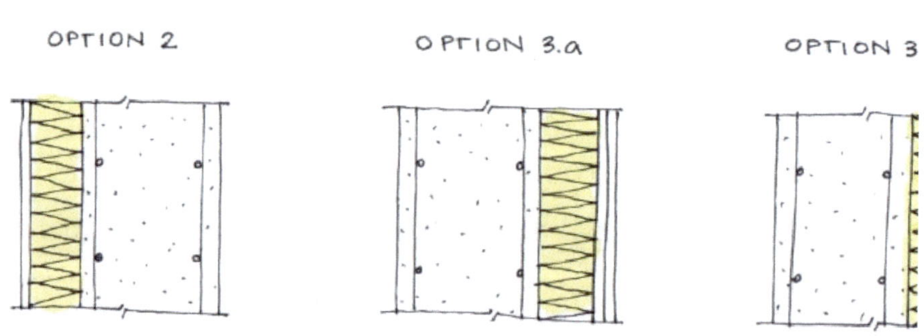


**Figure 3. Insulation options and their transmittance values**

## 5. DISCUSSION

Analysing the current European energy regulations and their transposition into the member countries, one thing is absolutely clear, and that is the need to seek a solution to the dichotomy and which this legislation is unable to resolve: the necessary energy refurbishment in protected buildings without losing their heritage value. A possible solution is proposed in the Theory of Energy Intervention in the Built Heritage or TEIBH [12]. This theory advocates a progressive intervention based on the predetermined heritage values of the original building (Fig. 4). In other words, a balance must be found between improving protected buildings in terms of energy efficiency and preserving their heritage value [13].

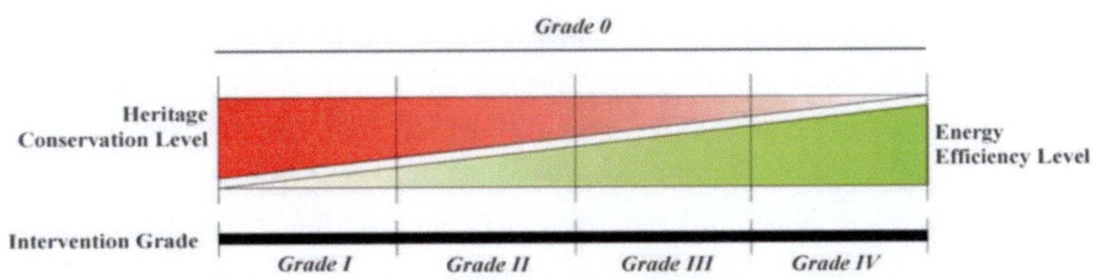


**Figure 4. Scheme of the Theory of Energy Intervention in the Built Heritage or TEIBH**

In the case of the brutalist buildings, first of all we find that many of them do not have any official protection, as it is the case of the three buildings in San Sebastian. This means that nowadays they can be intervened on at any time without taking into account their historical value. But on the other hand, if in the near future they are listed and officially protected by heritage legislation, they may never be improved energetically. Other brutalist buildings have such protection, and in some cases are being refurbished, but not from an energy point of view, but from a restoration point of view. For this reason, the previous analysis of each one of them is fundamental, both for their energy intervention and for their restoration. In this way, buildings will be adapted to the needs of comfort and to the reduction of energy consumption that is required today, beyond mere conservation.

For the three brutalist buildings in San Sebastian, the TEIBH should be applied individually, taking into account each one's heritage values and also considering what energy improvements can be made. For instance, energy requirements in a school may not be the same as in a sports facility. As a consequence, the use of each building and its energy needs for its users comfort should always be considered.

Of the three intervention options proposed, Option 1 is the one that best preserves the heritage value of the building but does not improve its energy performance in any way. This would be the general case of the restorations conducted so far on the brutalist buildings. Option 2 and Option 3 are similar from the point of view of energy improvement, but the former can lead to the loss of heritage values, by intervening on the façade from the outside. Option 3 is the one that best suits the TEIBH as it improves the energy performance without spoiling the materiality of the concrete façade. Within Option 3, there are two other ways to isolate the wall: Option 3.a, where the wall is simply isolated from the inside and Option 3.b where the existing air cavity is insulated. In the second case, as the cavity has a predetermined dimension, the insulation may not achieve the values established by current regulation, while in the first case, the insulation thickness may be exactly the accurate.

## 6. CONCLUSIONS

Energy efficiency improvement should become the central axis of any integrating approach to the refurbishment. This means that officially protected buildings that are part of the built heritage, should also try to achieve an improvement in their energy performance. This energy intervention must be adapted to the reality of the building, and energy improvement should not prevail over heritage value, so a balance must be sought between these two objectives: energy improvement and heritage conservation. With regard to brutalist buildings, their historical value must be protected, especially in the cases where this has not been done. But at the same time, a possible energy improvement intervention must be foreseen. Each building is different and the energy refurbishment must be undertaken individually. In the case of brutalist buildings, the construction solution of the concrete wall is similar in almost all buildings, so energy interventions shown before can serve as example for other cases. Thereby, it has been concluded that the option that can be best adapted, a priori, is the Option 3 with insulation on the inside of the façade. In any case, this option also presents problems that should be solved taking into account the reality of each building. The objective of reducing energy consumption and $CO_2$ emissions to the atmosphere while maintaining the values of the built heritage is a present challenge that must be adequately unravelled in order to enjoy historical buildings in the future.

## REFERENCES


[1] Council of European Union. "Directive 2002/91/CE of the European Parliament and of the Council, 16 December 2002, on the energy performance of buildings", OJ L1, of 4.1.2003, Brussels: CE, 2002.

[2] Council of European Union. "Directive 2010/31/EU of the European Parliament and of the Council, 19 May 2010, on the energy performance of buildings (recast)", O J L153/13 of 18.6.2010, Brussels: EU, 2010.

[3] Council of European Union. "Directive 2012/27/EU of the European Parliament and of the Council, 25 October 2012, on energy efficiency, amending Directives 2009/125/CE and 2010/30/EU, and repealing Directives 2004/8/CE and 2006/32/CE (Text with EEA relevance)", O J L315/1 of 14.11.2012, Strasbourg: EU, 2012.

[4] R. Banham. "The New Brutalism". *The Architectural Review*, vol. 118, pp. 354–361, Dec 1955.

[5] M. T. Valcarce Labrador. "El Nuevo Brutalismo: una aproximación y una bibliografía". *Cuaderno de Notas ETSAM – UPM*, vol.7, pp. 131–144, 1999.

[6] M. Oriol and G. Lafuente. "Construcción de Colegio, Iglesia y Convento en el Ensanche de Amara", 1965, Sig. A-02315-01, San Sebastián City Council Municipal Archive.

[7] L. Laorga and J.L. López Zanón. "Proyecto de Escuela Oficial de Náutica en Pasajes", Madrid, 1965.

[8] L. J. Arizmendi, "Construcción de los frontones Carmelo Balda", 1969, Sig. A-00182-01, San Sebastián City Council Municipal Archive.

[9] Spanish Government, MFOM. "Royal Decree 732/2019, of December 20, which modifies the Technical Building Code, approved by Royal Decree 314/2006, of March 17", BOE 311 of 27.12.2019, Madrid, 2019, updated DB-HE "Energy Saving" document, HE1, pp. 140488–140674.

[10] C. Croft, S. Macdonald and G. Ostergren, Eds. *Concrete: Case Studies in Conservation Practice*. Los Angeles: Getty Conservation Institute, 2018.

[11] N. V. Baker. "Walls", in *The handbook of Sustainable Refurbishment: Non-Domestic Buildings*, London: The RIBA Publishing, 2009, pp. 31–37.

[12] E.J. Uranga. "La Intervención Energética en el Patrimonio edificado Residencial. Análisis del barrio de Gros en Donostia/San Sebastian", Ph.D. dissertation, Dept. Architecture, UPV/EHU Univ., San Sebastián, 2017.

[13] E.J. Uranga, L. Etxepare, I. Lizundia and M. Sagarna. "Energy Efficiency Intervention and Preservation in Residential Built Heritage: Analysis, Proposal ad Results in the Gros District of San Sebastian", in *International Conference on Energy Efficiency in Historic Buildings*, Uppsala University Campus Gotland, Visby, Sweden, 2018.

# Action within the Flemish climate fund to reduce $CO_2$ emission of protected buildings in Flanders

N.Vernimme[1], Y. Vanhellemont[2], M. de Bouw[2]

[1] Dept. Research and Protection, Flanders Heritage Agency, (FHA), Brussels, Belgium. https://www.onroerenderfgoed.be/, (nathalie.vernimme@vlaanderen.be)

[2] Laboratory Retrofitting and Heritage, Belgian Building Research Institute (BBRI-WTCB), Brussels, Belgium. https://www.bbri.be/, (yves.vanhellemont@bbri.be, michael.de.bouw@bbri.be)

**Abstract –** In 2014 the Flemish government approved to finance (363,750€) a 7-year lasting research project for the field of cultural heritage in the framework of the Flemish Climate Policy plan 2013–20. The budget came from the Flemish Climate Fund, originating from the auction of European Emission rights. The scope of the project was defined by the Flanders Heritage Agency and consisted of the development and installation of a structural system of specialized energy consultants for cultural heritage in the Flemish Region, the organization of a question-portal for heritage professionals that retrofit protected monuments, the monitoring of cases and indicators of $CO_2$ reduction in registered protected monuments. The project was executed by the Belgian Building Research Institute (BBRI). The reduction potential of $CO_2$ for protected monuments in Flanders, outgoing from a reduction of 50% per monument on short term (timeframe of the project 2015-2021) was estimated, on the long run 288,000 ton $CO_2$-eq.

**Keywords** – regional climate policy – retrofit protected monuments – specialized energy consultants for heritage – specialized training for restoration architects.

## 1. CONTEXT: FLANDERS AND ITS CLIMATE POLICY

### 1.1 CLIMATE IN FLANDERS

Flanders is a highly urbanised region in Belgium at the North Sea coast. The climate is maritime temperate, with significant precipitation in all seasons. Changes in the climate have already and continue to have important effects such as (urban)heat stress, sea level rise, floods and drought. All climate scenarios for Flanders indicate for the end of this century a rise in the environmental temperature (+1.5 °C to +4.4 °C in the winter; +2.4 °C to +7.2 °C in the summer). There will be higher evaporation levels during winter and summer, and more precipitation during winter by 2100. The sea level at the Flemish coast could rise this century between 60 and 200 cm. Most climate scenarios show a drop in average summer precipitation and an increase in the number of extreme summer thunderstorms [1].

These effects will have an enormous impact on people, landscape, food ... and buildings, so also on our cultural heritage. Rising sea levels could threaten in the future hundreds of monuments along our coastlines. Already now heavier rainfall and changing humidity levels take a heavy toll from our cultural treasures. The lower humidity during the summer increases the amount of salt deposits and the risk of salt crystallization in monuments with porous stone. Increases in storminess and wind gusts lead to structural damage and sometimes even the collapse of historic buildings. Timber and other organic building materials are more often attacked by insects, moulds, fungi and invasive species such as termites.

### 1.2 THIRD FLEMISH CLIMATE POLICY PLAN 2013-2020

Since 2010, the countries under the United Nations Framework Convention on Climate Change have adopted the objective of not increasing the global average temperature by more than 2°C with respect to pre-industrial times. Flanders was committed to this objective, via its Flemish Climate Policy Plan 2013-2020 [2]. The Flemish Climate Policy Plan 2013-2020 used a dual approach in order to achieve a low carbon society in Flanders. The plan consisted of a general framework and 2 sub-plans: on the one hand the Flemish Mitigation Plan which had the purpose to reduce emissions of greenhouse gasses in Flanders between 2013 and 2020 as a means of combatting climate change. On the other hand the Flemish Adaptation Plan that aimed to understand the Flemish vulnerability to climate change and improve its ability to defend against its effects. The policy measures were implemented or carried out by various policy areas such as energy, transport, agriculture, industry, housing, government buildings, etc.

### 1.3 THE FLEMISH CLIMATE FUND

To support the measures mentioned in the Flemish Climate Policy Plan 2013-2020, the Flemish Government set up the Flemish Climate Fund. The Fund provides a financial framework for the long term climate policy and is financed with the returns from the auction of EU emissions certificates in the period 2008-2012 and the following years. The Climate Fund is in the first place an instrument to realise cost-effective Flemish measures to fight climate change. Next to this the Fund is used to buy emission credits, in case internal measures prove insufficient to reach the target.

In 2012 the Flemish ministers were asked to define extra proposals for their policy areas to help realize the foreseen 15 % reduction of non-ETS greenhouse gas emission for the period 2013-2020. Thirty three internal mitigation measures with potential for co-financing through the Flemish Climate Fund were proposed. The proposals were tested against an assessment framework to guarantee that the most prior and cost-efficient measures were put forward to start in the period 2013-2014. The principal assessment criteria were additionality (added value compared to existing policy), sustainability (side effects on environment, economy and income distribution), implementation trajectory (how quickly it leads to reductions) and cost efficiency (ratio of cost of measure/impact on emission reduction, or euros per ton of $CO_2$-reduction). On the basis of this assessment, 14 proposals were selected as priority; one of these was the proposal "specialised energy consultants for heritage properties". All the measures were included under the sectoral chapters of the Flemish Mitigation Plan.

## 2. THE PROJECT "SPECIALISED ENERGY CONSULTANTS FOR HERITAGE PROPERTY"

### 2.1 DEFINING AN APPROPRIATE MEASURE TO LOWER THE C02-EMISSION IN THE HERITAGE SECTOR

In 2010 the buildings sector was responsible for the emission of 18.9 Mton $CO_2$-eq or 38% of the overall Flemish non-ETS greenhouse gas emissions. The share of households was 14.4%. Greenhouse gas emissions from households are mostly related to the heating of spaces and the production of hot water. 2,184,307 buildings in Flanders were used for housing in 2012 [3]. Some 2 to 3 % of these (or 60,000 houses) have heritage values (listed or in the inventory of buildings with heritage value).

Houses with heritage value are mostly older than 75 years and have a large potential for improvement of insulation and other measures to reduce $CO_2$-emissions. Additional efforts are needed to improve these buildings, but the measures should be adapted in order not to harm the heritage values. The project "specialised energy consultants for heritage property" was defined by the Flanders Heritage Agency, the government agency of the minister responsible for immovable heritage. The scope was that 1/4 of the owners of houses with heritage value would order a specialized energy scan for their house and that 75% of them (or 11,250 owners) would implement adapted retrofit measures. The improved energy efficiency of the heritage buildings portfolio will yield added socio-economic benefits (reduction of energy bills for owners) and environmental benefits (reduction of air pollutants). The reduction potential of $CO_2$ for protected monuments in Flanders, outgoing from a reduction of 50% per monument on short term (timeframe of the project 2015-2020) was estimated 8242 ton $CO_2$-eq. On the long run the estimation is 288 kton $CO_2$-eq.

**Table 1. Estimation of reduction potential for proposal "specialised energy consultants for heritage property"**

| **Calculation reduction potential** | **"specialised energy consultants for heritage property"** | | |
|---|---|---|---|
| *$CO_2$-emission/ heritage building* | 6.7 | ton CO2 | |
| *Potential per scan 50%* | 3.3 | ton CO2 | |
| *Reduction/scan (75% implementation degree)* | 3300 | scans | 300/yr in 2016 -2017; 900/yr in 2018-20 |
| *Cycle life reduction potential* | 164,835<br>412,087.5<br>288,461.3 | ton $CO_2$ | 20 year<br>50 year<br>35 year |

The project consists of three main results: (1) the development and installation of a structural system of specialized energy consultants for cultural heritage in the Flemish Region, (2) the organization of a question-portal for heritage professionals that retrofit protected monuments, (3) the monitoring of cases and indicators of $CO_2$ reduction in registered protected monuments.

## 2.2 REALISATION OF THE PROJECT BY THE BELGIAN BUILDING RESEARCH INSTITUTE

The project consisted of three main axes:

- The development of a series of training packages for architects, in which all relevant topics, such as building physics and comfort, thermal insulation and ventilation, are treated. About 70 restoration architects already followed this training.
- The development of a question-portal for heritage professionals needing help to retrofit listed buildings.
- In order to better quantify the possible and real reductions in CO2-emission, a series of buildings are studied in more detail. The aim is to estimate to which measure the interventions, aiming to reduce the energy consumption of these buildings, are effective. As an extension to this, a survey is carried out in which we will estimate how

many heritage buildings are already equipped with energy-efficient techniques or materials. Finally, this information will be translated in reductions of $CO_2$-emissions in the built heritage in Flanders.

The results from the project are published on the project website, www.erfgoedenergieloket.be

### 2.2.1 *Development and installation of a structural system of specialized energy consultants for cultural heritage in the Flemish Region: the development of training packages.*

The basic starting point of the training packages is the conservation of heritage buildings, and the aim to protect their heritage values, in relation to the thermal optimisation of such buildings. The reduction of the energy consumption, at the core of the project, is itself an important mean for the conservation of heritage buildings: an optimal climate inside a building contributes to the preservation of building materials and possible valuable objects. But it also stimulates people to work and live in heritage buildings. Even if the reduction in energy consumption remains limited, the thermal comfort inside the building should be optimized as much as possible.

The following heritage principles are woven throughout the training packages:

- Respect for heritage values, even though it is usually not easy to concretize this in guidelines: the heritage values are different for each heritage building.
- Reversibility of interventions.
- The impact of energy- and insulation-interventions on building materials. These interventions should not have any negative impact on building materials and their durability.

The recommendations of the Flanders Heritage Agency, regarding roof insulation, insulation of windows and glazing, and the application of photovoltaic cells in heritage buildings, have been taken into account. These recommendations are structured in three Assessment Frameworks [4]. New Assessment Frameworks, regarding the insulation of floors and facades, are in preparation.

More specifically, the training packages treat the following topics:

- General aspects about durability and legislation.
- An introduction to building physics, with an emphasis on the behaviour of humidity and heat in building materials and buildings.
- The energetic performance of buildings and thermal comfort. With a strong emphasis on airtightness, as this is a major influential factor, connected to about every other insulation intervention in buildings.
- Diagnostics of the building envelope. The methods to investigate the thermal performance of the envelope. But vice versa, the properties of the envelope, the quality of the building materials, and possible degradation mechanisms, will strongly influence choices for interventions to improve the performance of the building. Therefore diagnostics of the building envelope is approached as a holistic topic, where both energy- and non-energy linked subjects are treated.
- Humidity in buildings. Humidity has a negative impact on the energy consumption and thermal comfort of a building, and determines to a large degree which types of interventions are possible.

- The treatment of roofs, both flat and inclined roofs. A large portion of the energy losses in buildings happen through the roofs. Specific heritage issues, such as detailing, interaction with wooden constructions etc. are taken into account.
- The insulation of facades. The interaction with humidity problems in facades is here of the utmost importance, as well as the interaction with several details: connection with roof insulation, the danger of creating thermal bridges, details regarding adjacent walls, floor constructions, airtightness, ...
    - The insulation of cavity walls.
    - Exterior façade insulation.
    - Interior façade insulation.
- Windows, glazing and blinds.
    - Discussion on how to improve the thermal performance of historic windows and glazing, while preserving as much as possible of the original materials and elements.
    - Protection against the sun becomes more important in a heating climate, especially when it is expected that long dry and sunny periods will be more frequent in the future. Floors and basements. Energy losses through floors and basements are usually less important. Non-insulated floors may even have a positive effect on the indoor climate during summer. Nevertheless, it may be better to insulate floors, which might be risky in some cases, due to humidity problems and the presence of wooden or steel floor constructions.
- Ventilation. Insulation and airtightness always come together, thus increasing the need for ventilation, adapted to the characteristics of each building.
- Heating. This topic deals with both heat production and heat distribution and strategies in historic buildings, taking into account high spaces (and problems with temperature gradients) and possibilities of compartmentalization of older, more spacious, buildings.
- Illumination and the application of photovoltaic cells in heritage buildings. Both energy consumption, efficient illumination and illumination techniques that are compatible to valuable interiors, are treated.
- Finally a series of practical cases is presented.

The training packages are published (in Dutch) on the project website [5] and will be made available in French in the near future.

#### *2.2.2 The question portal*

The project website also contains a question portal. Building professionals can ask for technical advice for their restoration projects, on the condition that the question is linked to the thermal optimization of a building with heritage value, including increasing the thermal comfort. The project deals mainly with individual housing, but occasionally we have been asked advice for other types of buildings.

The questions are treated by the BBRI, that has over 60 years of experience in helping contractors and other building professionals with technical problems on building sites.

The nature of the questions (and answers) is threefold:

- Short general questions, that can be answered in a more general and brief way. For instance regarding literature, references, or simple technical questions that are treated in publications, or where we can answer a question based on pictures and plans.
- Longer and more detailed questions, always regarding specific cases, where more time has to be foreseen, in order to discuss on site, together with the building owner, architect, contractor and representatives of the Heritage Agency.
    - Example: the discussion regarding the insulation of a roof of a building in the beguinage of Diest. Most houses on this site date back to the 17th century. The house still has its original wooden roof structure. For evident reasons, this structure had to remain visible, almost imposing a sarking-structure. Limited space around details (facades, corniches, dormers) oblige to use thin thermal insulation, for which aerogel insulation has been applied. The advice given was about the details of the insulation, and the application of vapour- and airtight protections in the roof structure.
- In some cases, a more extensive study had to be carried out, including tests.
    - Example: the restoration and thermal insulation of the facades of the American Hospital in Wervik (constructed 1919-1921). Problems were linked to the fact that the walls are very thin and suffer from rainwater infiltrations. The basic question was whether a water repellent agent could be applied to efficiently protect the facades and make thermal insulation possible. A simple question, from which arise several more questions. In this specific case, the efficiency and second order effects (influence on the drying rate) of a water repellent agent have been tested, together with effects regarding salts. Finally, the results of the test have led the architect into an alternative direction. In short: because of the impossibility to resolve the humidity problems completely, and because of the severe salt problems in the walls, a solution of the type box-in-box is applied, where the outer facades are kept as dry as possible by treating them against rising damp, and by ventilating the cavity between the old and new construction, taking into account the risk of augmented salt crystallisation in the cavity.

#### *2.2.3 Case-studies*

During the project, about 10 case-studies were followed [6]. The aim is to quantify to what extent one is really able to reduce the $CO_2$-emission of a building while respecting their heritage values. In practice, the following tests were applied to buildings, even though not all tests were performed on all buildings.

- Measurement of U-values (mostly facades or windows)
- Infrared thermography, both inside and outside buildings, revealing thermal bridges, air leaks or other important heat leaks. Mapping of temperature distribution over the outer surfaces, using drones.
- Monitoring of the inner climate (relative humidity and temperature)
- Air tightness (blowerdoor method)
- Comfort measurements (air temperature, radiation temperature, draft and $CO_2$-content of the air)
- This linked with the measured energy consumption of the building.

Evidently not all of these measurements are useful in all cases. In many cases, the buildings were not in use before the retrofit and windows or doors were missing, making it useless and/or impossible to quantify the above mentioned parameters.

The case-studies were selected based on their representativity, but also on the 'degree of invasiveness' of the planned interventions. The full list of cases is mentioned on the project website, but in order to give an impression of the diversity in studied building typologies, three examples are mentioned here:

**Figure 1. Three case-studies: 19th century mansion (Diest), House 'Billiet' (Bruges), Residence 'Duinpark' (Koksijde)**

- 19th century mansion in Diest: this building is not protected, but it figures on the inventory of built cultural heritage. The building was in a really bad condition, so during renovation only the walls, basement and the main entrance corridor have been preserved, together with some interior elements (i.e. the main staircase). BBRI gave advice regarding the humidity problems in the building, and will perform measurements after the building is being used again (since late autumn 2020, after being empty for several years).
- House 'Billiet' in Bruges: contrary to the previous example, this building has still preserved the most part of its valuable exterior and interior elements (including doors, tiles, floors, stained glass, ...). The building is therefore being treated in a less invasive manner. The building was constructed in 1928, in modernist style, by the architect Huib Hoste. Adjacent to the house, there is a workshop (same era) for diamond cutting, which is also listed and nowadays transformed into habitation. This part requires and allows a more invasive approach, since it is constructed in reinforced concrete with large windows and single glazing, which suffered severely from wood rot, degradation of the concrete, and which has evidently an extremely bad thermal behaviour.
- Residence 'Duinpark', Koksijde. This is a somewhat odd example, because it is an apartement building, constructed in the late 1950-ies. It is not a protected monument, but it is being treated as such, on the demand of the owners. Therefore we can consider it as a restoration project that might form a representative example of how to treat more recent heritage. The building suffers from severe material degradation (masonry and concrete) partially due to its vicinity to the coast. Moreover, the constant wind is a challenge to energy losses and a reduced thermal comfort through various air leaks. During renovation, the roof and facades will be partially removed and reconstructed, which opens the possibility to apply thermal insulation on facades and roofs. This building is, during this project, a subject to a novel way of measuring the energy consumption of a building, the so-called co-heating test. Rather simplifying put,

this test measures exactly how much energy is required to keep the inside of the building on a constant temperature. By measuring the temperatures of the adjacent spaces, one is able to determine very precisely the heating requirements of the building.

The results of the follow-up of the case-studies will be linked to a more general survey that is sent to restoration architects who participated in the training course and have restoration projects in the Flemish Region. With this survey, we aim to get more quantitative information on the thermal optimisations that are carried out in heritage buildings in the Flemish Region, in order to estimate the $CO_2$-reductions that are obtained, and to extrapolate towards $CO_2$-reductions that might be obtained on the long run.

## 3. CONCLUSION

The scope of the 'SPECIALISED ENERGY CONSULTANTS FOR HERITAGE PROPERTY' was the development and installation of a structural system of specialized energy consultants for cultural heritage in the Flemish Region, to retrofit heritage buildings and reduce their energy consumption (and therefore $CO_2$-emission) as much as possible, while obtaining a more comfortable inner climate.

For this purpose a series of training packages have been developed, to train architects so that they perform retrofits that reduce $CO_2$-emissions, while fully respecting the heritage values of protected buildings. A question portal, through which building professionals may ask for technical advice, turns out to be a fruitful mean of interchanging ideas, resulting in more qualitative restorations.

The effect of these actions is monitored throughout a series of case-studies that are followed from nearby, combined with a more general survey. The results from these actions will be available at the end of the project in December 2021. They will give an insight in the quantitative possibilities (or impossibilities?) of making heritage buildings energy-efficient.

The experiences obtained in this project have already contributed to the ongoing development of a specific method for obtaining an EPC-rating for heritage buildings.

## 4. REFERENCES


[1] Flemish Government, Environment, Nature and Energy Department (2013) "The Flemish Climate Policy Plan. 2013-2020, a summary."[online], pp. 62. Available: https://www.vlaanderen.be/publicaties/the-flemish-climate-policy-plan-2013-2020-a-summary . [Dec. 2, 2020]

[2] Flemish Government, Environment, Nature and Energy Department (2013) "The Flemish Climate Policy Plan. 2013-2020, a summary."[online]. Available: https://www.vlaanderen.be/publicaties/the-flemish-climate-policy-plan-2013-2020-a-summary . [Dec. 2, 2020]

[3] Flemish Government, Environment, Nature and Energy Department (2013) "The Flemish Climate Policy Plan. 2013-2020, a summary."[online], pp. 25. Available: https://www.vlaanderen.be/publicaties/the-flemish-climate-policy-plan-2013-2020-a-summary . [Dec. 2, 2020]

[4] Flanders Heritage Agency, Assessment frameworks [online]. Available: https://oar.onroerenderfgoed.be/publicaties/AKOE/3/AKOE003-001.pdf ; https://oar.onroerenderfgoed.be/publicaties/AKOE/4/AKOE004-001.pdf ; https://oar.onroerenderfgoed.be/publicaties/AKOE/5/AKOE005-001.pdf. [Dec. 2, 2020]

[5] BBRI, Training packages [online]. Available: https://www.erfgoedenergieloket.be/databank/downloadformulier/. [Nov. 30, 2018]

[6] BBRI, Case-studies [online]. Available: https://www.erfgoedenergieloket.be/databank/cases/. [Oct. 18, 2019]

# Barriers to reducing carbon from heritage buildings: residents' views from Cumbria and the English Lake District World Heritage Site

F.Wise, A.Moncaster, D.Jones

School of Engineering and Innovation, The Open University, Milton Keynes, United Kingdom, (freya.wise@open.ac.uk, alice.moncaster@open.ac.uk, derek.jones@open.ac.uk)

**Abstract** – Carbon emissions from the built environment must be urgently reduced to mitigate climate change. Retrofit of existing buildings is key. However, heritage buildings pose a particular challenge, particularly domestic buildings where retrofit is mainly resident-driven.

This paper analyses a survey of residents of pre-1940 buildings, exploring attitudes to carbon reduction, access to information sources, and planning policies. The study found that residents strongly desire to reduce carbon emissions from their older buildings but considered this to be difficult. Costs and knowledge were key challenges, but heritage values, planning barriers, and lack of tradespeople, were also important.

Most felt local planning policies were appropriate but potentially inconsistent. Residents accessed multiple sources for carbon reduction advice but preferred local and informal over statutory sources. The majority were unsure what information heritage organisations held. The study therefore highlighted the need to make information on carbon reduction more visible and targeted to residents.

**Keywords** – Carbon; Energy; Heritage buildings; Users; Retrofit.

## 1. INTRODUCTION

The built environment is responsible for significant energy use and resultant carbon emissions which must be urgently reduced to help mitigate climate change [1]. In Europe the building stock replacement rate is only around 1% per year [2], so carbon reduction through retrofit of existing buildings is a key strategy. In the UK however, up to 20% of these existing buildings have heritage value [3]. These buildings help shape the character of urban and rural landscapes and have a wide range of values [4]. As a result of these values and their traditional, often locally specific, construction techniques and materials, heritage buildings are challenging to retrofit sensitively.

The UK has some of the oldest housing stock in Europe with around 35% (10 million) built before 1944, and around 20% (almost 6 million) built before 1919 [5]. However, there are no clearly agreed definitions of what constitutes a heritage building. It is estimated that around 1-2% of homes are individually listed, the highest level of heritage designation in the UK [6]. There are also area listings such as conservation areas, national parks and World Heritage Sites (WHS) [3]. Additionally, it is recognised that many older buildings with no official designation have important heritage values [7].

The majority of residential heritage buildings are privately owned, and therefore decisions on retrofitting for energy reduction devolve to homeowners. This paper uses the results of a survey of residents of pre-1940 buildings in Cumbria to explore the range of barriers they perceive in balancing heritage preservation and carbon reduction [8]. Two of these barriers - the information sources that residents engage with, and residents' attitudes to local planning policies - are then explored in more detail. Finally, the implications of these findings for supporting decision-making on sensitive retrofitting for heritage buildings, and therefore overall carbon reduction, are discussed.

## 2. BACKGROUND LITERATURE

Various factors have been identified as barriers for residents making retrofitting decisions. Access to appropriate information is commonly identified as a key barrier, with an emphasis on the need for knowledge to come from trusted sources and be relevant to residents' specific situations [9], [10]. Several studies have found that residents are often part of 'knowledge networks', where they seek advice on retrofitting from friends, family and colleagues, and that these unofficial information sources significantly affect decision making [11]. Advice from personal contacts was also seen to increase understanding and retrofit adoption rates [12], [13]. This highlights that homeowners are part of a social-technical system, rather than isolated actors and that retrofit decisions are much more complex than simple cost benefit analysis [11].

These issues have been found to be particularly relevant in heritage buildings, where residents have to negotiate between competing demands while taking heritage values into account [7]. Residents have been shown to hold individually specific values in their heritage buildings -including in buildings with no official heritage designation- which influence their retrofitting decisions [14]. Indeed residents are often actively engaged in modifying standard solutions to their specific contexts and values, rather than being passive recipients of existing solutions [13]. Locally applicable, independent advice is therefore seen as a key requirement to effective retrofitting [10], with information barriers considered more problematic than a lack of technical solutions [7]. Policy responses such as Energy Performance Certificates (EPCs), often have mixed results, failing to achieve increased positive energy retrofitting activities [11]. These have been shown to have significant inaccuracies for heritage buildings [15], as acknowledged by the European Standard on improving the energy performance of historic buildings [16], which identifies that standard calculations are often inappropriate for heritage buildings and recommends a tailored approach to energy modelling. The application of planning policy to heritage buildings has also been identified as an important barrier [17]. The sometimes conflicting views of sustainability officers and conservation officers [18], and the lack of consistency between, and in some cases even within, different planning authorities [17], [19], all add to the barriers faced by homeowners. Meanwhile cost is frequently considered a key issue and the need for financial incentives is often identified [11], particularly for the upfront capital investment needed to enact retrofits [12]. Finally the need for educated and experienced tradespeople is key, especially in relation to heritage buildings and the particular, and locally contextualised, challenges that their traditional construction presents [9], [13].

In summary, there exist a range of barriers or inhibitors faced by residents making energy retrofitting decisions in heritage buildings: a lack of access to trusted and specific information about suitable and heritage sensitive options; planning inconsistency; financial costs; and lack of appropriately knowledgeable tradespeople. Residents negotiating these factors face making complex decisions, balancing a number of often competing issues.

## 3. METHOD

The county of Cumbria in the UK was the chosen setting for this study. Cumbria includes the Lake District National Park (LDNP) which has significant development restrictions above the UK's national and local planning frameworks. It has also been recently inscribed as a Cultural Landscape WHS by UNESCO [20]. This paper will focus on part of a survey which explored the carbon reduction views, heritage values and energy behaviours of residents of pre-1940 buildings in Cumbria. Based on the areas identified in the literature: residents' views of carbon reduction responsibility; the challenges, opportunities and barriers to carbon reduction; their attitudes to planning and their use of retrofit information sources will be examined.

The survey was informed by both the literature and by interviews with Cumbrian sustainability and conservation professionals, and was piloted with a number of heritage buildings residents before launching. It was distributed via the email lists of local sustainability and conservation organisations and over 750 leaflets were also hand delivered to older houses across Cumbria. The survey ran from the 31st of October 2019 to the 10th of January 2020. 484 people looked at the first page online and 185 started the survey, 37 of which did not submit their response and 1 failed to confirm their consent. 147 responses in total were therefore analysed.

Descriptive statistics and cross tabulations were assessed in SPSS [21]. Five respondents asked that their comments not be published, and this has been honoured. Respondents were representative for Cumbria's rural/urban division and there was a range of housing types, although data was skewed towards detached houses with very few flats/apartments. 97% of respondents were owner occupiers and respondents' buildings ranged from Grade I Listed (the highest UK heritage designation category) through to buildings with no official heritage designation. Buildings in conservation areas (37%), undesignated buildings (34%) and those in the LDNP (27%) made up the majority of buildings. Buildings were aged 1400-1938, with around half dating from 1801-1900 corresponding to a major construction increase in Cumbria at that time.

## 4. RESULTS AND DISCUSSION

### 4.1 RESPONSIBILITY AND ATTITUDES TO EMISSIONS REDUCTION

The vast majority (82%) of residents felt Governments had the highest levels of responsibility to reduce carbon from heritage buildings, although homeowners (74%), building professionals (71%), energy companies (65%) and local authorities (56%), were also thought to share responsibility. This may suggest that residents are aware of the need for collective action to tackle climate change. Only 31% of residents thought Historic England had significant responsibility, despite their statutory advice role for alterations to designated heritage buildings.

95% of residents felt that it was generally much harder to reduce carbon emissions from heritage buildings compared to modern buildings. The majority (49%) felt the reduction potential from their own buildings was limited, whilst 38% felt it was moderate and only 13% substantial. Despite the perception of difficulties, however, most residents were either quite (36%) or very (50%) motivated to reduce carbon from their buildings. This suggests that residents desire to reduce emissions if barriers can be overcome.

### 4.2 BARRIERS TO RETROFIT

Respondents were asked to rate the importance of a range of barriers, identified from the literature (Fig.1). Cost, knowledge of suitable options and planning restrictions were considered the most important by the majority which is consistent with findings from other studies. A lack of heritage sensitive options, and the availability of tradespeople, were also important barriers, but time commitment, disruption, and things already achieved, were not considered that important by most residents. This may provide further evidence of residents' desire to reduce carbon, as they generally seem prepared to invest time in, and accept the potential disruption of, carbon reduction measures. A number of free text comments expanded on these results ('Availability of options suited to our property' 'Can't find tradespeople qualified in sympathetic alterations' 'well in excess of £30k to insulate the walls and floor. Not affordable.' 'Keeping the feel of the building even on the parts which are not listed.').

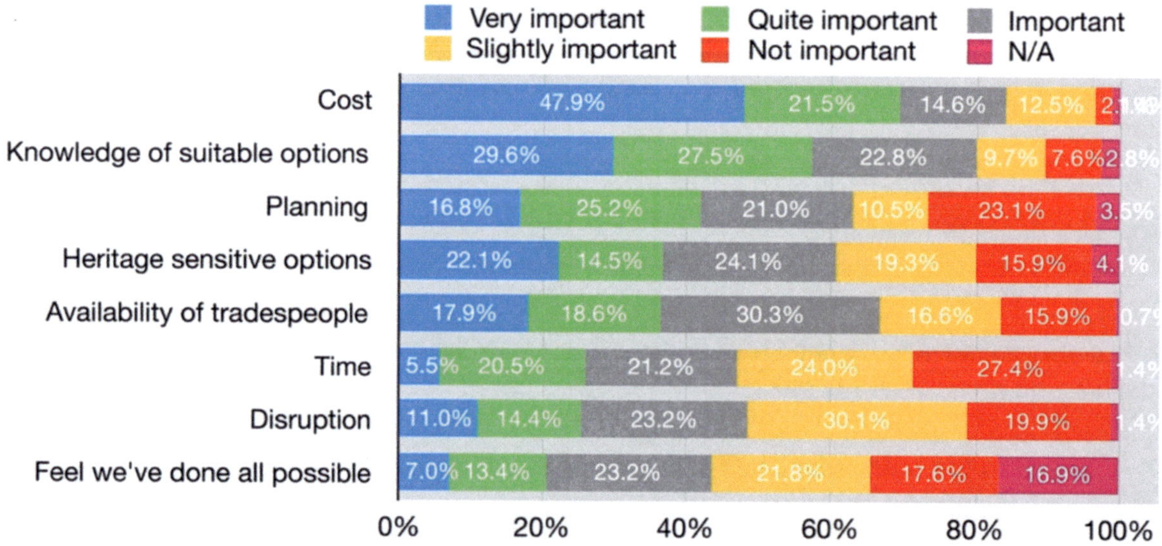


**Figure 1. Importance of barriers to reducing carbon emissions to survey respondents**

Two of these barriers will be explored further below. Although cost was found to be the most important barrier, there is significant research on this topic and it is very geographically, and socially, context-dependent, set in a rapidly changing political landscape. The focus of this paper was therefore the next two most important barriers, knowledge of suitable options and planning.

### 4.3 ACCEPTABILITY OF PLANNING REGULATIONS TO RESIDENTS

Residents felt that planning restriction levels were generally acceptable (Fig.2) but there was concern about the consistency with which policies were applied (Fig.3). Comments revealed that this was mainly related to the enforcement of regulations, consistency between planning officers, and perceived bias to certain projects types (Table 1). One of the reasons for this inconsistency may be that expertise is divided between sustainability and conservation branches of planning departments, meaning that a holistic view is not taken [18].

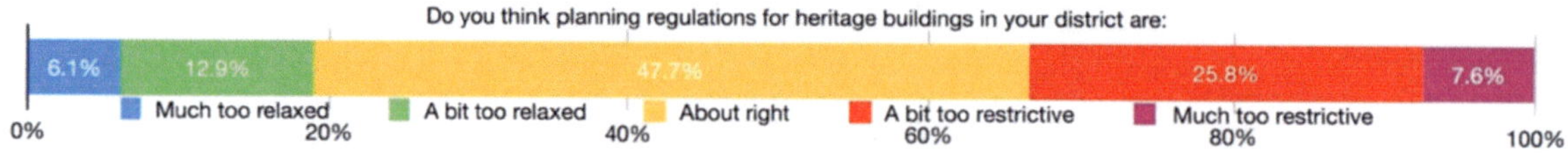


**Figure 2. Attitudes to planning levels for heritage buildings**

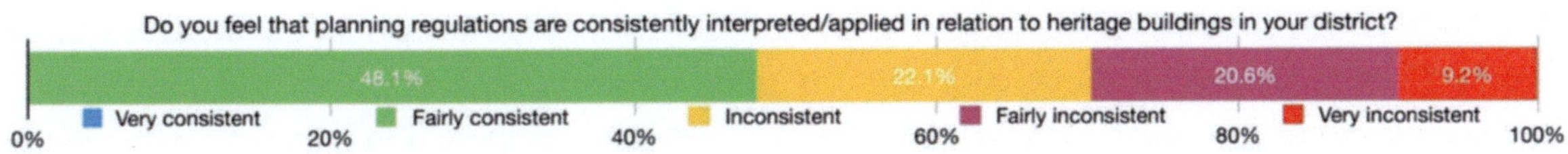


**Figure 3. Consistency of planning regulations**

**Table 1. Selection of respondents' free text comments on planning**

| |
|---|
| 'We had local slate when we re-slated- two doors down on the same terrace had some awful Spanish slate that is inconsistent across the terrace and spoils the look, yet no action from planning.' |
| 'Seem to change a lot over time - e.g., window replacements - UPVC along the road but strict moulding in wood applied to us at separate time.' |
| '...Seems to be a constant problem with individual interpretation of the rules and regulations, e.g., double glazing, changes to the building fabric.' |
| 'Depends on the quality of the relevant heritage planner & their knowledge: sometimes they can be a "mixed bag".' |
| 'Ordinary people have trouble getting permission to make changes; but some local rich influential people seem to get permission for anything.' |
| 'I'm pretty sure that we would be turned down for solar panels on our Victorian terrace, but wealthy landowners get away with developments as far from the local vernacular as you can get,' |

Planning can be an emotive and contentious issue [17], so it is positive that residents generally report it to be fairly appropriate. However, inconsistency between different areas of planning, and between judgements was found repeatedly in the literature as well as this study, and is likely to erode residents' faith in planning systems. This is therefore an important area for policy makers to improve.

## 4.4 INFORMATION SOURCES, BOTH POTENTIAL AND ACCESSED

As identified above, access to appropriate information sources is important to enable residents to make informed decisions about suitable carbon reduction strategies for their buildings. Fig.4 shows where respondents would seek information on carbon reduction from buildings.

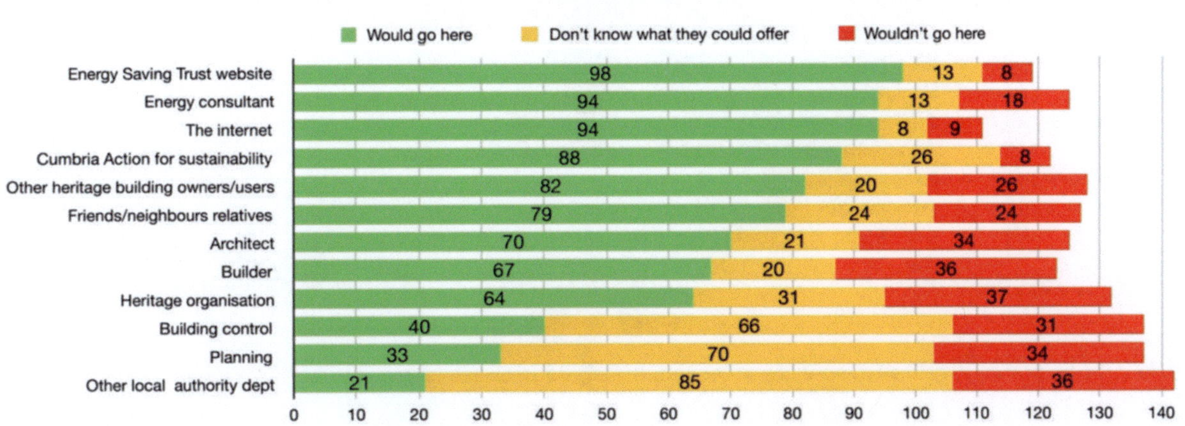


**Figure 4. Places respondents would go to find information (number of respondents)**

It can be seen that the top potential information source is online (The Energy Saving Trust has a website offering generic carbon reduction advice) [22]. In addition, five of the top six potential information sources, with the exception of 'energy consultant' are free, suggesting that people may be more likely to investigate these first. The popularity of sources such as other heritage building owners and social networks (friends/neighbours/relatives) confirm other findings that point to the importance of informal information networks [12]. Cumbria Action for Sustainability (CAfS) is an active sustainability charity which provides information, events and initiatives across Cumbria [23]. One of the survey distribution channels was through their email list however, so respondents may have more familiarity with their resources than average.

It is noticeable that statutory information sources such as planning departments and building control are the least likely to be accessed and that a significant proportion of residents appear uncertain of what these sources could offer. Importantly, it is also clear that less than half of residents would seek information from heritage organisations, despite these organisations providing significant and relevant resources. This could be linked to confusion on the remit of heritage organisations and a perception that they only cover significant, or at most, listed buildings, with residents not recognising their wider historic environment remit. Residents in undesignated buildings are likely to find similar retrofits unacceptable to those in designated buildings [14] and building construction techniques are likely to be similar, so this information would be highly relevant. This highlights a need for heritage organisations to publicise their resources more widely and promote targeted information to undesignated residential heritage buildings, especially around the importance of maintenance and simple changes such as draught stripping, or the benefits of adding shutters to sash windows [24].

Fig.5 shows whether those who had actually accessed the information sources had found them satisfactory or not. Information from planning, builders and energy consultants was the least satisfactory, perhaps linking back to the frustration at inconsistency amongst planning officers identified above. Residents considered other heritage building owners, CAfS, the internet and social networks the most satisfactory. This could imply that trusted, local and context specific information sources are amongst the most useful for residents, whilst the breadth of knowledge on the internet and its ease of access could be a factor in this source's popularity.

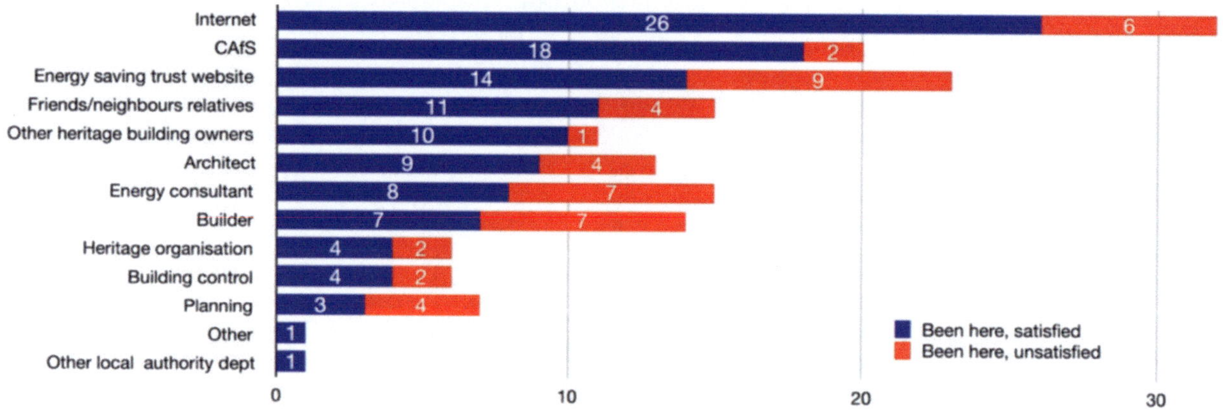


**Figure 5. Satisfaction levels with sources accessed, arranged in order of most use**

Residents' desire to access peer advice indicates a potential to form local support networks for heritage building residents to share information and develop communities of knowledge which could provide practical help about 'what actually works' in local heritage buildings [13]. This type of network could also provide links and recommendations to suitably experienced tradespeople, another identified challenge. Some of this role is fulfilled by CAfS, with open building visits, speakers and advice services during their annual 'Green buildings week', but this focusses on generic building sustainability and is not specific to heritage buildings. Therefore, a gap exists between heritage specific information which many residents would not access, and generic sustainability information which may not be relevant to heritage buildings. There is currently no mechanism for this through either planning or the energy assessor and EPC scheme which has proved largely inappropriate for heritage buildings [15]. There is therefore a need for information to be targeted to heritage building residents, and especially those in undesignated buildings, in a way that is easily accessible, context specific and trustworthy.

## 5. CONCLUSION

This paper has identified key barriers to retrofitting and explored two of them in more detail in the context of pre-1940 buildings in Cumbria. It has shown that heritage residents desire to reduce carbon despite feeling that it is challenging. It has confirmed findings from other studies identifying cost and knowledge barriers as the most important. Planning is also an important barrier and is seen as mainly appropriate but with perceptions of inconsistency and bias in the system, which is unhelpful for residents struggling to navigate complex decisions and processes.

One limitation of the study is the focus on a specific geographic area, hence the applicability of the results in other contexts requires further study. The initial results however demonstrate that a focussed and contextualised understanding of residents' attitudes may be critical to support decision making for sustainable retrofit choices. Further work is required to confirm these findings in other contexts and to expand understanding of the barriers that residents face in their heritage retrofitting in greater depth, to better examine the 'boundaries' of resident decision making.

Residents appear to struggle to know what options are available, with information barriers second only to cost. This may be especially true for residents in undesignated buildings who may not look at heritage information sources but who still have strong feelings about their buildings' heritage values and are concerned about the sensitivity of retrofits. Information from local groups has been shown to be valued by residents, and social networks were also identified as important. Facilitating local networks of heritage building residents to offer mutual support and provide appropriate information could help to overcome this informational barrier, and local information tailored to older buildings would be a valuable addition to support residents' decision making. Importantly, any resources need to be visible to heritage residents and especially to those in undesignated buildings who make up a significant proportion of UK building owners. Improving the visibility and relevance of information sources is therefore a key strategy to help support residents of heritage buildings in negotiating sensitive carbon reduction decisions and thereby help to mitigate climate change.

## 6. REFERENCES


[1] Global Alliance for Buildings and Construction, International Energy Agency, and United Nations Environment Program, "2019 Global Status Report for Buildings and Construction: Towards a zero-emission, efficient and resilient buildings and construction sector," p. 41, 2019.

[2] M. Almeida, M. Ferreira, and R. Barbosa, "Relevance of Embodied Energy and Carbon Emissions on Assessing Cost Effectiveness in Building Renovation—Contribution from the Analysis of Case Studies in Six European Countries," *Buildings*, vol. 8, no. 8, p. 103, Aug. 2018, doi: 10.3390/buildings8080103.

[3] D. Pickles and I. McCaig, "Energy Efficiency & Historic Buildings: Application of Part L of the Building Regulations to Historic & Traditionally Constructed Buildings," Historic England, 2017. Accessed: Apr. 05,

2019. [Online]. Available: https://historicengland.org.uk/images-books/publications/energy-efficiency-historic-buildings-ptl/heag014-energy-efficiency-partII/

[4] L. Mazzarella, "Energy retrofit of historic and existing buildings. The legislative and regulatory point of view," *Energy Build.*, vol. 95, pp. 23–31, May 2015, doi: 10.1016/j.enbuild.2014.10.073.

[5] J. Piddington, S. Nicol, H. Garrett, and M. Custard, "The Housing Stock of the United Kingdom," BRE Trust, Watford, Feb. 2020. [Online]. Available: https://files.bregroup.com/bretrust/The-Housing-Stock-of-the-United-Kingdom_Report_BRE-Trust.pdf

[6] Historic England, "Listed Buildings | Historic England," *Historic England*, 2019. https://historicengland.org.uk/listing/what-is-designation/listed-buildings/ (accessed Jun. 17, 2019).

[7] D. Herrera-Avellanosa *et al.*, "Deep renovation of historic buildings: The IEA-SHC Task 59 path towards the lowest possible energy demand and CO2 emissions," *Int. J. Build. Pathol. Adapt.*, vol. 38, no. 4, pp. 539–553, Jan. 2019, doi: 10.1108/IJBPA-12-2018-0102.

[8] S. Nicol, M. Roys, D. Ormandy, and V. Ezratty, "The cost of poor housing in the European Union," BRE, Watford, 2015. [Online]. Available: https://www.bre.co.uk/filelibrary/Briefing%20papers/92993_BRE_Poor-Housing_in_-Europe.pdf

[9] K. Gram-Hanssen, J. O. Jensen, and F. Friis, "Local strategies to promote energy retrofitting of single-family houses," *Energy Effic.*, vol. 11, no. 8, pp. 1955–1970, Dec. 2018, doi: 10.1007/s12053-018-9653-5.

[10] C. Maby and A. Owen, "Installer Power: The key to unlocking low carbon retrofit in private housing," Severn Wye Energy Agency, Glouster, Sep. 2015. Accessed: Jul. 08, 2020. [Online]. Available: https://www.linktoenergy.org.uk/wordpress/wp-content/uploads/2015/11/SWEA-Installer-Power-Executive-Summary.pdf

[11] F. Bartiaux, K. Gram-Hanssen, P. Fonseca, L. Ozoliņa, and T. H. Christensen, "A practice–theory approach to homeowners' energy retrofits in four European areas," *Build. Res. Inf.*, vol. 42, no. 4, pp. 525–538, Jul. 2014, doi: 10.1080/09613218.2014.900253.

[12] N. Hrovatin and J. Zorić, "Determinants of energy-efficient home retrofits in Slovenia: The role of information sources," *Energy Build.*, vol. 180, pp. 42–50, Dec. 2018, doi: 10.1016/j.enbuild.2018.09.029.

[13] R. Galvin and M. Sunikka-Blank, "The UK homeowner-retrofitter as an innovator in a socio-technical system," *Energy Policy*, vol. 74, pp. 655–662, Nov. 2014, doi: 10.1016/j.enpol.2014.08.013.

[14] M. Sunikka-Blank and R. Galvin, "Irrational homeowners? How aesthetics and heritage values influence thermal retrofit decisions in the United Kingdom," *Energy Res. Soc. Sci.*, vol. 11, pp. 97–108, Jan. 2016, doi: 10.1016/j.erss.2015.09.004.

[15] D. Pickles and C. Cattini, "Energy Efficiency and Historic Buildings: Energy Performance Certificate," Historic England, 2015. [Online]. Available: https://historicengland.org.uk/images-books/publications/eehb-energy-performance-certificates/

[16] CEN, "EN 16883:2017 Conservation of cultural heritage - guidelines for improving the energy performance of historic buildings." British Standards Institute (BSI), 2017. [Online]. Available: https://shop.bsigroup.com/en/ProductDetail/?pid=000000000030322690

[17] C. M. Stuart, "Managing or Driving Change? Establishing Consensus of Opinion on Improving the Energy Efficiency of Historic Buildings," *Hist. Environ. Policy Pract.*, vol. 5, no. 2, pp. 182–195, Jul. 2014, doi: 10.1179/1756750514Z.00000000048.

[18] K. Friedman and D. A. Cooke, "Is UK Planning a barrier to energy efficient heritage retrofit: a comparative analysis of a selection of London Boroughs," in *Proceedings Retrofit 2012*, Salford, 2012, p. 13.

[19] N. Ginks and B. Painter, "Energy retrofit interventions in historic buildings: Exploring guidance and attitudes of conservation professionals to slim double glazing in the UK," *Energy Build.*, vol. 149, pp. 391–399, Aug. 2017, doi: 10.1016/j.enbuild.2017.05.039.

[20] Lake District National Park Authority (LDNPA), "English Lake District World Heritage Site," *Statement of Outstanding Universal Value*, 2020. http://lakesworldheritage.co.uk/world-heritage/souv/ (accessed Jun. 02, 2020).

[21] IBM Corp, *SPSS Statistics for Macintosh, Version 25.0.* Armonk, NY: IBM Corp, 2017.

[22] Energy Saving Trust, "Energy Saving Trust," 2021. https://energysavingtrust.org.uk (accessed Jun. 04, 2021).

[23] Cumbria Action for Sustainibility (CAfS), "Cumbria Action for Sustainability," 2021. https://cafs.org.uk (accessed Jun. 04, 2021).

[24] C. Wood, B. Bordass, and P. Baker, "RESEARCH INTO THE THERMAL PERFORMANCE OF TRADITIONAL WINDOWS: TIMBER SASH WINDOWS," English Heritage, 2009.

# Link Historic Buildings to Cloud with Internet of Things and Digital Twins

Zhongjun Ni, Yu Liu, Magnus Karlsson, Shaofang Gong

Department of Science and Technology, Linköping University, Norrköping, Sweden.
(zhongjun.ni@liu.se, yu.a.liu@liu.se, magnus.b.karlsson@liu.se, shaofang.gong@liu.se)

**Abstract –** Information and communication technologies (ICTs) help preserve historic buildings and optimize energy efficiency. This study proposes a digitalization framework for historic buildings by utilizing ICTs, such as Internet of Things (IoT), digital twins, and cloud computing. A digital twin is a digital representation of physical world assets that genuinely reflects the properties of real-world objects and processes. In this study, historic buildings are modeled with cloud-based digital twins. Indoor environmental data are collected with locally deployed sensors and ingested to a digital twin in real-time. The digital twin enables decision-makers to remotely monitor the indoor environment of a historic building and actively manipulate actuators to perform maintenance. Empowered by data analytics and artificial intelligence (AI), a digital twin can further simulate and predict state changes in a historic building to reach desired autonomous maintenance and energy saving.

**Keywords –** Internet of Things; digital twins; cloud computing; historic buildings; energy efficiency

## 1. INTRODUCTION

Preservation of cultural heritage and historic buildings requires preventive approaches to maintain the materials, structure, appearance, and environmental conditions while considering energy efficiency and human comfort [1], [2]. Several factors, such as ambient climate, building characteristics, and occupants' behaviour, can affect energy consumption of a building [3]-[5]. To achieve human comfort, it is necessary to reasonably control heating, ventilation, and air conditioning (HVAC) systems [6], [7]. Traditional approaches to improve indoor environmental quality (IEQ) in historic buildings highly rely on artificial manipulation of HVAC systems [8].

Assisted by evolving information and communication technologies (ICTs), represented by Internet of Things (IoT), historic building preservation has undergone continuous development. For instance, lots of studies have demonstrated deploying various types of sensors in historic buildings to achieve real-time monitoring of indoor environment [9], surface conditions [10], and structural health [11]. Our previous research [12] has developed a remote monitoring and control system for cultural and historic buildings. Wireless sensor networks based on the Zigbee protocol are deployed to heritage buildings, which enables sensing and manipulating of indoor climate from a centralized server. The results have led to the CultureBee system [13] that has been adopted by many Swedish churches. Based on the result, further studies [14], [15] proposed a universal framework for IEQ monitoring and management that leverages public cloud platform to enhance the system reliability and scalability while lowing the deployment cost.

As a continuation, this study aims to develop a cloud-based digitalization framework that integrates IoT and digital twins for energy efficiency optimization and smart maintenance of historic buildings. Three historic buildings located in Norrköping, Sweden, as shown in Fig. 1, are selected as case studies where environmental sensors are installed on site and corresponding digital twins are created and stored in the cloud. Based on continuously collected sensing data, the digital twins can genuinely reflect real-time operating conditions and predict future states of historic buildings. According to learned occupant and operating activities, digital twins also facilitate decision-making of maintenance by exploiting novel artificial intelligence to realize energy saving.

(a) (b) (c)

**Figure 1. Three historic buildings are selected for case studies. (a) Östgötateatern, (b) Stadsmuseum, and (c) Hörsalen.**

## 2. SYSTEM OVERVIEW

Fig. 2 depicts the proposed digitalization framework, which comprises five layers, namely perception layer, transmission layer, storage layer, analysis layer and application layer, following a bottom-up order. These layers can be grouped into two main parts: the local part and the cloud part. In the perception layer, sensors and actuators are deployed to collect data of indoor environment, ambient climate, facility operation status and occupants' behaviour about a building. The transmission layer is a bridge between the local part and the cloud part. Real-time data collected from the local part and control commands issued from the cloud part are exchanged through the transmission layer. In the storage layer, data from heterogeneous sources are pre-processed and stored in relational or non-relational databases. Digital twin instances are also created and stored in this layer. Once data are ready, evaluation methods and AI models deployed in the analysis layer can utilize them to achieve different goals. The application layer provides interactive functions to users, such as data visualization, remote control, and energy consumption prediction.

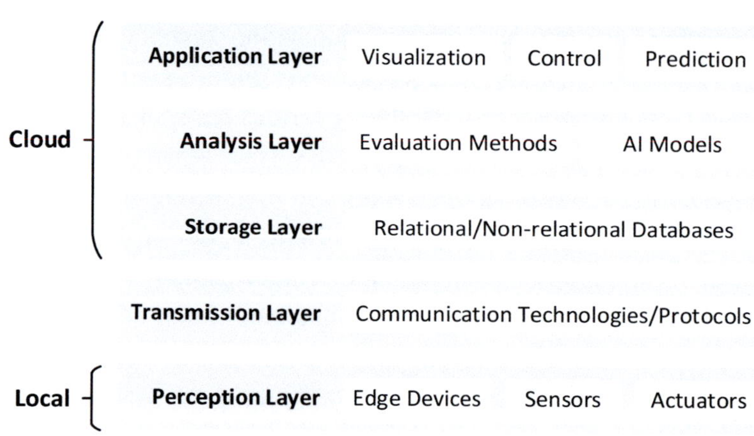


**Figure 2. Layered architecture of the cloud-based digitalization framework for historic buildings.**

## 3. SYSTEM DESIGN AND IMPLEMENTATION

The detailed design and implementation are presented in Fig. 3. In the local part, the core module is an edge device composed of a microprocessor and a microcontroller. The edge device aims for data collection, control forwarding, and edge computing. The cloud part integrates a series of services provided by Microsoft Azure. Components of the cloud part are driven by carefully designed data streams and event streams to provide applications ultimately. Detailed implementation of each part is discussed in the subsequent subsections.

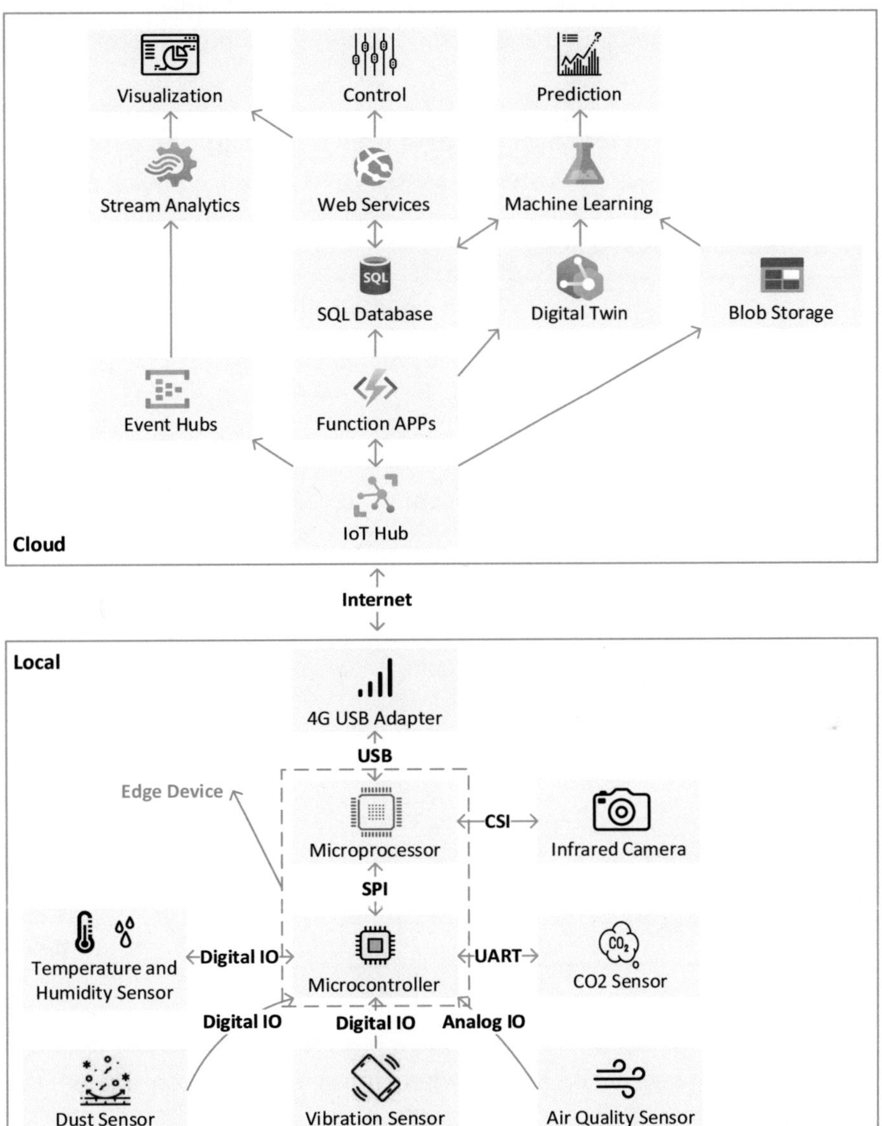


**Figure 3. Design and implementation of the cloud-based digitalization framework for historic buildings.**

### 3.1 THE LOCAL PART

The implementation of the local part is presented from aspects of hardware and software.

#### *3.1.1 Hardware*

In the edge device, a Raspberry Pi Compute Module 3+ (CM3+) is used to provide the function of the microprocessor. The CM3+ contains a 32 GB embedded multimedia card (eMMC) flash memory and an ARMv8 system-on-chip (SoC), which has 1 GB random-access memory (RAM) and runs at 1.2 GHz. The CM3+ works with a board hosting 120 general-purpose input/output (GPIO) pins, a universal serial bus (USB) port, and two camera ports. The microcontroller used in the edge device is an Arduino Uno board, which is based on the ATmega328 chip. Each of the 14 digital pins on the board can be used as an input or output (IO). In addition to rich IOs, the board supports multiple communication buses, such as universal asynchronous receiver-transmitter (UART) and serial peripheral interface (SPI). The microprocessor and the microcontroller communicate via the SPI bus.

The edge device accesses the Internet through a 4G USB adapter. The adapter model is ZTE MF833V. According to testing, the upload and download speeds of the adapter are around 12 Mbit/s and 38 Mbit/s, respectively.

Six types of sensors are used for collecting data from a building. The detailed specifications are summarized in Table 1. The connections between sensors and the edge device are shown in Fig. 3.

**Table 1. Detailed specifications of used sensors.**

| **Sensor Types** | **Specifications** | | | | | |
|---|---|---|---|---|---|---|
| | *Measuring object* | *Unit* | *Range* | *Accuracy* | *Resolution* | *Model* |
| *Temperature and humidity sensor* | Temperature | °C | -40 - 80 | ±0.5 | 0.1 | DHT22 |
| | Relative humidity | / | 5 - 99% | ±2% | 0.1% | |
| *$CO_2$ sensor* | $CO_2$ concentration | ppm | 0 - 2000 | 200 | 1 | MH-Z16 |
| *Dust sensor* | Concentration of small particles (diameter > 1μm) in the air | pcs/0.01cf | 0 - 8000 | / | / | Grove dust sensor |
| *Air quality sensor* | $NH_3$, NOx, benzene, smoke, and other harmful gases | ppm | 0 - 2000 | / | / | MIKROE-1630 |
| *Vibration sensor* | Whether the object vibrates | / | True/False | / | / | Grove vibration sensor |
| *Camera* | Picture | pixel | / | / | 3280 × 2464 | Raspberry Pi NoIR v2.1 |
| | Video | | / | / | 1080p/720p/480p | |

A sensor box is presented in Fig. 4. A plastic box is used to pack the hardware. We fix the edge device on the bottom and the sensors on the lid. Many holes are drilled on the box surface to ensure enough air circulation.

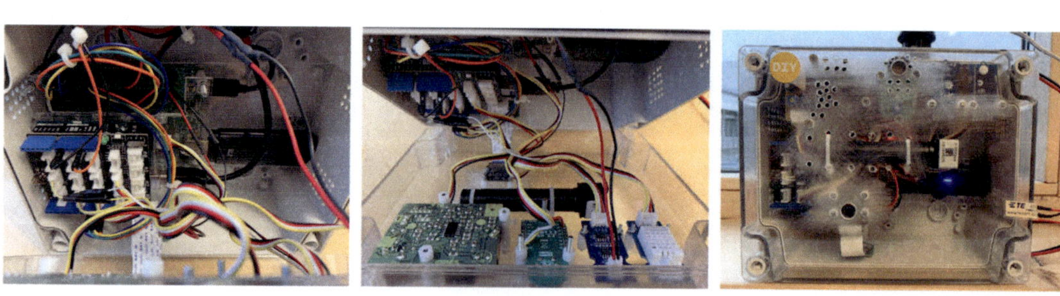

(a) (b) (c)

**Figure 4. Hardware setup and deployment. (a) Placement of the edge device, (b) placement of the sensors, and (c) a sensor box in operation.**

### *3.1.2 Software*

The microcontroller performs two tasks. One is to periodically obtain readings from sensors, and the other is to return the latest sensor readings when receiving a query from the microprocessor. The main modules of the microcontroller program are shown in Fig. 5. The sensor manager uses sensor drivers to obtain latest readings and stores these readings. The SPI slave handles queries from the microprocessor. The watchdog timer (WDT) ensures the long-term stable operation of the program.

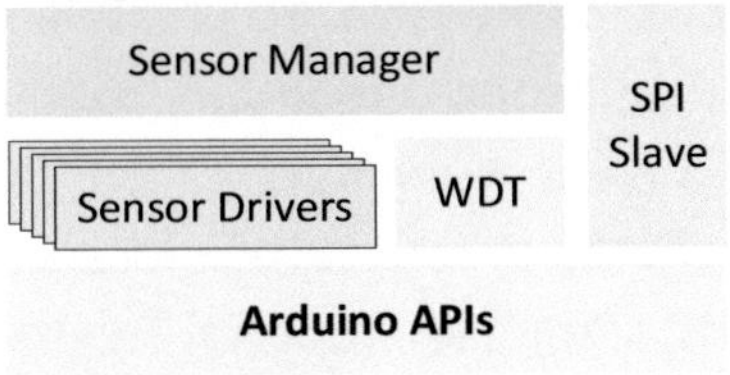


**Figure 5. Program architecture of the microcontroller.**

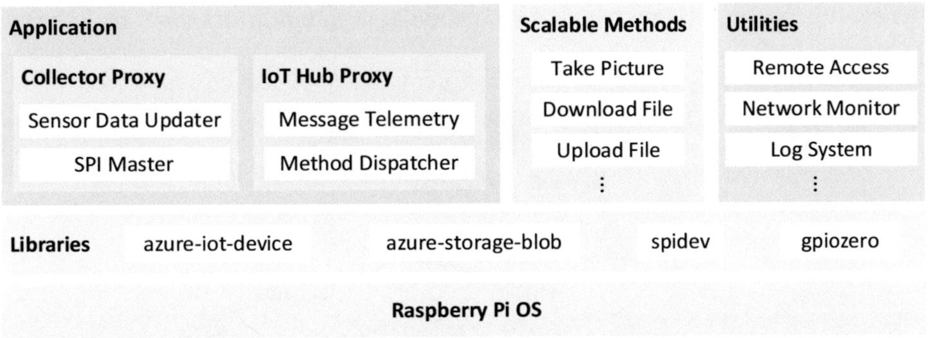


**Figure 6. Program architecture of the microprocessor.**

The core function of the microprocessor is to exchange data with the cloud. On the one hand, it periodically uploads sensor readings and actuator status to the cloud. On the other hand, it takes corresponding actions when receiving instructions issued by the cloud. The main modules of the microprocessor program are shown in Fig. 6. In the application part, the collector proxy periodically queries sensor readings from the microcontroller via SPI Master. The IoT hub proxy interacts with the cloud, i.e., to package sensor readings and actuator status into messages and telemetry to the cloud. When receiving an instruction from the cloud, the method dispatcher parses the instruction and executes the corresponding method, such as taking pictures, downloading files, and uploading files. A series of utilities ensure the long-term operation of the program. For instance, remote access allows researchers to log in to the edge device for management, and the network monitor maximizes the network availability and prevents the device from being offline.

## 3.2 THE CLOUD PART

The implementation of the cloud part is explained from aspects of data routing and digital twins.

### *3.2.1 Data routing*

The IoT hub manages local edge devices and enables bidirectional communication between the cloud platform and the local edge devices. When receiving a message from an edge device, the IoT hub publishes an event. The function apps that subscribe to the event can parse the event and perform operations such as writing data to the database and updating status of digital twins. Based on these events, real-time stream analysis can also be performed. Once data are ready, we can apply different machine learning algorithms, deploy various web services, and finally provide users with applications such as data visualization, remote control, and energy consumption prediction.

### *3.2.2 Digital twins*

A digital twin instance of a building is created based on Azure Digital Twins, a platform as a service (PaaS) that enables creating knowledge graphs based on digital models. A digital model is an abstraction of real-world entities of the same type. To create a digital twin instance of a building, a series of digital models such as Building, Floor, and Room, need to be defined. As an example, Fig. 7 shows a digital twin of room TP6137 in a building called Täppan at Linköping University. The relationship between models is designated to reflect their interactions. Each digital model has several fields to reflect objects in the real world, such as Property, Telemetry, and Component. Properties are data fields that represent the state of a building entity such as area and height. Telemetry fields represent measurements or events and are often used to describe sensor readings. Components are used to represent a group of instances of other models, e.g., actuators such as lighting devices and heating equipment.

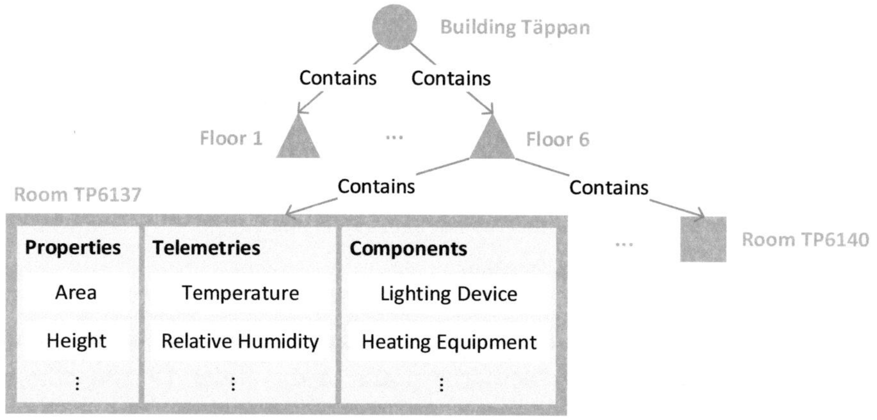


**Figure 7. Digital twin of Room TP6137, Building Täppan, Campus Norrköping, Linköping University.**

## 4. RESULTS AND DISCUSSION

To preliminarily test and verify the functionalities of the framework, we deployed a sensor box (see Fig. 4c) in a room at Linköping University. Environmental parameters were collected, transmitted to the cloud platform, and ingested to the digital twin instance through the aforementioned approach.

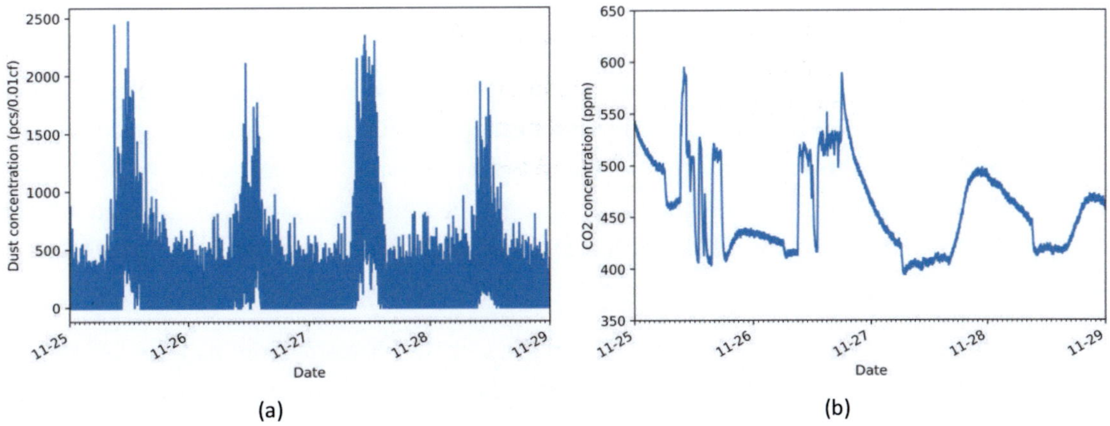


**Figure 8. Historical data of sensors from November 25 to 28, 2020. (a) Dust sensing, and (b) $CO_2$ sensing.**

Fig. 8 depicts historical data from two sensors. It is seen in Fig. 8a that the dust concentration was high from 9 am to 3 pm in these four days. After investigation, we find that the scheduled indoor ventilation in the building during that period enhanced the fluidity of indoor air and drove the movement of small particles, which led to the increase of dust concentration in the room. Simultaneously, the correlation between $CO_2$ concentration and room occupancy is seen in Fig. 8b; details are described in Fig. 9.

After comparing the $CO_2$ concentration changes on November 26 (one person in the room) and November 27 (no one in the room), we find an occupant's presence can significantly affect the $CO_2$ concentration. The detailed explanation of $CO_2$ concentration changes near the three marked points in Fig. 9a is as follows. 1) The occupant entered the room and started working, which caused the $CO_2$ concentration to increase. 2) The occupant went out for lunch and returned after an hour, which caused the $CO_2$ concentration to decrease first and then increase. 3) The occupant left the room after getting off work, which caused $CO_2$ concentration to decrease.

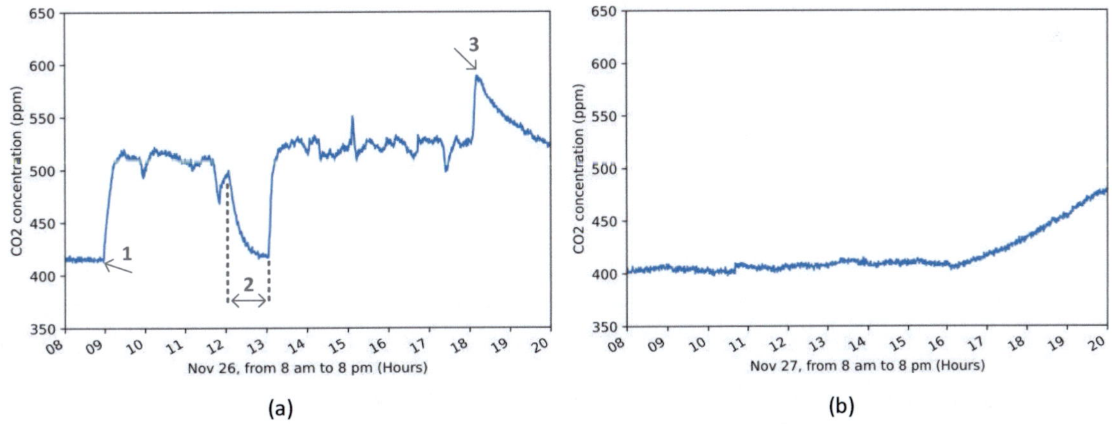


**Figure 9. Occupant's presence can affect $CO_2$ concentration. (a) One person in the room, and (b) no one in the room.**

Obviously, this correlation can be used to predict the occupancy to realize autonomous control of HVAC systems and lighting. As for the future, with continuously collected sensor data from the three historic buildings, more correlations between sensor readings and occupants' behaviour can be investigated to enrich more applications and finally achieve energy efficiency optimization and intelligent maintenance.

## 5. CONCLUSION

This study proposes a cloud-based digitalization framework by utilizing IoT and digital twins for historic buildings, which is to be deployed to three heritage buildings that are chosen for case study. The detailed design and implementation in terms of both hardware and software are presented. The experimental deployment results show that data can be collected, transmitted, and stored in the cloud while digital twins that reflect the latest status of a building can be created and fed with real-time sensor data. A sample digital twin instance is given to illustrate the building models.

In the future, after deploying sensor boxes to the selected historic buildings, we will focus more on digging information from collected data. More specific AI models can be trained to provide plenty of applications, such as anomaly detection, energy consumption prediction, and autonomous control for historic buildings to reach energy efficiency optimization.

## 6. ACKNOWLEDGMENT

We thank Gustav Knutsson for his assistance in ordering electronic components and experimental materials. This study has been financially supported by the Swedish Energy Agency within the program of Spara och Bevara. Norrevo Fastigheter AB in Norrköping is acknowledged for their support and providing access to the historic buildings.

## 7. REFERENCES

[1] K. Van Balen, "Preventive Conservation of Historic Buildings," Restor. Build. Monum., vol. 21, no. 2–3, pp. 99–104, 2015, doi: 10.1515/rbm-2015-0008.

[2] A. Martínez-Molina, I. Tort-Ausina, S. Cho, and J. L. Vivancos, "Energy efficiency and thermal comfort in historic buildings: A review," Renew. Sustain. Energy Rev., vol. 61, pp. 70–85, 2016, doi: 10.1016/j.rser.2016.03.018.

[3] A. D'Alessandro et al., "Real-time urban seismic network and structural monitoring by means of accelerometric sensors: Application to the historic buildings of Catania (Italy)," 2018 IEEE Int. Conf. Environ. Eng. EE 2018 - Proc., pp. 1–5, 2018, doi: 10.1109/EE1.2018.8385254.

[4] P. Bacher and H. Madsen, "Identifying suitable models for the heat dynamics of buildings," Energy Build., vol. 43, no. 7, pp. 1511–1522, 2011, doi: 10.1016/j.enbuild.2011.02.005.

[5] P. Rocha, A. Siddiqui, and M. Stadler, "Improving energy efficiency via smart building energy management systems: A comparison with policy measures," Energy Build., vol. 88, pp. 203–213, 2015, doi: 10.1016/j.enbuild.2014.11.077.

[6] Z. Yu, B. C. M. Fung, F. Haghighat, H. Yoshino, and E. Morofsky, "A systematic procedure to study the influence of occupant behavior on building energy consumption," Energy Build., vol. 43, no. 6, pp. 1409–1417, 2011, doi: 10.1016/j.enbuild.2011.02.002.

[7] A. Boodi, K. Beddiar, M. Benamour, Y. Amirat, and M. Benbouzid, "Intelligent systems for building energy and occupant comfort optimization: A state of the art review and recommendations," Energies, vol. 11, no. 10, 2018, doi: 10.3390/en11102604.

[8] P. H. Shaikh, N. B. M. Nor, P. Nallagownden, I. Elamvazuthi, and T. Ibrahim, "Intelligent multi-objective control and management for smart energy efficient buildings," Int. J. Electr. Power Energy Syst., vol. 74, pp. 403–409, 2016, doi: 10.1016/j.ijepes.2015.08.006.

[9] A. Kupczak, B. Pretzel, and M. Strojecki, "Different HVAC systems in historical buildings to meet collection demands," pp. 337–344, 2018.

[10] R. Walker and S. Pavía, "Thermal and moisture monitoring of an internally insulated historic brick wall," Build. Environ., vol. 133, no. February, pp. 178–186, 2018, doi: 10.1016/j.buildenv.2018.02.020.

[11] S. Raffler, S. Bichlmair, and R. Kilian, "Mounting of sensors on surfaces in historic buildings," Energy Build., vol. 95, pp. 92–97, 2015, doi: 10.1016/j.enbuild.2014.11.054.

[12] J. Zhang, "A Remote Monitoring and Control System for Cultural Heritage Buildings Utilizing Wireless Sensor Networks," Ph.D. dissertation, Dept. of Sc. and Tech., Linköping Univ., Norrköping, 2013. Accessed on: December 06, 2020. [Online]. Available: http://liu.diva-portal.org/smash/get/diva2:679457/FULLTEXT02.pdf.
[13] "CultureBee - wireless monitoring of cultural heritage," Culturebee.se. [Online]. Available: http://culturebee.se/. [Accessed: 06-Dec-2020].
[14] Y. Liu, K. Akram Hassan, M. Karlsson, O. Weister, and S. Gong, "Active Plant Wall for Green Indoor Climate Based on Cloud and Internet of Things," IEEE Access, vol. 6, pp. 33631–33644, 2018, doi: 10.1109/ACCESS.2018.2847440.
[15] Y. Liu, K. Akram Hassan, M. Karlsson, Z. Pang, and S. Gong, "A data-centric internet of things framework based on azure cloud," IEEE Access, vol. 7, pp. 53839–53858, 2019, doi: 10.1109/ACCESS.2019.2913224.